# THE JEWS
# OF SOUTH-WEST ENGLAND

# THE JEWS OF SOUTH-WEST ENGLAND

## The Rise and Decline of their Medieval and Modern Communities

by

BERNARD SUSSER

UNIVERSITY
*of*
EXETER
PRESS

First published 1993 by
University of Exeter Press
Reed Hall, Streatham Drive
Exeter, Devon EX4 4QR
UK



**British Library Cataloguing in Publication Data**
A catalogue record of this book is
available from the British Library

ISBN 0 85989 366 9

Typeset in 10/12pt Palatino
by Colin Bakké Typesetting, Exeter

Printed and bound in Great Britain
by Short Run Press Ltd, Exeter

*To*

*my friend, guide and mentor*

# Arthur Goldberg ז"ל

*A proud Jew*

*and*

*a patriotic citizen*

Remember the days of old,
Consider the years of many generations;
Ask thy father, and he will declare unto thee,
Thine elders, and they will tell thee.

(*Deuteronomy* 32:7)

My hopes are with the Dead; anon
My place with them will be,
And I with them shall travel on
Through all Futurity;
Yet leaving here a name, I trust,
That will not perish in the dust.

(Robert Southey, *His Books*)

# CONTENTS

# ACKNOWLEDGEMENTS

This book is based on a thesis submitted in June 1977 to the University of Exeter, England, for the degree of PhD in the faculty of Social Studies.

Many people have helped me in the course of my research, and any merit in it is in large measure due to them. First and foremost, I gladly acknowledge my debt to Dr H.E.S. Fisher of the University of Exeter who taught me all I know of historical method and the art of marshalling evidence. Without his constant friendship, encouragement and advice over many years neither the original thesis nor this book would ever have been completed.

The late Dr Cecil Roth initiated me into Anglo-Jewish historical research and most generously encouraged me by allowing me access to his entire collection. The late Dr Vivian Lipman pointed me in the direction of the Decennial Census returns, and Mr A. Schischa taught me the Yiddish I needed to decipher the Plymouth Congregation's Yiddish records, as well as helping me with a number of difficulties.

The late Sir Israel Brodie and Lord Jakobovits allowed me to use the archives of the Chief Rabbi's Office. The late Mr Wilfred Jessop, Chicago, generously loaned me his unpublished work, 'A Coat of Many Colours'.

I am grateful to a host of correspondents whose help is acknowledged in the notes, but in particular to the late Mr L. Berlin, son of the Revd M. Berlin who was minister to the Plymouth Congregation 1896–1906, and Mr L. Norman, son of the secretary of the Plymouth Congregation in 1858 (!), because they gave me much information as well as a large number of documents relating to the Congregation. Mr Percy Aloof searched the microfilms (an eye-straining job) and transcribed the 1871 and part of the 1881 Plymouth census returns for me until ill-health prevented him from continuing.

Nearly thirty years have elapsed since I began my research and many who gave me access to their offices and libraries and showed me how to use their treasures are now in the World of Truth. I thank them for the help they gave me; all my doubts, queries, suppositions and mistakes are known to them. I will be glad to hear from any reader who can throw any more light on any aspect of this book so that errors can be corrected.

# ACKNOWLEDGEMENTS

My thanks are due to Channa and Jacob. When they were little they allowed me to do the necessary research, and when they grew up they read the many draft manuscripts, checked them and pointed out passages which I had not made clear.

Mr Martin Gilbert generously allowed me to reproduce maps from his *Jewish History Atlas*, and I am grateful to him.

The late Dr Richard Barnett gave me a number of photographs of Jews in the South West of England. Some of them are reproduced.

Mr Simon Baker, of the University of Exeter Press, has seen this work through to publication. His meticulous attention to detail has saved it from many an error. I thank him and the Press for their help and co-operation.

The generosity of the family of the late Mr Arthur Goldberg ז"ל has made the publication of this book possible: the reading public and I are greatly indebted to them.

According to the Jewish proverb, 'The last, Ah!, the last is the most precious'. In respect of my wife's contribution to this work I can only echo Rabbi Akiva's words to his disciples: My knowledge and what you learn from me is all due to her.

Dated on the seven hundred and second anniversary of the Edict of Expulsion of the Jews from England.

*Jerusalem, 18 July 1992*
*Bernard Susser*

# GLOSSARY

The following terms are used in this book in a particular sense.

**Congregation:** When capitalized, the term implies an assembly of Jews with a corporate identity.

**Jew:** One who considers himself or is or was considered by others to be Jewish; the child of a Jewess.

**South West:** The English counties of Cornwall and Devon.

The transliteration of Hebrew words generally follows the Sephardi pronunciation, except where a word is widely used in its Ashkenasi form.

**Adar** Month in the lunar Jewish calendar, generally corresponding to February/March.

**Adar Sheni** Month in the leap lunar Jewish calendar, generally corresponding to March.

**Aliyah** Emigration to Israel.

**Aliyah (pl. Aliyot)** Being called to the Torah.

**Ashkenasim** Jews emanating from Central or Eastern Europe, cf. Sephardim.

**Av** Month in the lunar Jewish calendar, generally corresponding to August.

**Baal Habayit (pl. Baalei Batim)** (Vestry) member of a Congregation.

**Baal Tefillah** Prayer leader.

**Bar Mitzvah** When a boy becomes thirteen years and one day old.

**Bat Mitzvah** When a girl becomes twelve years and one day old.

**Bentsch Gomel** To recite a thanksgiving blessing when called to the Torah after deliverance from peril.

**Beth Din** Jewish court of law.

**Bimah** Central platform from which the cantor conducts the prayers, and from which the Torah is read aloud to the congregation.

**Brit Milah** The Covenant of Circumcision.

**Chazan / Sheni** Cantor or second cantor.

**Cheder** A school for teaching children the elements of the Jewish religion, approximating to 'Sunday school'.

**Chuppah** Marriage canopy, and hence the marriage ceremony.

**Cohen (pl. Cohenim)** A priest of the tribe of Levi.

**Ellul** Month in the lunar Jewish calendar, generally corresponding to September.

**Etrog (pl. Etrogim)** Citrus fruit used in the ritual at the festival of Tabernacles.

**Gabbai** Treasurer.

**Gabbai Zedakah** Charity treasurer.

**Gaon** Rabbinical leader of a community; any outstanding Torah scholar.

**Haftarah** Portion from the Prophets read in the synagogue on Sabbath morning after the reading of the weekly pericope.

**HaLevi** A Levite.

**Halitzah** See *Shetar Halitzah*, below.

**Hamets** Fermented grain flour whose use or even possession is forbidden on Passover.

**Hanukah** Eight day Festival of Lights commemorating the re-dedication of the Temple by the Hasmoneans in 165 BC.

**Hasid** A follower of the Jewish religious and social movement founded by Israel Baal Shem Tov in the mid-eighteenth century.

**Hatan Bereishit** The first man called to the Torah at the beginning of the annual cycle of the Pentateuchal reading.

**Hatan Torah** The last man called to the Torah at the end of the annual cycle of the Pentateuchal reading.

**Havdalah** Ceremony involving the use of wine, spices and lights, marking the end of the Sabbath.

**Hebrah Kaddishah** Society for the burial of the dead.

**Herem** Decree of excommunication.

**Heshvan** Month in the lunar Jewish calendar, generally corresponding to November.

**Hezkat Hakehilla** Vestry rights.

**High Holydays** The festivals of the New Year and Day of Atonement.

**Hoshanna Rabbah** The last intermediate day of Tabernacles; a minor festival in its own right.

**Iyyar** Month in the lunar Jewish calendar, generally corresponding to May.

**Kaddish** Doxology recited for eleven months after, or on the anniversary of, the death of one's parents or other near relatives.

**Kahal (or Kohol)** The governing body of the Congregation; the vestry.

**Kehillah** Community.

**Kislev** Month in the lunar Jewish calendar, generally corresponding to December.

**Kosher (or Kasher)** Proper, ritually fit. Usually used in connection with food.

**KZ** The initial letters of the phrase *'Kohen Zedek'* (= righteous priest), often appended to the Jewish name of a priest, giving rise to the surname 'Katz'.

**Matnat Yad** A Festival day on which *yizkor* (see below) is said.

**Matzah (pl. Matzot)** Unleavened bread eaten at Passover.

**Mikveh** A bath constructed in accordance with Jewish law, in which (a) a woman must immerse monthly before resuming marital relations, (b) a convert must immerse to become a Jew, and (c) many men immerse, particularly on the eves of Sabbaths and Festivals.

**Milah** Circumcision.

**Minhag** Custom or tradition, often with the force of Jewish law.

**Mi Sheberach** A prayer in the name of one called to the Torah invoking blessing on him, his family, friends and communal leaders.

**Mitzvah** A religious obligation; an honour, such as an aliyah, bestowed on a worshipper.

**Mohel** Layman or doctor who performs the religious Covenant of Circumcision.

**Nisan** Month in the lunar Jewish calendar, generally corresponding to April.

**Ohel** Chapel in the cemetery.

**Omer** Period of forty-nine days between Passover and Pentecost.

**Orach (pl. Orchim)** Visitors; worshippers in the synagogue who do not pay a membership fee to the Congregation.

**Parnas** Chairman or president of the Congregation.

**Pesach** Passover.

**Pinkes** A ledger; a book containing communal records.

**Porger** One who removes forbidden fat and sinews from meat.

**Purim** Festival occurring in March commemorating the events recorded in the book of Esther.

**Rav** Rabbi.

**Reb** Honorific title corresponding to 'Mr'.

**Rosh Hodesh** The first day or days of the Jewish month, a minor festival.

**Segan** One who bestows aliyot.

**Sephardim** A Jew of Spanish or Portuguese descent.

**SGL** The initial letters of the phrase *segan leviyah* (= Levitical excellence), often appended to the Jewish name of a levite, and giving rise to the surname 'Segal'.

**Shabbat** The Sabbath.

**Shavuot** Jewish festival of Pentecost, occurring in May.

**Shechitah** Jewish method of slaughtering food animals.

**Shetar Halitzah** Document obliging brother to contract levirate marriage or its substitute.

**Sheva Berachot** 'Seven Blessings', recited at the wedding service, or at meals in the first week of the marriage.

**Shevat** Month in the lunar Jewish calendar, generally corresponding to January/February.

**Shochet (pl. Shochetim)** Slaughterer of animals for food using the method of shechitah.

**Shool** Synagogue.

**Sivan** Month in the lunar Jewish calendar, generally corresponding to June.

**Succah** Tabernacle; a room or hut with a roof of leaves and branches inhabited during the festival of Tabernacles.

**Succot** Festival of Tabernacles, occurring in October.

**Takkanot** Regulations.

**Tallit** Shawl worn by Jewish men during prayers.

**Tammuz** Month in the lunar Jewish calendar, generally corresponding to July.

**Tannaim** Conditions; contract; engagement to marry.

**Tevet** Month in the lunar Jewish calendar, generally corresponding to January.

**Tishri** Month in the lunar Jewish calendar, generally corresponding to October.

**Torah** Parchment scroll of the Pentateuch (particularly *Sefer Torah*); the corpus of Jewish teachings.

**Tosafist** Medieval French and German Talmudic commentators.

**Toshav (pl. Toshavim)** One who pays a membership fee for a seat in the synagogue.

**Treifah** Ritually unfit, not kosher (usually used in connection with food).

**Vacher** One who watches over a corpse before its burial.

**Yahrzeit** Anniversary of death.

**Yizkor** Commemorative prayer.

**YIA** Initial letters of the phrase *Yibaneh Ireinu, Amen* (= May our city [i.e. Jerusalem] be rebuilt, Amen), appended after the name of any city in the Diaspora.

**YZV** Initial letters of the phrase *Yishmereihu Zuro Ve-goalo* (= May his Rock and Redeemer protect him), appended to the Jewish name of a living Jew.

**ZL** Initial letters of the phrase *Zichrono Liveracha* (= May his memory be for a blessing), appended to the name of a dead Jew.

# ABBREVIATIONS

The following abbreviations are used in this book.

| | |
|---|---|
| A/c. | Account |
| Berlin, PHC tomb. | Tombstone in the Jewish Cemetery, The Hoe, Plymouth, as recorded by the Reverend M. Berlin |
| Circum. Reg. | Joseph Joseph's Circumcision Register |
| EHC | Exeter Hebrew Congregation |
| EHC tomb. | Tombstone inscription in the Jewish Cemetery, Exeter as recorded by B. Susser |
| FHC | Falmouth Hebrew Congregation |
| FHC tomb. | Tombstone inscription in the Jewish Cemetery, Falmouth, as recorded by B. Susser |
| *Gent. Mag.* | *Gentleman's Magazine* |
| H. in H. | Hand in Hand Jewish Friendly Society, Plymouth |
| HO | Home Office |
| *JC* | *Jewish Chronicle* |
| *Jew. Encycl.* | *Jewish Encyclopaedia* |
| *Jew. Year Bk* | *Jewish Year Book* |
| Mesh. Nefesh | Meshivat Nefesh Society, Plymouth |
| Min. Bk | Minute Book |
| *MJHSE* | *Miscellanies of the Jewish Historical Society of England* |
| *PAJHS* | *Proceedings of the American Jewish Historical Society* |
| PCC | Probate Court of Canterbury |
| PenHC | Penzance Hebrew Congregation |
| PenHC tomb. | Tombstone in the Jewish Cemetery, Penzance as recorded by B. Susser |
| PHC | Plymouth Hebrew Congregation |
| PHC Bk of Records | Plymouth Hebrew Congregation's Book of Records |
| PHC Inscrip. | Inscription at the Synagogue, Plymouth |
| *Pig. Direct.* | *Piggott's Directory* |
| *Ply. Direct.* | *Plymouth Directory* |
| Ply. Syn. Cat. | Catalogue of leases and documents relating to the Plymouth Hebrew Congregation |
| PHC tomb. | Tombstone inscription in the Jewish Cemetery, The Hoe, Plymouth as recorded by B. Susser |
| PRO | Public Record Office, London |
| *TJHSE* | *Transactions of the Jewish Historical Society of England* |
| *Trew. Flying Post* | *Trewman's Exeter Flying Post* |
| *VJ* | *Voice of Jacob* |

# LIST OF MAPS

# LIST OF TABLES

# LIST OF ILLUSTRATIONS

# Preface

Jewish settlement in Devon and Cornwall, the two most south-westerly counties of Great Britain, began in the remote mists of Biblical and Roman times. There was an officially recognized Jewish community in medieval Exeter until 1290 which is well-documented. Then the pages of Jewish history in the South West are virtually blank for the next four centuries until individual Jews once again settled in sufficient numbers to coalesce and form independent Jewish Congregations, in Devon in the mid-eighteenth century and in Cornwall in the latter part of that century.

The Congregations were formed in the mould of contemporary English institutions, and thus were an early initiation for the ethno-centric immigrant on the road to acculturation and eventual assimilation. By and large, the indigenous population of the South West accepted and even welcomed the immigrants—certainly the English-born Jew merged almost indistinguishably into the wider social scene.

Two factors mainly determined the rise and decline of the South-West Congregations: economic and religious. On the economic front, Jews went to the South West when they could earn a better living there than elsewhere; in their heyday these four Congregations formed a significant proportion of Anglo-Jewry. When better opportunities arose elsewhere they emigrated to new pastures. At the same time religious considerations played their part. Jews who wished to abandon the observances of their faith, did so either formally by conversion to Christianity or, more usually, informally by simply living like their neighbours. In either case, they generally moved away from the four centres (Plymouth, Exeter, Penzance and Falmouth) with established Congregations and synagogues. Jews who wished to retain their ancestral faith and its observances stayed close to the synagogue and their fellow Jews. When a Congregation became attenuated through emigration for economic reasons, religious factors came into play. If the Congregation could no longer provide a proper Jewish education for children, a Jewish social life for the young unmarrieds, regular synagogal services and a full provision of *kasher* food, observant Jews together with their families moved to the larger Anglo-Jewish communities where they could find these facilities.

Jews who were prepared to compromise with their religious observances and stayed behind rapidly became fully assimilated, and they or their children disappeared from the Jewish community.

To examine this rise and decline of Jewish communities in the South West it will be necessary to investigate the areas of town settled by Jews, their numbers, demographic data, occupations, communal organisation, as well as their religious and cultural life. The slow process of assimilation will be examined in detail, describing the stagnation of the Exeter Congregations in the late nineteenth century and the steady decline of the Plymouth Congregation in the late twentieth century. Many provincial Anglo-Jewish Congregations in the nineteenth century and even more in the twentieth century have followed in the footsteps of the four South-West Congregations. The chapters which follow are a memorial to Anglo-Jewish life in the provinces which is fast vanishing from all but a handful of cities.

CHAPTER ONE

# The Early Settlement of Jews in Devon and Cornwall

## *Part 1*

### *Ancient traces of the Jews in Devon and Cornwall*

No decisive evidence has been adduced to show the presence of organized Jewish communities in England before AD 1070, but there is some varied evidence worthy of consideration indicative of the presence of Jews in Britain before this date, and especially in Devon and Cornwall. Writers on Anglo-Jewish history from the seventeenth to the nineteenth centuries have suggested that Jews first visited England in company with the Phoenicians about the time of King Solomon. This suggestion was based on the links between the Kingdoms of Judah and Tyre.[1] Ancient historians referred to Phoenician voyages to the Cassiterides, later identified as Britain, in search of tin and lead, and it was thought likely that Jews had accompanied them.[2] That there may have been some connection between the inhabitants of Devon and Cornwall and the dwellers on the Palestinian coastline is shown by food habits which they still hold in common. Both areas use saffron in cooking, particularly in the baking of cakes.[3] In these two regions as well as in Brittany, which was also under Celtic influence, clotted cream is manufactured.[4] A further indication of some degree of intercourse between the ancient Israelites and Celts is said to be the similarity in sound and meaning of words and phrases in the Hebrew and Celtic languages.[5] So much so, that in 1827 the British and Foreign Bible Society distributed Hebrew Bibles among the Cornish as being nearest the vernacular.[6]

The presence of smelting ovens in Cornwall and Devon which are called 'Jews' Houses' or 'Jew's Houses'[7] may point to early Jewish participation in the mining industry. The earliest method of smelting was a simple pit in which the ore was burnt and the metal subsequently collected from the ashes. This method was in use until the third or second century BC. An improvement on this primitive method involved the building of a furnace made of hard clay in the shape of an inverted cone about 3 feet across and about the same height. A blast of air from a bellows to the lower part of the furnace served to produce an intense heat, and the molten tin was discharged from a small opening at the bottom. This type of oven was in use from the second century BC until about AD 1350 and was called by eighteenth-century tinners 'a Jew's House'.[8]

The tin from a Jew's House was known as 'Jew's House tin',[9] and it is somewhat suggestive that a farm on which such a Jew's House was discovered in 1826 was locally known as 'Landjew'.[10]

Jews may have had at least one well established trading centre in Cornwall in the pre-Roman period, as the town Marazion was anciently known as Market-Jew, and the main street of Penzance which leads to it is even today called Market-Jew Street.[11] Nor is this the only town in Cornwall whose name is said to be Hebraic in its origin. There is also the village of Menheniot, which name, a correspondent to the *Jewish Chronicle* suggested, is derived from the two Hebrew words, *min oniyot*, which mean 'from ships'.[12] The current pronounciation of the name of the Cornish town of Mousehole as 'Muzzle' might also be influenced by Hebrew, as 'Muzzle' is the homonym of the Hebrew word meaning 'luck'. It might be objected that the apparent Hebrew origins of the names of these towns is due to mere coincidence. It is known, however, that in the nineteenth century the cryptic Hebrew expression *Makom Lamed* (= 'L(ondon) place') coined by local Jews when referring to London, passed into general Cornish usage.[13]

It is worth noting that much of the evidence which points to Jewish settlement or influence in Britain during the pre-Roman period, relates in the main to Devon and Cornwall.

Were there Jews in Roman Britain? This question has been considered by Dr Applebaum and largely on the basis of archaeological and literary evidence, he suggests a positive answer.[14] It is highly likely, he says, that there were at least a few Jewish soldiers in oriental units of the Roman army which served in Britain. It is also possible that there were some Jewish traders who were connected with the import of pottery, glass and oriental wares. They may even have formed small communities at Colchester, York, Corbridge and London. Moreover, there is a distinct likelihood that some Jewish slaves were brought to England after the Bar Kochba uprising in AD 135. Indeed, Carew supposed that Jews were sent as slaves by one of the Flavian Emperors to work the mines of Cornwall, but the only evidence for his view appears to be a coin of Domition found in an old mine gallery.[15]

The archaeological evidence relates to finds of coins and pottery. According to Dr Applebaum, Near-Eastern coins of the Roman period found in Dorset and Devon show an early connection between those areas. A close analysis of these coins indicates that Exeter was one of the first ports of call for sea-traffic coming from the Mediterranean up the Channel. Analysis of the coins also shows that they mainly originate from Antioch, Chalcis, Cyrrhus, Hierapolis, Edessa, Samosata, Zengma and Singara, all of them towns with a high percentage of Jews in their population.[16]

The particularly strong link between Exeter and the Near East makes it likely that there were some early Jewish associations with that city. Dr Applebaum suggests that archaeological finds in Exeter should be closely examined as some of them might indicate that Jews passed through or settled in Exeter in Roman times.[17] Following this suggestion the discovery in Exeter of a sherd of carinated bowl with an incised graffito of the second or third century AD was noted.[18] Lady Fox considers the inscription 'to represent an amalgamation of a trident with a conventional palm leaf, bordered by incised lines; there seems to be the loop of a cursive letter above the "trident"'. There is a possibility, and it cannot be put any higher than that, that the 'loop of a cursive letter' is the figure of a citron. If so, this would indicate Jewish associations with the bowl, as the palm and citron were used extensively on Jewish coins and burial caves.[19] The 'trident' could be the Hebrew letter *Shin*[20] and if this identification is correct it heightens the likelihood of the figures on the bowl being palm and citron and strengthens the assumption that there were Jewish connections with Exeter at this early period.

On the other hand it must be allowed that the references to Britain in *Midrashic* literature of the Mishnaic and Talmudic periods do not appear to relate to the South West.[21]

A persistent legend also refers to the presence of at least one Jew in England at the beginning of the Christian era. He was Joseph of Arimathaea, a wealthy Essene Jew who, it is said, out of sympathy with Jesus gave him burial in a rock tomb near Jerusalem.[22] According to legend he came to England as one of the Seventy Apostles to erect the first oratory; and out of the staff which stuck in the ground at Glastonbury as he stopped to rest himself, there grew a miraculous thorn, said still to blossom every Christmas Day.[23] A variant of the legend makes Joseph travel through Cornwall accompanied by Jesus.[24] This legend may be the folk memory of some ancient time when one or more notable Jews visited England.

In the Saxon period individual Jews may have visited England but the evidence formerly used to support the hypothesis that there were organized Jewish communities has been largely rebutted.[25] It has also been suggested that the prevalence of Biblical names in Cornwall such as Benjamin, David, Isaac, Joseph, Samuel, and Solomon, during the Saxon period indicates some intercourse between Jews and Cornwall,[26] the more so as these names were not used in other parts of the country, not even at Exeter, which is barely 40 miles from the Cornish border.[27] But there is always the possibility that these names came to Cornwall with the spread of Christianity to that county, or that they were a legacy from Jews who were there in the Roman or pre-Roman periods. The latter suggestion would account for a mid-fourth century Duke of Cornwall who became a Christian when already an adult, and yet was called Solomon even before his baptism.[28]

## *Part 2*

## *Jews in Medieval Devon and Cornwall*

There are no known authentic references to Jews in England during the reign of William the Conqueror, other than an incidental remark by William of Malmesbury, the medieval chronicler, that the Conqueror had brought Jews with him from Rouen. Of William Rufus it is related that he favoured the Jews of London when they were involved in a religious discussion with bishops and churchmen. It is not likely that there was any settled and relatively numerous Anglo-Jewish community until after the massacre of the Jews of Rouen by Crusading knights in 1096.[29]

Little more is known of the Jews in England until Henry I issued a charter of uncertain date giving protection to some individuals.[30] Among other privileges they were guaranteed freedom of movement throughout the country, relief from ordinary tolls, free recourse to royal justice, and permission to retain land taken in pledge as security. Aided by these privileges, English Jewry slowly gathered strength and, under Stephen's continuation of Henry's protective policy, Jewish communities developed at Norwich, Lincoln, Winchester, Cambridge, Thetford, Northampton, Bungay, Oxford, Gloucester, Bristol and York. There were also isolated Jewish families at Worcester and Leicester.[31] But beyond the bare knowledge that these communities existed and made greater or lesser contributions to various tallages, little else is known.[32] Jews may have journeyed into Devon and Cornwall but there is no evidence demonstrating the settled presence of Jews in these counties until the closing years of Henry II's reign.

Henry II not only confirmed but even extended the privileges which his father had granted to the Jews, and they flourished under his protection amid the general peace which prevailed in the realm. These favourable conditions led to an influx of Jews from Europe, an immigration which was reinforced by Jewish refugees expelled from the Île de France in 1182.[33] Its increased numbers enabled the Anglo-Jewish community to consolidate itself, and it now attained the zenith of its prosperity in medieval times.

Another factor made it easier for Jews to create new communities outside London. Hitherto, the only consecrated burial ground available to the Jews of England had been in London, and great inconvenience and even distress was experienced by relatives who had the unpleasant task of escorting a corpse along the roads of twelfth-century England. In 1177, each community was allowed to buy land outside its city walls for use as a cemetery. This new concession encouraged Jews to settle in towns remote from the capital. Thus by 1189, the end of Henry II's reign, besides the communities in the towns which have already been enumerated, there were further groups of Jews established at Stamford, Lynn, Bury, Bedford, Ipswich, Canterbury, Hereford, Dunstable, Chichester, Devizes and, in Devon, at Exeter (Map 1).[34]

So far as Devon was concerned, it was not only that it was easier for a Jewish community to maintain itself but, from the middle of the twelfth century, there was also a positive attraction for Jews to settle there. Tin mining was in operation in Devon in 1156[35] and it is possible that rich Jews from already established centres provided some of the capital for one of Britain's first capitalist industries,[36] at the same time sending their agents to safeguard their interests. If so, these agents probably formed the nucleus of the subsequent medieval Jewish community in Exeter. One may gauge the extent of the involvement of medieval English Jews with tin mining in Devon by the steep decline in the Devon output of tin from 87 thousand weight in 1291 to 38 thousand weight in 1296, a decline which has been attributed to the expulsion of the Jews in 1290.[37] Furthermore, the name of at least one mine owner, Abraham the Tinner, who owned a number of stream works in 1342 and employed several hundred men, suggests that he was of Jewish origin.[38]

The first mention of a resident Jew in Exeter was in 1181 when Piers Deulesalt[39] paid 10 marks[40] that the king might take care of his bonds,[41] and by 1188 there were enough Jews to form a distinct community. The earliest recorded act of the new community was to pay a fine of one gold mark[42] that its members might be allowed to set up a *Beth Din* to try 'pleas which were between them in common'.[43]

How did the members of this community earn their daily bread? Without doubt, their main means of livelihood was the interest they received from money which they advanced on the security of lands, rents and chattels. Hundreds of documents survive relating to such transactions and innumerable references in the published volumes of the Plea Rolls of the Exchequer of the Jews. Only a double standard of ethics on

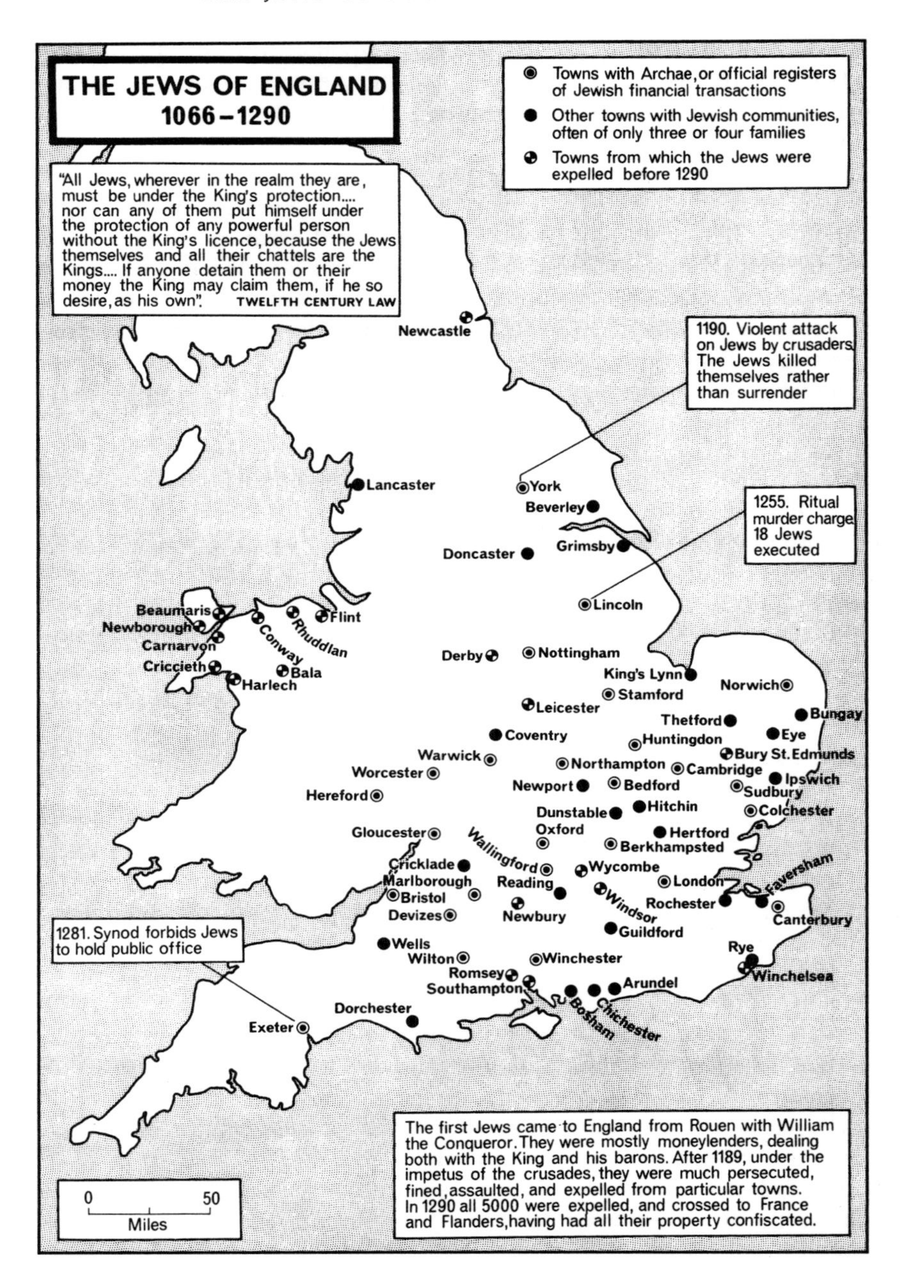


*Map 1*

The Jews of England, 1066–1290.

the part of both Jews and Christians made this possible. The Church regarded usury as reprehensible and forbade its adherents to take interest. In principle, Jews were similarly inhibited, but the Church generally turned a blind eye to their activities—their spiritual welfare being of no concern to the Church. So far as Jews were concerned they could not take interest from one another, but it was not generally regarded as blameworthy to take interest from a non-Jew.[44]

It has been strongly urged that Jews in medieval England, particularly those in the small towns and villages, must have engaged in other occupations and trades.[45] The proposition is logically acceptable though the evidence put forward to demonstrate it is sometimes a little forced. There is no reference in the Plea Rolls or any of the other published sources to suggest that the Jews of Exeter did anything other than advance money. Even in this, their most certain occupation, a degree of doubt remains. Did they act on their own behalf or were they acting as agents for the rich Jews of other towns? It is tantalizing to read of the rural activities of the Exeter financier, Jacob Copin,[46] who was assaulted whilst transacting business in Newton, and not to know what took him there, or on whose behalf he went.

A spontaneous popular outbreak against the Jews of England occurred in London in 1189 at the coronation of Richard I. This was followed by attacks on the Jews in almost all the towns in which they resided.[47] The riots caused a heavy loss to the Exchequer both by the impoverishment of the Jews who survived and the despoliation of those who were killed, part at least of whose property should have escheated to the Crown on their death. Provision was made in 1194 to safeguard the royal rights in the future. Six or seven cities[48] were designated as centres through which all business transactions had to be channelled, and in those cities records were kept of all Jewish possessions and credits. In each of the designated towns two Jews and two Christian clerks, called chirographers, were appointed to safeguard the royal interest under the supervision of a representative from the newly established central authority. Orders were given that all deeds and contracts (chirographs) were to be drawn up in duplicate and the counterparts deposited in a chest (*archa*) secured by three locks. Ultimately, chirograph-chests were set up in each of the principal Jewish centres in the country, some twenty-seven in all.[49] In the thirteenth century, they were established in the West Country at Bristol, Devizes, Exeter and Winchester.[50] As an *archa* was sometimes established for the sake of a single Jewish financier,[51] the presence of an *archa* in a town does not necessarily indicate a sizeable Jewish community. This may be illustrated by reference to the situation at Exeter where in 1276 there seem to have been only two Jews actively engaged in money-lending: Auntéra widow of Samuel son of Moses, and Isaac son of Moses, apparently her brother-in-law. In 1290, there was hardly a shadow of the formerly prosperous community of Exeter, the sole representative being a Jewess named Comitissa. Besides her house in the High Street, no other house was either owned or leased by Jews in Exeter, nor was there a synagogue. Yet, though the Jews were so few in number, at the Expulsion there were actually two *archae*, an old one for bonds executed up to 1275 and a new *archa* which had been opened in 1283.[52]

The chirographers, both Christian and Jewish, were paid for their responsibilities, receiving one penny between them each time a chirograph or other instrument was removed from the *archa*.[53] They had to deposit a pledge against the proper performance of their duties.[54] Their office left them open to charges of incompetence,[55] malpractice, for which they were collectively fined,[56] and fraud.[57] The chirographers, Christian and Jewish, were put to the trouble of travelling to London to give evidence in a case where the amount involved was forty shillings.[58] There is no evidence, however, to suggest that the office was either courted or shunned. Occasionally it passed from father to son.[59] The chirographers of the Exeter *archa*, so far as they have been identified, are listed in Table 1.

Eventually the administrative system of the *archae* and chirographers became an effective instrument to raise money for the crown, but before then England had to pay a ransom of £100,000 to the Emperor Henry VI for Richard the Lionheart. The contribution of English Jewry to this sum was assessed at 5,000 marks, a disproportionate amount, being three times as much as was demanded from the burghers of London, by far the richest city in the kingdom. Representatives of every Anglo-Jewish community were summoned to a 'Parliament' at Northampton on 30 March 1194, to decide the amount each community should pay towards the required sum. Payments to the *Northampton Donum*, as the tallage is called, reveal the presence of some twenty major Jewish communities as well as a number of minor ones scattered throughout the country. The most important centres were London, Lincoln, Canterbury, Northampton, and Gloucester, each with from twenty to forty contributors.[60] The Jews of Exeter at this time were neither numerous nor affluent as only one man, Amiot,[61] contributed to the *Donum*, and his mite, £1. 3*s*. 3*d*., was less than 2 out of the 5,000 marks demanded.[62]

The seventeen years during which John was King of England, 1199–1216, marked a turning point in the history of England as well as in the affairs of its Jewry. Hitherto, England had been closely connected with northern France and from thence had come the main body of the Jewish settlers in England. From its inception the Anglo-Jewish community had been French in character, and it maintained its links with France until the expulsion from England in 1290. The Jews of England spoke French,[63]

*Table 1: The Chirographers of Exeter, 1224–1290*

| *Jewish chirographers* | *First and last known dates of office* | *Christian chirographers* | *First and last known dates of office* |
|---|---|---|---|
| Moses le Turk | 1224–33[a] | Lawrence Cissore and | 1244 |
| Ursell, son-in-law of Amiot | 1224[b] | Henry Picot | 1224–3[i] |
| Hak (Isaac) son of Deudoné | before 1244[c] | Richard Bollock | 1266–77[j] |
| Josce Crespin and | 1224–66 | Thomas de Langedon | 1266[k] |
| Bonenfant son of Leo | 1244[d] | David Taylor | 1277[l] |
| Lumbard Episcopus | 1260–66[e] | | |
| Leo of Burg' | before 1266[f] | | |
| Jacob Copin | 1266–80[g] | | |
| Jacob Crespin | 1275–90[h] | | |

(a) Adler, 'Medieval Jews', pp. 227, 239.
(b) Ibid. p. 239.
(c) Ibid. p. 229.
(d) Ibid. p. 228. Josce Crespin and Bonenfant acted together. Josce of Exeter, chirographer in 1266, was probably Josce Crespin (Rigg, *Plea Rolls*, I, 132).
(e) Adler, 'Medieval Jews', p. 229 and Rigg, *Plea Rolls*, I, 135.
(f) Rigg, *Plea Rolls*, I, 135. Adler, 'Medieval Jews', p. 231, identifies him with Lumbard Episcopus.
(g) Adler, 'Medieval Jews', pp. 231, 232.
(h) Ibid. p. 232. Another Exeter Jew, Deulegard, became chirographer of Winchester in 1253 (ibid. p. 229).
(i) Ibid. p. 239. They acted together.
(j) Ibid. p. 233.
(k) Rigg, *Plea Rolls*, I, 132.
(l) Adler, 'Medieval Jews', p. 233.

bore French names, and regarded France as their haven of refuge in time of trouble. The Jews of England looked to the French rabbis for religious guidance.[64] English Jews shared with English nobles and the higher and more learned of the English clergy a common language and secular culture.[65] Their 'Frenchness' was characteristic not only of the Jews of London but also of those living in the provinces. Exeter Jews, even as late as the Expulsion, bore French names, such as Amiot, Bonenfant, Bonefey, Deudoné, Deulecresse, and Deulegard; whilst the women were graced with names such as Amité, Nona, Chère, Ivoté, and Juetta.[66]

A further indication of the close political ties and the similarity of administrative functions in the two countries are the Charters of Privilege granted by successive sovereigns from Henry I onward which were issued to the Jews of Normandy as well as of England. Once Normandy was lost, England again became, politically as well as geographically, an island. The Jews of England, too, were cut off from the great Continental centres of Jewry. The influx from abroad which had formerly provided the numbers necessary for the expansion and consolidation of provincial communities was checked. Had John decided to leave Normandy to the Normans the process of isolation may have done no harm and might even have strengthened the Jewish community by making it culturally and intellectually self-supporting. Unfortunately for England in general and its Jewry in particular, John's endeavours to recover Normandy led him to impose crippling taxation. It was this heavy burden of taxation which fell principally upon the Jews that hastened the decline of medieval Anglo-Jewry.[67]

The first attempt to satisfy the King's rapacity was made in Bristol in 1210.[68] John arrived there after a campaign in Ireland, and he ordered all the wealthier Jews of England to appear before him for a scrutiny of their resources. Great cruelties were perpetrated on the assembled Jews to force them to reveal their wealth—the story of the Jew of Bristol whose teeth were extracted one by one became proverbial. As a result of the investigation, they were tallaged for 66,000 marks and kept in prison until the sum was paid or satisfactory guarantees given.[69] Eleven years later there were still debts outstanding from this tallage. The Exeter Jews who had not satisfied their obligation were Samuel of Wilton who had died, his widow, Iveta, Deodatus son of Amiot, Jacob of Gloucester, Samuel Episcopus,[70] and Sampson *cum ore* (with the mouth).[71] Clearly the Exeter community had grown in number and in financial importance since the *Northampton Donum* of 1194 to which only one individual contributed. The growth of Exeter Jewry at this period is indicated not only by its contribution to the Bristol Tallage but also by a number of references in the *Fine and Oblate Rolls, 1204–5*, to a further six members of the community. They were Samuel of Exeter and his wife Juetta,[72] Deulecresse le Evesque,[73] and Jacob son of Yveliny together with his brother Deulecres and his sister Sarra.[74] Possibly the provincial communities had been reinforced by Jews from London who were anxious to escape the close scrutiny of King John's officials, for his exactions pressed close one upon the other until the once wealthy Jews of London were so reduced that 'they prowled about the city like dogs'.[75]

The outbreak of civil war rendered the position of Jews throughout England insecure. The barons saw the Jews not only as creditors but also as royal agents, for the Jews were used by the King 'like a sponge, sucking up the floating capital of the country, to be squeezed from time to time into the Treasury'.[76] To mitigate the disposessing effect of this method of raising money for the King, the barons caused the tenth and eleventh clauses to be written into Magna Carta. These stipulated that debts due to Jews or other usurers should bear no interest during the minority of the heirs of a deceased debtor, and that if the debts fell into the hands of the King then only the capital need be paid. Similarly, a widow's dowry and the support of her minor children were to be the first charge on the estate and all debts to be paid out of the residue. These clauses bear witness to the animosity of the Jew's everyday clients, and it has been said that only John's death in 1216 saved the Jews from further despoliation by him and from attacks by the populace.[77]

When the infant Henry III came to the throne, England was torn by the strife engendered by the conflict of John and the barons. But the Jews suffered not only from the troubled political scene but also from the deteriorating religious climate. In 1218, Stephen Langton, Archbishop of Canterbury, gave effect to the discriminatory decrees of the fourth Lateran Council.[78] Popular feeling was aroused against the Jews, and in Exeter, at least, this must have run high against them, and they needed special protection. This may be deduced from a writ addressed to the Sheriff of Devon in 1218 couched in the following terms:[79]

> We command you that you have our Jews of Exeter in ward and countenance, neither doing, nor suffering to be done, to them, any mischief or molestation, and that, if any offend against them in any wise, you cause reparation to be made them without delay. We command you, likewise, that you neither lay, nor suffer be laid, hand . . . on their chattels, and that if any Jew offend in aught for which he deserve to be put by gage and pledge, you attach him by view of Deulecresse Episcopus, our bailiff in those parts, to be before our Justices assigned to the custody of the Jews at Westminster at a convenient term to answer thereof, and in the attachments do the said Justices to wit of the offence of the said Jew and the term which you have appointed him, and that you also have a care that, if any Jew or Jewess fall into our mercy, you may not by the Assize of our realm exact from such Jew or Jewess more than 20*d*. only. We therefore command you that on account of any amercement that concerns the Sheriff or Constable you exact not more than 20*d*. only. You are also, as you receive mandate from Us, to distrain their debtors within your bailiwick to pay their debts, that the debts which they owe Us may not remain unpaid by your default.

The measures taken by the Sheriff in pursuance of the writ presumably took effect, for there is no record of any violence or outbreaks against the Jews at this period.

During the royal minority, successive regents set about the task of rebuilding the shattered Jewish community. Apart from protection of the person secured by writs similar to that just quoted, financial relief was also granted. Tallages, at least until 1227, were light. The first of these was in 1221, when seventeen Jewish centres contributed £564 towards a dowry for Princess Joan, sister of Henry III.[80] To this dowry Exeter Jewry contributed £8 5*s*. 8*d*. The Exeter Jews who paid most towards this sum were Jacob of Gloucester,[81] £3 11*s*. 8*d*., and Deulecresse le Eveske, £2 10*s*. 0*d*., followed by Ursell who gave 18*s*., Ursell, son-in-law of Amiot, 15*s*., Moses le Turk, 6*s*., and Moses of Exeter, 5*s*.[82] The Jewish community of Exeter throve under the rule of the regents, for only two years later, in 1223, fifteen Exonian Jews and Jewesses were able to pay £78 10*s*. 6*d*. to a royal tax which brought in about £1,680 from the Jews of the whole country. Three of the Exeter contributors to this tallage were women.[83] Bona daughter of Abraham gave £1 10*s*. 0*d*., and Chère together with Hanot between them paid £1 11*s*. 8*d*. The men mostly gave larger amounts. Jacob of Gloucester headed the list with a payment of £17 and 14 marks, Deulecresse comes next with £13 18*s*. 10*d*., Moses le Turk gave £9 6*s*. 8*d*., Bonefey son of Isaac £4, Ursell £4, Ursell, son-in-law of Amiot, £3 19*s*. 8*d*., Sampson £1 7*s*. 4*d*., and Moses son of Solomon 13*s*. 4*d*. New Exeter residents mentioned for the first time are Solomon of Dorchester who together with his son-in-law Deulecresse subscribed £2 10*s*. 0*d*., Jacob of Norwich who gave 5 marks and Lumbard son of Deulecresse Episcopus who gave 13*s*.[84] The last tallage imposed on English Jewry during Henry's minority was one of 4,000 marks, but, unless the records are incomplete, Exeter Jewry did not contribute towards it.[85]

The first quarter of the thirteenth century marks the zenith of Anglo-Jewry's, as well as Exeter Jewry's, prosperity in the medieval period. Under the personal rule of Henry III, however, the Anglo-Jewish community was taxed so heavily and in such arbitrary fashion that it began to diminish in numbers and financial importance. In the first decade of Henry's personal rule onerous tallages were imposed in 1229, 1230, 1231, 1233, 1234, 1236, and 1237 of 6,000, 9,000, 15,000, 35,000, 750, 10,000 and 3,000 marks respectively.[86] The 3,000 marks in 1237 were a gift to Richard of Cornwall, for his intended crusade. Detailed accounts of the collection of these imposts have not survived and it is therefore not possible to assess the relative importance of Exeter Jewry's contribution to the national total, but presumably they had to pay their share.[87] Nor was there any relief from the high level of taxation in the succeeding years of the reign. In February 1241, 109 Jews from the twenty-one communities of the realm then recognized were summoned to Worcester to a so-called 'Parliament of Jews'.[88] They were the delegates appointed to consider ways and means of raising a new tallage of 20,000 marks, and they themselves were to act as assessors and collectors to aid the Sheriffs in collecting the tax from their fellow Jews. The larger Jewries such as London, York, Cambridge and others sent six delegates, whilst Exeter Jewry, never very numerous, sent only four. They were Jacob of Gloucester, Deulecresse Episcopus, Bonenfant son of Judah, and Josce Crespin son of Abraham.[89] When the Exeter Jews did not pay their share of this Worcester tallage it was Joseph Crespin, by then one of the Jewish chirographers of Exeter, who became responsible for the outstanding balance of £31 1*s*. 4*d*.[90]

The royal exactions continued unabated until the Jews of England had been stripped of all their immediately realizable possessions. But even when the Jews themselves were no longer able to provide the money which Henry always needed they still represented a marketable asset. Accordingly, in 1251, the King mortgaged them for two years as security for a loan of 5,000 marks to Richard, Earl of Cornwall, who 'was thus permitted to disembowel those whom the King had flayed'.[91] At the end of this period Exeter Jewry paid £3 15*s*. 0*d*. towards a total collection of £320 and the following year, in 1254, the communal fund of the Jewry in Exeter gave £2 towards a tallage of 1,000 marks. Besides the small communal donation two individuals were also assessed for this tallage, one an unnamed local debtor of Aaron son of Abraham of London for £10, and Bonenfant the chirographer for 12 marks.[92]

Between 1254 and 1260 the Jews of England were tallaged for 32,000 marks. Most of this large sum was paid by rich individuals who were often eventually reduced to penury.[93] Such a one was the formerly wealthy Exeter Jew, Jacob of Gloucester. In 1221, and again in 1223, he had been the largest Exeter contributor to tallages imposed on Jews in those years. And yet, in the closing years of his life[94] he was only able to give 13*s*. to the tallage demanded in 1260. Nor were his fellow Jews in Exeter by then in much better case. Lumbard Episcopus[95] paid 6*s*. 8*d*., Bonenfant four marks, his brother Samuel two and a half marks and Joseph son of Moses gave 4*s*., the total amounting to £8 6*s*. 4*d*. This modest sum seems to represent the total realizable wealth of Exeter's Jews as they were unable to pay anything at all to a further tax in that same year.[96]

When the Jews were squeezed by the King, they in turn had to press their Gentile debtors for prompt repayment. In the circumstances there were many disputes which were brought to court. In the Exeter courts, at least, the Jew was fairly treated. An interesting record of the procedure adopted in Exeter has survived. It is contained in a description of local customs and business regulations current in Exeter just after the middle of the thirteenth century. The description was published under the title of *An Anglo-Norman Custumal of Exeter* by Professor J.W.

Schopp and Miss R.C. Easterling.[97] The reference to the Jews reads as follows:

> A plea of a Christian against a Jew could not be heard without Jew and Christian nor the plea of a Jew against a Christian without a Christian and a Jew, and the Jew must give wage and pledge that he will pursue his plea and the Christian likewise that he will stand to the law, and if he cannot find a pledge, he, the Christian, must be plevied by affiance [i.e. he must be bailed on his solemn promise].[98]

The procedure outlined in the custumal was used in actual practice.[99]

The Jews formed one of the objects of dispute in the constitutional struggle between Henry III and his barons. The lesser baronage, in particular, was affected by financial dealings with the Jews. One of the complaints specifically aired at the Parliament of Oxford in 1258 was that the Jews sold lands taken in pledge by them to the great magnates, who then refused to accept repayment of the outstanding debt so that they could retain possession of the lands. In the revolt of the barons led by Simon de Montfort from 1262 to 1267, much havoc was wrought to the Jewries of London, Canterbury, Bristol and other cities and in each place the *archa* was either burned or carried off.[100] The revolt did not spread to Devonshire and the small Jewry of Exeter was left unharmed by the general disorders. Indeed, it was regarded as a place of safety as six deeds were sent from London for safe deposit in the Exeter *archa* when the seal of the Exchequer of the Jews was stolen during the uprising in London.[101] It is therefore not surprising to read that the Exeter chirographers, both Christian and Jewish, were reluctant to leave the security of Devon and go to London in 1266.[102] Their fears were justified as a Jew, Jacob Baszyn, formerly resident in Exeter, was murdered in Oxford in 1286, probably in the disturbances of the 'Disinherited Knights' following de Montfort's fall.[103]

The last years of Henry's long reign marked an effort to strengthen the position of the Jews. Those who had emigrated were encouraged to return. Jews were permitted to claim their debts, notwithstanding the destruction of the *archae*, if reasonable proof of them could be produced. Citizens were nominated to protect the Jews in towns where they had been particularly hard used.[104] It was, perhaps, these improved conditions which encouraged Jacob Copin, Jewish chirographer in Exeter, to seek redress when he was assaulted. The incident occurred in 1270 whilst he was transacting business in the vill of Newton.[105] He was assaulted by one Robert of Buleshill, Christiana his wife and William le Layte. When Copin brought an action against them they absconded, and the Sheriff was ordered to arrest them and bring them to justice.[106] The provisions to strengthen the Jewries of the realm, successfully enabled the Jews to recuperate. Whereas in 1267 the Jews of Exeter had been totally denuded of their ready cash, by 1272 Jacob Copin was able to pay £20 to Henry's last tallage. This comparatively large amount does not necessarily indicate that Copin himself had become a wealthy man, but rather that he had collected it from his fellow Jews in Exeter and paid it on their behalf.[107]

When Edward I came to the throne of England (1272–1307) he showed himself to be no less extravagant than his predecessors and he sought every opportunity to raise revenue. A memoranda roll of Edward I's reign records that a fortnight after Easter, 1272, the Sheriff of Devon was to render an account of his tallage. Apparently, he promptly did as he was ordered as there is no record of any failure to do so.[108] In 1277, on the other hand, John Wygger, the Sheriff of Devon, was fined for the late return of writs concerning Jews, though he was later pardoned because he was dead![109] Late in 1273, a tallage was levied on the Jews assessed at one-third of all their movable goods, that is, their bonds and valuables. Once again the effect was to throw the burden upon Christian debtors who were pressed to repay so that the Jews might be quit of their own obligations. Again Christian debtors protested against the effects of Jewish usury but this time their protest was reinforced with the authority of the Church. In 1274, the Council of Lyons, acting under the spur of Pope Gregory X, urged the Christian world to greater efforts against the sin of usury. Encouraged by the Church and Laity the *Statutum de Judeismo* was issued at Worcester in 1275 which absolutely forbade Jews to lend money at interest. Outstanding transactions had to be wound up as soon as possible. The right of distraint on land was severely curtailed and the alienation of real estate by Jews was forbidden without special permission.[110]

The crippling provisions of the *Statutum,* if carried out to the letter, would have left the Jews with hardly any means of livelihood. It is true that some, particularly in the provincial centres, traded in corn and wool, as appears from a summary of the bonds for money, corn and wool held by Jews throughout England at the Expulsion.

The round figures in which the amounts of corn are expressed and the absence of any qualitative differentiation suggests that dealings in corn 'conceal clandestine moneylending operations'.[111] But the specification of quality[112] and details of time and place of delivery[113] as well as external evidence[114] seems to indicate that the transactions were genuine. Exeter's main import at this time was wine[115] which the Jews also imported[116] so the Exeter Jews may have participated in the wine trade. But in the main, the Jews of Wiltshire, Exeter and Bristol left bonds which related to monetary transactions,[117] and must have found themselves hard pressed financially. Although the Jews of England were able to eke out only a bare existence after the passing of the *Statutum* in 1275, they were nevertheless still expected to make large contributions to the royal purse. A tallage was imposed in 1275 which most were unable to pay. Those who were still in this unfortunate position by the following year were imprisoned, their chattels confiscated, and their wives and children deported. Debts owing to Jews became part and parcel of the commercial life of the wider community. Thus a Plea Roll of 1277 records[118] that a further tallage of 3,000 marks was imposed in 1277–8 notwithstanding the difficulty experienced in collecting the earlier tallage.[119]

With their old sources of livelihood cut off, the poorer Jews sought other means, often illegal, to keep themselves from starvation. Some sought refuge in apostasy, the numbers in the *Domus Conversorum*[120] rising to nearly one hundred. Others are said to have taken to highway robbery. Most carried on their old profession, making use of the devices invented by Christian usurers to evade canon law.[121] Yet another way of augmenting their income, coin-clipping, was practised by some Jews. They filed the edges of the coins in which they dealt and melted the 'clippings' into bullion. The evil was no new one; early in John's reign an Assize of Money had taken place to enquire into the debasement of the coinage. Not only the Jews but also the Cahorsins, the Flemish wool traders, and indeed everyone who handled coins in large numbers was suspected, and with some reason, of making illicit profits from coin-clipping.[122] In December 1276, it was found that the coinage was in a bad shape, so justices were appointed to try accusations of coin-clipping made against both Jews and Christians.

At this stage Christians were suspected as the principal offenders; it was only later, almost perhaps by accident, it was realized that in these trials lay a golden opportunity to extract further revenue from the Jews. In November 1278, all the Jews of the kingdom were arrested and a house to house search made. Six hundred and eighty men and women against whom evidence was found or could be manufactured were sent for trial to London, and 293 of them were hanged.[123] The King profited considerably from the Jews' troubles. So much property fell into his hands as the result of the forfeitures and escheats consequent upon the executions that a special department was set up to deal with it. The Exeter Jewry did not pass unscathed through these perilous times. Eleven Exonians including one non-Jew were arrested. They were Blakeman son of Jacob Copin, Aaron of Caerleon, Deulecresse le Chapleyn, Leo and Copin sons of Lumbard, Solomon the son of Solomon, Benedict of Wilton, Ursell, Isaac Ericun, Aaron of Dorchester, Jorin son of Isaac, and a non-Jew, James de Fenys. The defendants were allowed bail, but there is no record of what finally happened.[124] Jacob Copin, father of the accused Blakeman was hanged about this time,[125] probably on a charge of tampering with the coinage.[126] Ursell may have followed his or a namesake's earlier example and fled the country.[127]

The *Statutum de Judeismo* was unworkable in the context of medieval English society. The severity of its provisions was soon ameliorated and a modified form of money-lending was authorized. However, the local Jewry never really recovered from the setback it received after the Act was passed and from the troubles that followed the charges of coin-clipping. The Jews who were left in Exeter resumed their former occupation but they lent only small sums.[128]

The hostility of the clergy towards the Jews was not peculiar to England. In 1096, there were popular outbreaks stimulated by centuries of clerical animosity against the Jews in all the cities of northern and central Europe.[129] The legislation covering Jews which was passed by various synods in England reflected, and was sometimes prompted by, the pronouncements of successive popes and European Church Councils.[130] The Fourth Lateran Council of 1215, for example, was concerned at the rapid spread of heresy in Europe, and decided that too free intercourse with Jews was largely responsible for it.[131] The Council passed certain measures, designed to check Jewish influence among Christians, which were introduced into England in 1218,[132] and were renewed and reinforced at synods at Worcester in 1240, at Chichester some six years later, at Salisbury in about 1256, and at Exeter in 1287.[133] The anti-Jewish code thus promulgated contained a number of humiliating and degrading conditions. All Jews had to wear a distinguishing badge—'ostensibly to prevent the scandal of unwitting sexual intimacy between unbelievers and the faithful'.[134] Jews were forbidden to employ Christian servants, to enter churches or keep their property in them. Jews were not allowed to build new synagogues. They had to acknowledge the local priest as their overlord by paying church tithes based not only on their real estate but even on the usurious profits which they were supposed not to make.[135] Life was only made bearable by royal influence and also by Time—the great healer. The renewal of the regulations at successive synods was necessary because their full force became blunted, for with the passage of time many of the regulations lapsed into desuetude.[136]

One result of religious fanaticism on the one hand and ignorant credulity on the other was the invention in England of the infamous Ritual Murder accusation. This was first made on Easter Eve 1144, when the dead body of William, a young apprentice, was found in a wood near Norwich. Rumours spread that he was a victim of the Jews who had crucified him on their Passover, but no Jew was tried or punished for the alleged crime.[137] Further accusations of similar crimes followed at Gloucester in 1168, Bury St Edmunds in 1181, and Bristol in 1183. No trials were held after these latter charges either, but popular rumour was sufficient to establish the martyrdom of the children involved.[138] These child martyrs attracted considerable numbers of pilgrims, no doubt with economic benefit, to the cathedrals and abbeys where their relics were housed.

The Church was not the sole beneficiary of the new hagiolatry; the King, too, required compensation for the supposed loss of one of his subjects. In 1239, an alleged murder in London was punished by the confiscation of one-third of the ten richest Jews' property.[139] Further charges accompanied by heavy fines for the Jewish community at large and fatal consequences for the individuals concerned were made in London in 1244, Lincoln[140] in 1255, and Northampton in 1277.[141]

Converts both to and from the dominant faith frequently caused trouble. At Lynn in Norfolk in 1190, for example, some Jews followed a former coreligionist into a church where he had taken refuge to escape their insults. An uproar followed which soon turned into a riot.[142] Exeter Jews were probably involved when many of the leading members of the Oxford Jewry were imprisoned on a charge of rescuing a boy who had become a Christian, the lost infant being traced to Exeter in 1236.[143] But the greatest damage was caused by conversions to Judaism. In 1222, a deacon was burnt at Oxford for becoming a Jewish proselyte. After his conversion, the Archbishop of Canterbury, zealously supported by the Bishops of Norwich and Lincoln, threatened with excommunication even those who sold to the Jews the basic necessities of life. Fortunately for the Jews this last enactment was countermanded by the King, else they would have starved.[144] Another proselyte is said to have been one of the direct causes of the decision to expel the Jews form England. He was Robert of Reading, a Dominican friar who some years before the Expulsion embraced Judaism, adopted the name of Haggai and married a Jewess.[145] But even converts from Judaism to Christianity proved to be a source of vexation to the Church. For whereas Christians apostatized at the risk of their lives and demonstrated depth of character and courage in so doing, most of the Jews who left their community had much to gain. The wealthy Jew was deterred from accepting Christianity as, at his baptism, he had to sacrifice the greater part of his wealth to the King, who was deprived henceforth from deriving a profit from 'his' Jew. So it was those who had little or nothing to lose materially who joined the Church, and these converts often proved to be little loss to the faith they left and little gain to the one they joined.[146] But the Christian clergy was not only irritated by the qualitative but also by the quantitative gain,[147] which remained comparatively small.[148] Six known converts came from Exeter, four men and two women, though there may have been more.[149] The men were the brothers Samuel and Solomon, sons of Leo,[150] who were converted before 1266 and 1270 respectively,[151] Henry of Exeter, a clerk, and Richard the elder of the same city, a tailor.[152] One of the women was called Alice of Exeter[153] and the other was Claricia,[154] daughter of Jacob Copin who was hanged about 1280.[155] Claricia entered the *Domus* in 1280, probably at a tender age because she died there seventy-six years later, by which time she had become the sole resident.[156] These six Exeter Jews who converted to Christianity represent about 6.7 per cent of Jews who are known to have resided in Exeter from 1180 until 1290.[157] If this proportion was typical of Anglo-Jewry as a whole, then the frustration of Christian clerics 'to whom the unconverted Jew was a standing reproach' is, perhaps, understandable.[158]

In 1286, Pope Honorius addressed a letter to the Archbishops of Canterbury and York in which he called upon them to reaffirm the decisions of the Lateran Councils. This letter had a baleful influence on the Synod of Exeter a year later in 1287 under Peter Quivil,[159] when 'all the ancient canonical strictures against the Jews were reinforced with a severity seldom paralleled'.[160] Special stress was laid on the following enactments:

> No Christian should take medicine from a Jewish doctor;
> Jews had to pay taxes to parish clergy and wear a distinguishing badge when out on the streets;
> They were forbidden to appear on the streets or even to have their windows open at Easter;
> Jews and Christians were not to visit each other or join in any festivities;[161]
> They were not to enter churches, or to build new synagogues.[162]

These enactments must have been strictly enforced in the city of their promulgation, and for a time, at least, a certain degree of unpleasantness must have ensued for the Jews of the city. This Synod was probably the cause of the rapid dwindling in the number of Exeter Jews shortly before the Expulsion. For whereas there seem to have been at least forty Jews actively engaged in money-lending in the years immediately before 1290,[163] only one, a Jewess called Comitissa, had a house there when the final blow fell.[164]

## *The Expulsion, 1290*

A combination of crippling taxation and repressive legislation had brought the fortunes of Anglo-Jewry low. As providers of capital they had been superseded by Christian moneylenders.[165] They did not fit into feudal society nor were the leaders of that society sufficiently advanced in their political thinking to accept on equal terms people of different religious faith.[166] On this count alone, Edward could justify his decision to expel a group which refused to be assimilated. But there was an even more attractive advantage to be gained by a general expulsion of the Jews—a financial one. This had already been demonstrated when Edward expelled the Jews from his Gascon dominions in 1289 and confiscated their property.[167] The confiscated property proved a welcome addition to his depleted funds, but it did not amount to a very great deal, and Edward's needs were both great and pressing, as he had entered into heavy commitments to ransom his cousin, Charles of Salerno.[168] When he turned to his English Jewry, he discovered that it produced very little in terms of day-to-day revenue.[169] Its assets in bonds and immovables, however, were substantial. Lists of all the bonds and obligations on account of money and produce owing at the time of the Expulsion to the Jews in eleven of the seventeen towns in which they resided are extant. The total value on record is about £9,100, the Exeter Jewry accounting for some £1,238 of this amount. There were 39 bondholders in the Exeter *archa* and five of them had bonds for more than two-thirds of Exeter's total (see Table 2).

All these assets were Edward's for the taking if he repeated his Gascony expedient. This he determined to do, and on 18 July 1290, writs were issued to the sheriffs of various English counties, informing them that all Jews were ordered, on pain of death, to leave the realm before 1 November 1290.[170] The edict was carried out with humanity, and the few who ill-treated the Jews as they left the country were punished. Provincial Jews made their way to London and from thence they sought refuge with their brethren overseas. The medieval Anglo-Jewry was no more.[171]

*Table 2: Bonds in the Exeter* archa, *1290*

| | *Money bonds* | | | *Corn bonds* | | |
|---|---|---|---|---|---|---|
| | £ | *s.* | *d.* | £ | *s.* | *d.* |
| 1. Deulecress, Chaplain | 75 | 16 | 4 | | | |
| 2. Lumbard, son of Deulecress | 1 | 0 | 0 | | | |
| 3. Lumbard, son of Solomon | 14 | 0 | 0 | | | |
| 4. Tercia, widow of Lumbard | 16 | 10 | 0 | | | |
| 5. Copin, son of Lumbard | 6 | 6 | 8 | | | |
| 6. Symme, son of Lumbard, and Joce fil Isaac | | | | 20 | 0 | 0 = 60 qrs. |

*Table 2 (continued)*

| | Money bonds | | | Corn bonds | | |
|---|---|---|---|---|---|---|
| | £ | s. | d. | £ | s. | d. |
| 7. Samuel, son of Moses | 7 | 13 | 4 | | | |
| 8. Amité, widow of Samuel son of Moses | 198 | 0 | 0 | 20 | 0 | 0 = 60 qrs. |
| 9. Dyaye, son of Samuel, son of Moses | 4 | 0 | 0 | | | |
| 10. Jacob, son of Samuel | 10 | 0 | 0 | | | |
| 11. Moses, son of Samuel | 10 | 0 | 0 | | | |
| 12. Moses, son of Josce | 5 | 0 | 0 | | | |
| 13. Isaac, son of Moses | 89 | 13 | 4 | | | |
| 14. Cok, son of Moses | 0 | 15 | 4 | 7 | 6 | 8 = 22 qrs. |
| 15. Cok | 1 | 0 | 8 | | | |
| 16. Aaron, son of Josce | 7 | 10 | 0 | | | |
| 17. Isaac, son of Josce | 1 | 2 | 10 | 6 | 13 | 4 = 20 qrs. |
| 18. Leonin, son of Josce | 0 | 7 | 0 | | | |
| 19. Aaron of Caerleon | 27 | 6 | 0 | | | |
| 20. Salomon, son of Aaron | 20 | 0 | 0 | | | |
| 21. Isaac, son of Solomon | 20 | 0 | 0 | | | |
| 22. Salaman, son of Salaman | 26 | 0 | 0 | | | |
| 23. Isaac | 1 | 3 | 6 | 6 | 13 | 4 =20 qrs. |
| 24. Isaac de Campeden | 4 | 13 | 0 | | | |
| 25. Abraham, son of Isaac | 1 | 12 | 6 | | | |
| 26. Abraham | 4 | 0 | 0 | 63 | 6 | 8 = 190 qrs. |
| 27. Abraham, son of Miles | 4 | 4 | 0 | | | |
| 28. Jacob, son of Peter | 0 | 6 | 8 | | | |
| 29. Jacob, son of Perez | | | | 6 | 13 | 4 = 20 qrs. |
| 30. Jacob Crispin | 114 | 13 | 4 | | | |
| 31. Abraham and Cok | 2 | 13 | 0 | 6 | 13 | 4 = 20 qrs. |
| 32. Amitecote and Abraham | 0 | 8 | 0 | | | |
| 33. 'A certain Jew' | 13 | 16 | 8 | | | |
| 34. Duntere | | | | 10 | 0 | 0 = 30 qrs. |
| 35. Antere and Leo | 3 | 0 | 0 | | | |
| 36. Ivoté, daughter of Benedict Bateman | 2 | 0 | 0 | | | |
| 37. Jacob Copin | 357 | 10 | 0 | | | |
| 38. Ursellus, son of Manser | 6 | 0 | 0 | | | |
| 39. Comitissa | 0 | 2 | 0 | 33 | 6 | 8 = 100 qrs. |
| Total | 1,058 | 4 | 2 | 180 | 13 | 4 |

(Source: *TJHSE*, II (1894), 91.)

Was there any organized community of Jews in Cornwall in the medieval period? There is no evidence to suggest that there was. Individual Jews certainly settled there[172] and may have participated in the tin trade as may be inferred from the *Liber Rubeus* of the Treasury from the *Capitula de Stannatoribus*, 9 Ric. I:

> Also neither man nor woman, Christian nor Jew, shall presume to buy or sell any tin of the first smelting, nor to give or remove any of the first smelting from the Stannary or out of the place appointed for weighing and stamping, until it shall be weighed and stamped in the presence of the keepers and clerks of the weight and stamp of the farm.
>
> Also neither man nor woman, Christian nor Jew, shall presume, in the Stannaries nor out of the Stannaries, to have in his or her possession any tin of the first smelting beyond a fortnight unless it be weighed and stamped.
>
> Also neither man nor woman, Christian nor Jew, in market towns and borough on sea or on land, shall presume to keep beyond thirteen weeks tin of the first smelting weighed and stamped, unless it be into the second smelting and the mark discharged.
>
> Also neither man nor woman, Christian nor Jew, shall presume in any manner to remove tin either by sea or by land, out of the counties of Devon and Cornwall unless he or she have the licence of the Chief of the Stanneries.[173]

It is possible, however, that the phrases 'neither man nor woman, Christian nor Jew' were a stereotyped legal term which meant 'everybody'. If this was so then no inference can be drawn from the *Liber Rubeus* about medieval Jewish tin trading. There is, however, a very explicit reference to such trading in Camden's *Brittania* (1586) where he says that in the time of King John the mines were farmed by the Jews for 100 marks.[174]

Some support for the assumption that medieval Jews actually worked underground is to be found in the folklore of Cornish legend. According to the tradition of Cornish miners, unidentifiable noises were thought to be caused by spirits called 'knockers'. These were supposed to be the spirits of Jews who were at work in the mines. So as not to offend them, nobody was allowed to make the sign of the cross underground.[175] Some miners thought that the knockers were the spirits of the Jews who were alleged to have crucified Jesus, and it was said that the knockers never worked on Christmas Day, the Jews' Sabbath, Easter Day and All Saints Day.[176]

A medieval relic which perhaps demonstrates the close connection between Jews and the mining industry has survived in the form of an

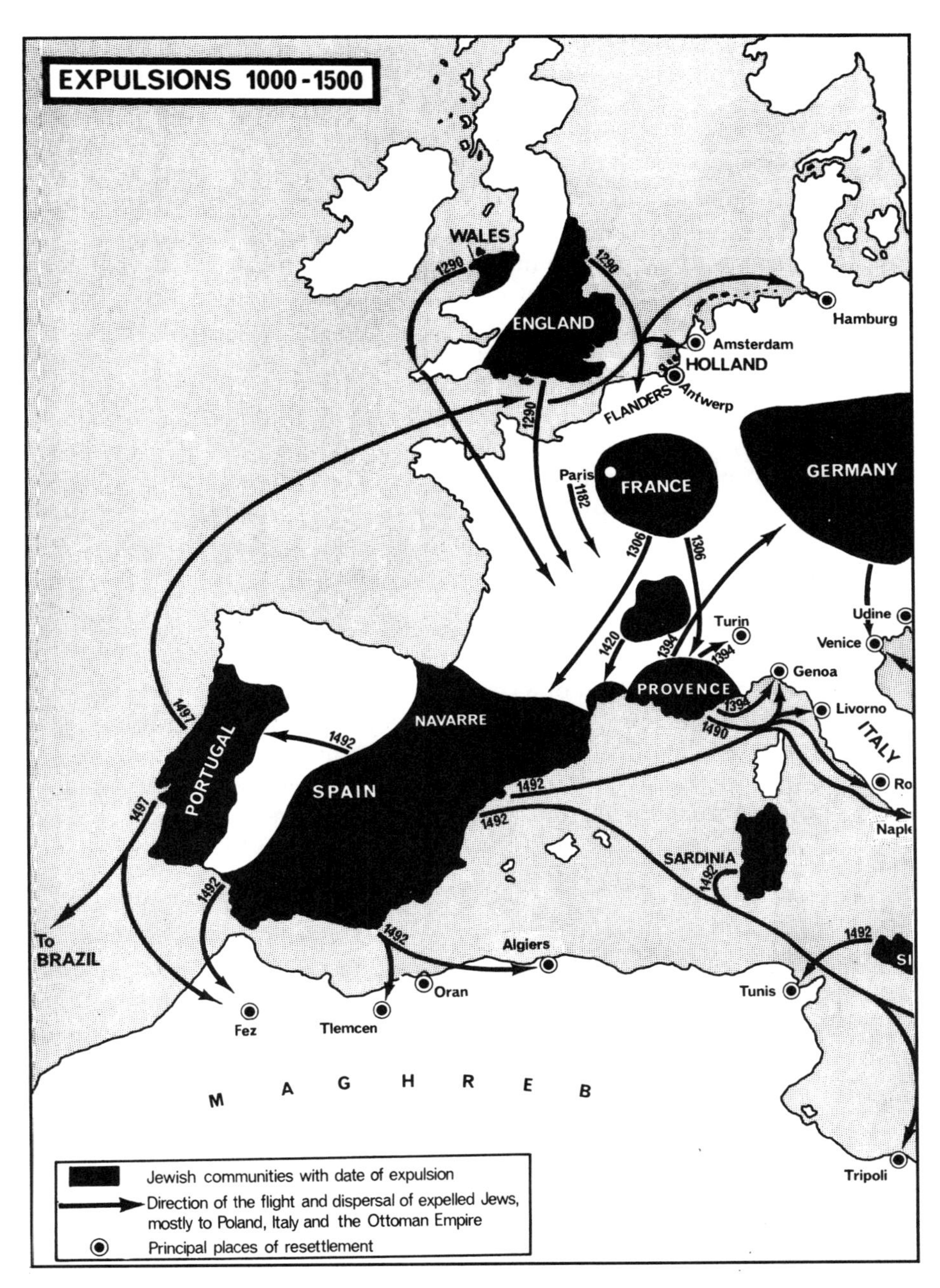


*Map 2*

Routes taken by the Jews of England and Wales on their expulsion.

alloy casting of a figure inscribed with four Hebrew letters. The figure was dug up on Bodmin Moor near Helmen Tor in the parish of Lanlivery, significantly, perhaps, near the site of a Jew's House.[177] It is at present housed in the collection of the Royal Institution of Cornwall, Truro. The figure is 5.5 inches high, 4.3 inches wide at its base, 3.6 inches deep and weighs 9 pounds 9 ounces. It clearly portrays a bearded man sitting on a high backed chair or throne. Until it was lost in a late nineteenth-century fire, there was a crown on the man's head. The man has tentatively been identified with Richard, Earl of Cornwall and the four Hebrew letters, *Nun, Resh, Shin, Mem,* have been said to be the initial letters of four Hebrew words referring to Richard as 'Rapacious Eagle—The Almighty is our King'.[178] As the letters are scattered over the figure this identification is most uncertain and it is difficult to do more than hazard a guess at the meaning of the letters and the purpose of the figure. One possibility which suggests itself is that the object was made by a Cornish miner who gave it as a pledge to a Jew. If this presumption is correct then the letters might represent its value and the time of its redemption. On the other hand, it is unlikely that the pledgee would mar a pledge with such large and deeply incised letters. Perhaps the object was used in pseudo-Cabbalistic or Black Magic rites and hence the Hebrew lettering,[179] or it may even have been used as a medieval chess piece.

There is also a suggestion of a connection between medieval Jewry and Cornwall which relies on the literal meaning of the Hebrew phrase used to designate England.[180] Some medieval Jewish scholars referred to England as *Ketzei HaAretz,*[181] a phrase which can be literally rendered as Land's End. According to one suggestion the Jews' first contact with England was at Land's End and they afterwards used the Hebrew name for Land's End to designate England as a whole.

Finally, it is difficult to imagine why a writ issued in 1283 to some twenty Sheriffs, all known to have Jews under their jurisdiction, to go in person to all the chests of the chirographers of the Jews in their bailiwicks, should also be sent to the Sheriff of Cornwall, unless there were Jews actually in his county;[182] however, no reference to an *archa* in Cornwall (and at least one *archa* was established elsewhere for a single Jew)[183] has come to light.

After their expulsion from England, the medieval Jews left behind only a slight impression of their two century sojourn. In Devon, one relic of their stay has not been noted in any of the standard works of Anglo-Jewish history. It is *Jews' Bridge,* the lowest bridge over the River Bovey, some two and a half miles south-west of Chudleigh,[184] which was built before 1399.[185] In 1406, the vicar of Buckfastleigh willed 12 pence each to the Totnes Bridge, the Dart Bridge and the *'ponti Judeorum'*.[186] In 1421, an indulgence of 40 days was granted to those who contributed to the upkeep of 'Jewysbrugge'.[187] It was repaired in 1643,[188] and widened in 1753, at which time it was called Judar's Bridge,[189] It was replaced in the early nineteenth century[190] and again in 1966, still keeping its name *Jews' Bridge*. The origin of the name is not now known, though the suggestion by William White[191] that it was so called because it was built by one of the Jewe family is untenable in view of the early reference to *ponti Judeorum*. Possibly, it was so called because the medieval Jews paid special tolls when they used it;[192] that they acquired the toll rights; that it was built with monies raised from a special tax levied on Jews;[193] that the burial ground for medieval Jewry was situated close by;[194] and sadly and most likely, that the bridge was the site of a tragedy involving Jews.[195]

Edward I's edict of 1290 expelled all Jews from the Kingdom on pain of death. During the course of the next three and a half centuries, before their presence was officially accepted, individual Jews entered England in spite of the severe penalty which they faced.[196] One such case occurred in 1409, when a Jewish woman, Johanna, and her daughter, Alice, were discovered at Dartmouth.[197] It is not known how they arrived, nor how long they had been in residence but upon discovery, perhaps to avoid fatal consequences to themselves, they intimated that they were prepared to become Christians. Henry IV ordered the Keeper of the *Domus Conversorum* 'to admit them for the term of their lives and grant them the usual wages of the converts, 1*d*. per day'.

A stronger personality and much more important visitor was Joachim Ganz or Gaunze who was largely responsible for the revival of copper mining in Keswick and Cornwall. He was a German-Jewish mining engineer who was invited by the Company for the Mines Royal to advise on methods of copper extraction. He visited the company's mines at Keswick and produced a report on the treatment of copper ores which is still of value. He spent three years from 1586 to 1589 in Cornwall ensuring a supply of copper ore for the rising metallurgical industries of Wales. He was expelled from England after a too forthright declaration of his Judaism which may well have influenced Francis Bacon in his literary and scientific work.[198]

A chance Jewish visitor to Devon was a privateer who was arrested in Plymouth in 1614. He was Samuel Palache who had received a commission from the King of Morocco to act against the Spaniards. Having taken two prizes he set sail for Holland but was forced by bad weather to take refuge in Plymouth, where he was arrested on a charge of piracy but was released the following year by an Act of the Privy Council.[199]

CHAPTER TWO

# Immigration and Emigration: The Jewish Communities of Devon and Cornwall after 1656

## *Part 1*

### *The Composition and Growth of Jewish Communities in Devon and Cornwall*

A small number, perhaps a dozen or so, Marrano[1] families settled in England about 1630, retaining their Catholic guise as it was thought that the edict of Edward I banning Jews from England was still in force. In 1656, Oliver Cromwell gave what appears to have been informal permission for Jews to reside and trade in England.[2] Gradually, they increased their numbers so that by the end of the seventeenth century there were about a hundred Sephardi[3] families whose economic importance to England was out of all proportion to their numbers.[4]

Alongside the Sephardi colony there grew up a settlement of Ashkenasi Jews which was eventually to outnumber the original Sephardi settlers. The new immigration was the overspill of German-Polish Jews who settled in Hamburg and Amsterdam and eventually made their way to England. In 1690 there were probably not more than two or three hundred of these Ashkenasi Jews in England, virtually all of them in London.[5] Their numbers were greatly swollen by Jewish refugees from persecution in Bohemia (1744–5), the Seven Years' War (1756–63), and the Haidamack massacres of 1768, so that by the end of the eighteenth century there were between 20,000 and 26,000 Jews in England, three-quarters of them in London, the rest mainly in the seaports.[6]

In the next half-century there was a steady increase in the Jewish population in England averaging about 300 per annum, so that there were in all some thirty-five or forty thousand Jews by about 1850, and of these not more than 3,000 were Sephardim.[7] In the three decades between 1850 and 1880, the Jewish population of Britain increased by about 70 per cent. In part, the increase was due to natural causes, but in the main it was due to an immigration from East Europe which eventually completely changed the composition of Anglo-Jewry. After 1880, prompted by pogroms and vicious discriminatory legislation, there was a mass emigration of more than a million Jews from Russia, Poland, Hungary (mainly Galicia) and Romania. Of these, 850,000 went to America, 100,000 to Britain and the rest went to Germany and elsewhere. The East European emigration increased the number of Jews in England, estimated in 1880 to be 60,000, to more than 180,000 in 1905 (see Map 3).[8]

Before discussing the effect of these influxes on the Jewish population of Devon and Cornwall, it is necessary to say a word about Jewish population estimates. It has never been possible to count accurately the number of Jews in England at any particular time as there are no precise data on which to work.[9] Furthermore, there are many definitions of a Jew, based on religious and racial considerations as well as on social and cultural associations. For the purpose of this work, a wide definition has been adopted, that is, those who regarded themselves or who were regarded by others as Jews, irrespective of religious practice or parentage. Although a broad definition has been adopted, in all but a handful of cases the discussion throughout this book relates to Jews who had some contact, even if tenuous, with the synagogal organizations.

The development of the Anglo-Jewish community as a whole in the eighteenth century is reflected in miniature in the South West of England. There is some reason to suppose that some Marranos settled in the South West in the sixteenth century, possibly in Exeter, but they certainly had some connections with Plymouth where the London Marranos stationed agents. These agents served a dual purpose. Their primary function was to promote trade for the Marrano mercantile houses which they represented. A secondary function was to inform Marranos who had fled from Spain and Portugal and who wished to practise Judaism openly again whether or not it was safe to proceed to the Low Countries to do so. A well-known example of this procedure is provided by the case of Gracia Mendes, 'the most adored Jewish woman of her age'. In 1536, she was on her way to Antwerp where she intended to revert to her ancestral faith. Her ship put into Plymouth and her firm's agents there warned her that the Inquisition was operating in Antwerp. She and her family accordingly broke their journey and stayed in London until it was safe to proceed.[10]

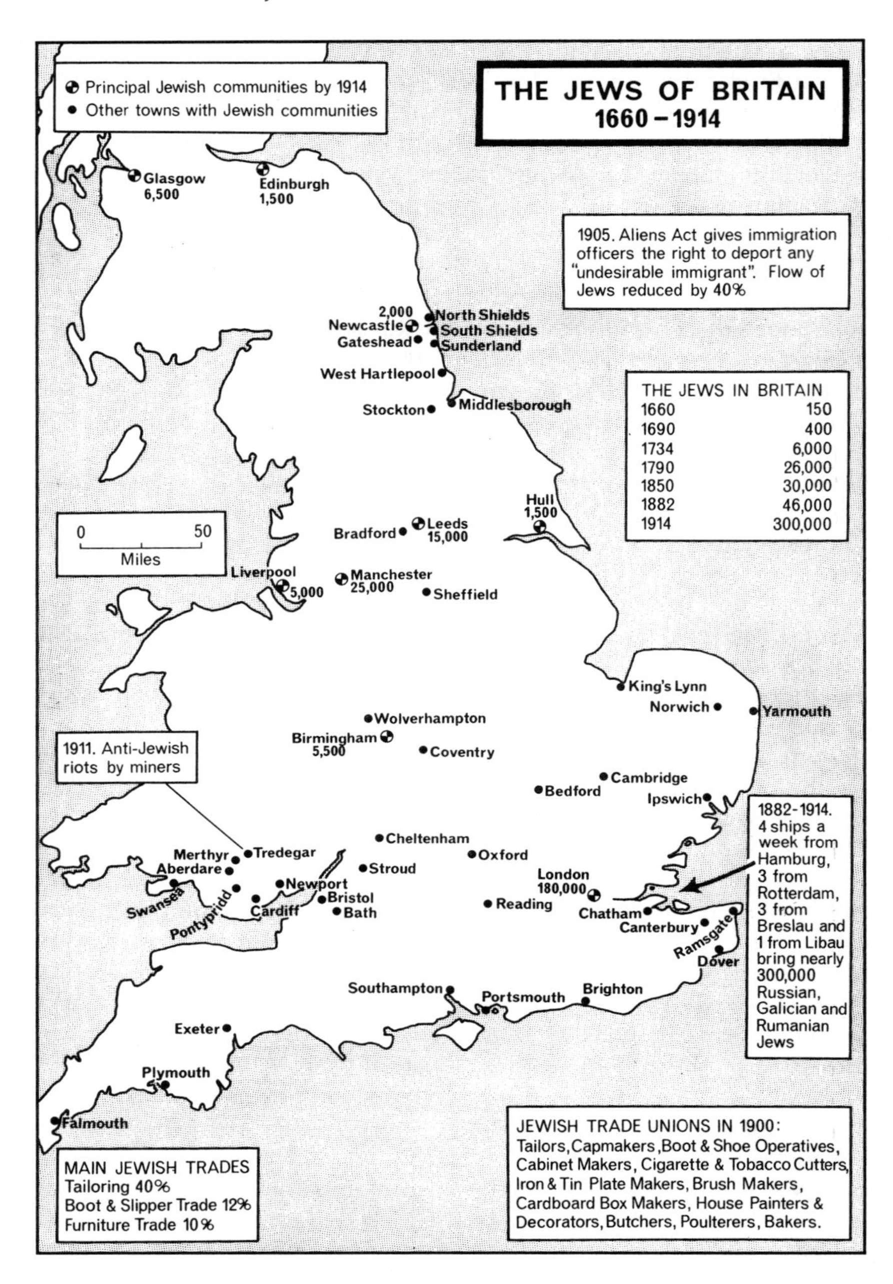


*Map 3*
The Jews of Britain, 1660–1914.

The mercantile connections established by Marranos in the South West during the sixteenth century were maintained in the seventeenth and form a direct link with the first known Jewish settlers in the eighteenth. In 1617, a Marrano, Antonio Dacosta Doliveira, was in Plymouth where he acted commercially on behalf of the Spanish Ambassador to London, Count Gondomar.[11] It appears that the Marranos in England made little secret of their religious beliefs and further that some of them may have settled in Exeter, as in 1600 the Bishop of Exeter complained of the prevalence of 'Jewism' in his diocese.[12] It is possible, however, that the Bishop might have been referring to heterodox beliefs held by some of his flock, and not to the presence of Jews or pseudo-Jews.[13]

There appears to have been a Jewish scholar in Plymouth in 1634, and if so it would further demonstrate the gradual penetration of the country outside London by Jews before their presence in England was officially countenanced. He was a Jno. Lawrenson, a 'Hebrew high Jerman who was mayntayned at the charity of the Town of Plymouth at the Universitye'.[14] It is not clear what the description 'Hebrew high Jerman' means; perhaps it designated a Jew from Germany proper as opposed to one who came from the Low Countries. Lawrenson is not a typically Jewish name though this alone would not necessarily preclude his Jewish origins. Nor need he have been a practising Jew, he might have been a convert and yet still be called a Jew.

Additional support for the view that Jews in the seventeenth century ranged as far afield as Cornwall is provided by a circumstance which, tragically, has added to the nomenclature of Cornish roads. The lane leading from the village of Herland Cross in the Parish of Breage to Pengilly Farm is called Jews' Lane. In this lane a Jew hanged himself after some outrage done to him by a Squire Sparnon. Soon after the Jew's suicide it was said that the lane was haunted. The Reverend Robert Jago was paid five guineas to exorcise the ghost, which he did by drawing a circle with a whiplash and uttering 'certain formulae'.[15] No date is given for this incident but it must have been between 1633, when Robert Jago was presented to the Vicarage of Wendron and Helston, and 1685, when he died.[16]

After 1656 Jews were free to travel about the country on their lawful business openly as Jews. An important part of their business was the West India trade which was partly conducted through the port of Falmouth.[17] It was through this trade that commercial and social contact was established between Cornishmen and Jews *qua* Jews for the first time since the Expulsion in 1290. This contact is graphically described in a pamphlet[18] published in London in 1685 by Samuel Hayne, Customs Officer, who alleged that the Jews conspired with Sir Peter Killigrew and Brian Rogers, a Falmouth merchant, to defraud the Customs. Their method, he claimed, was to send Dutch cargo to Falmouth in an English vessel so that it could be sent on to Barbados as an English export, thus avoiding the duty that Dutch goods would have attracted.[19] The local Cornish population was well disposed towards the Jewish merchants and at the trial of Gomasero and Losado, two London Jews, the jury found against Hayne's allegations, as he put it, 'because it was about the time that Ignoramus Juries were in fashion'.[20] The real reasons why the Cornish favoured the Jews' case are apparent from Hayne's pamphlet. The first, discreditable if true, was that the Jews bribed the jurymen and their families.[21] The second, and more important, was that the Cornish merchants were anxious about the effects of driving away Jewish capital and trade. Whenever Hayne attempted to tighten the administration of the Customs he was met, in his own words, with this objection:

> The Jews are a very Rich sort of People, their trade is very great, they employ many ships etc. and should that be cut off, abundance of people, both here and in the Plantations, would feel the want of them. Moreover, the King would be much lessened in the Customs by the breaking off of their trade.[22]

As a natural consequence of this Sephardi trade with the South West some Sephardim settled in Devon in the first half of the eighteenth century. In 1729 the public was advised that 'Jacob Monis, a learned Jew born at Padua' taught Hebrew, Italian, Spanish and Portuguese 'speaking, reading and writing grammatically' and could be contacted at Brices Printing House, Exeter.[23] He was followed by Gabriel Treves who settled in Exeter with his family and opened a snuff shop there about 1733.[24] He brought over his nephew, Joseph Ottelenghe, from Leghorn in 1734,[25] and the family quarrels between them flared up into a minor *cause célèbre*.[26] There was apparently a Gubby family in Axminster, and one Benjamin Gubby was apprenticed to a barber in Tiverton in 1738.[27]

From the beginning of the eighteenth century Ashkenasi Jews also made their way from London down to Cornwall, probably peddling. It was possibly two such as these who suffered some reversal of fortune and were given charity by the Borough of St Ives in 1703.[28] A number of Ashkenasi Jews were lodging in a public house in Exeter in the early 1730s. They were established there on a sufficiently permanent basis to hire Ottolenghe to slaughter fowls ritually for their food.[29] Within a few years people with Ashkenasi names settled in Exeter. There was a Simon Nathan who was declared bankrupt in 1741,[30] and a Mrs Sara Abrahams andAbraham Ezekiel who paid 1½*d*. and 2*d*. respectively each week to Exeter's Poor Rate from 1752.[31] About this time there was the nucleus of a community as Ezekiel obtained a lease for a Jewish cemetery in 1757,[32] and it is hardly likely that Ezekiel and his fellow Jews would have interred Jews there unless the community was established on a permanent basis.[33] It may therefore be surmised that in the early 1750s there were some twenty or thirty Ashkenasi families in Exeter.

A similar picture of a community comprised of Ashkenasi Jews emerges in Plymouth, but this time the evidence points to a Congregation organized about a decade earlier than the one in Exeter. Joseph Jacob Sherrenbeck and his wife Sarah were settled in Plymouth by 1744 when Sarah bought a piece of land on the Hoe.[34] But there must have been at least a dozen or more Jewish families in Plymouth in 1745 because services with readings from the Torah, which can only be done in the presence of ten male Jews aged 13 or over, took place. This may be inferred from an inscription on a silver pointer still in use at the Plymouth Synagogue. The inscription reads (in translation): 'This pointer belongs to Joseph ben Judah Jacob from Sherrenbeck, *PH* Plymouth, in the year 5505' (= 1745). The two Hebrew letters *PH* may represent the Hebrew word *po* which means 'here', and this would not be out of place in such an inscription. There is, however, a mark over the letters *PH* which elsewhere on the pointer is used to indicate an abbreviation, and although it is not a standard one, *PH* might well stand for *Parnas HaKahal*, President of the Congregation. If so, the inscription on this pointer is evidence of an early transitional stage when the community had gelled but its permanence was not yet assured, Sherrenbeck therefore retaining possession of the pointer and not presenting it at that time to the nascent Congregation. This inference from the inscription on the pointer is confirmed by the nineteenth-century Plymouth historian Worth who apparently had evidence that Jews held small but organized services in a rented room in Broad Hoe Lane in the early 1740s.[35]

It is worth noting that evidence of a few Jews settled in a town, however, is like an iceberg—a little shows above the water whilst the major part is concealed. In Plymouth, for example, extra-synagogal sources reveal the presence of only five Jews in the town before 1760.[36] There were, however, many more, as the inscription on the silver pointer indicates, and the fortunate discovery of a synagogue account book[37] reveals 52 male members in 1759. One more example may be given. In Barnstaple, the only known Jews were Abraham Ralph and his family. Yet, apparently, regular 'Synagogue assemblies were always held in his house'—implying again the quorum of ten adult male Jews—for an unknown number of years before Ralph's death in 1805.[38]

The influx of Jews to Cornwall came somewhat later and to a lesser extent than to Devon, and once again the Jews were Ashkenasim. It is not likely that there were sufficient male Jews to organize services on a regular basis until at least the late 1760s, and even then only in Falmouth. For in that period Alexander Moses, or Zender Falmouth as he is otherwise known, subsidized young Jews enabling them to act as pedlars around the neighbourhood of Falmouth on condition that they returned in time to form a quorum for Sabbath services.[39] Had a congregation been functioning at that time it is unlikely that he would have been obliged to make such a condition. There must have been a quorum of ten adult male Jews, residents or visitors, in Falmouth in 1763 as the marriage of Eleazar Hart, father of Lemon Hart the famous rum merchant, took place in Falmouth in December 1763.[40] By 1766, however, there was a sufficient number of Jewish settlers to buy a house on the sea-front, to be used as a synagogue.[41] So far as Penzance is concerned, other than Margoliouth's unsubstantiated statement, there is no evidence of any Jew there before 1781, when the name of Lemon Hart appears on a clock face, apparently sold by him.[42]

Of course, Jews settled throughout the South West, but in Exeter, Plymouth, Falmouth and Penzance they settled in sufficient numbers to form viable communities. The minor settlement elsewhere in Devon and Cornwall will be described in detail below.[43]

From whence did these Ashkenasim come? In the main they came from Central Europe, from Germany and from what was formerly Czechoslovakia. The synagogal ritual, so far as it has survived in the Plymouth Congregation, conforms to the German liturgical pattern rather than the Polish or Russian.[44] An additional pointer to a Central rather than an East European provenance of these Jews is provided by the type of Yiddish used in the Plymouth Congregation's minutes of 1779–1830, which is predominantly Central European in character. Furthermore, the place of origin of some of the 'founding fathers' is known. Joseph Jacob derived his surname Sherrenbeck from his town of origin in Germany.[45] Eleazar Hart records in his family bible in 1763 on his marriage in Falmouth that he came from Weinheim in Germany.[46] The progenitor of the Joseph family of Cornwall was a Joseph Joseph who left Mulhous, Alsace, for England in the early eighteenth century.[47] Of the 58 Jewish aliens in Plymouth in 1798–1803, eleven came from the Margravate of Ansbach, ten from in or near Mannheim, and a further twenty-seven from other places in Germany and Bohemia.[48]

The great incentive to emigration was clearly the persecution of the Jews in her dominions by the Empress Maria Theresa and the Seven Years' War with its resulting devastation in Central Europe. This may be seen from an analysis of the Plymouth Aliens List which records the names and date of arrival in England of the 55 Jewish aliens who were in Plymouth in 1799. It is convenient to tabulate the arrival figures in eleven-year periods, as is shown in Table 3.

*Table 3: Arrival in England, 1745–1799, of Jewish Immigrants present in Plymouth in 1799.*

| *Year of arrival* | *Number of immigrants* |
|---|---|
| 1745–55 | 6 |
| 1756–66 | 19 |
| 1767–77 | 11 |
| 1778–88 | 12 |
| 1789–99 | 7 |
| Total | 55 |

(Source: Lipman, 'Aliens List'.)

It should be noted that the number of immigrants given in Table 3 was not the number who actually arrived in Plymouth in any particular period. The figure represents only the number of immigrants who happened to be in Plymouth in 1799. But it does throw some light on the pattern of arrival, although it must be borne in mind that a certain number who arrived in the 1745–55 period would have died or left Plymouth by 1799. The large increase in the 1755–66 period is reflected in the rapid expansion of the Plymouth Congregation where more than 40 families joined the Congregation between 1756 and 1760.[49]

The immigration of Jews into the South West followed a discernible pattern. Immigrants tended to travel with relatives or friends from the same area, and when they were once settled they attracted further friends and relatives from their former localities. In the Plymouth Aliens List, 1798–1803, for example, it is possible to identify at least seven, and possibly eight, sets of brothers, representing sixteen or eighteen heads of families.[50]

Of the 53 Jewish families in Plymouth in 1821, at least 34 were interrelated, and eight men came from one town, Mannheim.[51] A further indication that the immigration was prompted by family or friendly connections is that a large proportion of immigrants made their way straight to Devon and Cornwall from their place of embarkation. Of the 58 names on the Plymouth Aliens List, 15 went to unknown destinations after landing. Of the remaining 43, 20 went straight to Devon and Cornwall and another 5 had arrived there within one year, 11 stayed in London for varying lengths of time, between two and twenty years, and 7 went straight to other towns. It is unlikely that the 25 who went directly or almost directly to the South West after landing at Harwich, London, Dover and Gravesend, would have done so unless there was a special reason, and that can hardly be any other than the presence of relatives and friends who had offered to help them. A further indication of this trend is that most immigrants to Plymouth after 1771 went directly there, whereas before 1770 they tended to go in roughly equal numbers to Plymouth and Cornwall, London and other towns, as Table 4 shows.

Although the communities of the South West were founded by Ashkenasi Jews who formed the overwhelming bulk of their membership from the time of their foundation onwards, a small number of Sephardim also settled in the area or passed through it from time to time in the second half of the eighteenth century. There were never sufficient of them to make even a little community of their own with their own synagogue for worship. On the other hand, if the experience of the early nineteenth century, from which there are many more records, was already anticipated in the middle of the eighteenth, then there must have been a constant flow of Sephardi visitors, either leaving or entering England. At least ten Sephardi Jews landed in Cornwall between 1757 and 1788, nine of them at Falmouth and one at Fowey.[52] The names of these ten have fortunately survived on a list of aliens associated with the Sephardi Bevis Marks Synagogue, London, in 1803. But there might very well have been others who had died or moved away by 1803, or for some other reason were not included on the list.

Among the Sephardi Jews who settled in Exeter in the latter part of the eighteenth century was a certain da Costa who ran a successful school for languages in Exeter from before 1772, when an Anthony Fiva disclaimed responsibility for an anonymous letter in Italian about him, until 1780.[53] There was also a Samuel Lopes who occupied 'The White House', Exeter, in 1797 at a rental of £6 7*s*. per annum.[54] These Sephardim may have been connected with the Ashkenasi community in Exeter, but as no minutes of

*Table 4: Jewish Emigration to the South West, 1745–1800.*

| | *The number of immigrant Jews who went* | | | |
|---|---|---|---|---|
| | *straight to South West* | *to South West within 1 year* | *to London first* | *to other towns first* |
| Before 1770 | 7 | 2 | 6 | 5 |
| 1771–80 | 3 | 2 | 1 | – |
| 1781–90 | 7 | – | 3 | 1 |
| 1791–1800 | 3 | 1 | 1 | 1 |
| Total | 20 | 5 | 11 | 7 |

(Source: Lipman, 'Aliens List'.)

the Exeter Congregation have survived from this period it is not possible to be sure.

The most important Sephardi settler in the South West, because of the part which he and his collateral descendants played in local and national affairs, was Manasseh Masseh Lopes. Lopes was born in Jamaica in 1755, where his father Mordecai Rodriguez Lopes had made a fortune from sugar plantations. He married Charlotte, only daughter of John Yeates of Monmouth, on 19 October 1795.[55] His father died six months later and left his fortune to Manasseh. He came to Devon in 1798, when he bought the manors of Maristow, which became his principal seat, Buckland Monachorum, Walkhampton, Shaugh Prior, and Bickleigh; and in 1808 he added the manor of Meavy, in all some 32,000 acres. Manasseh and Charlotte had an only daughter Esther who died in 1819, aged 23. Presumably she had never been expected to have children because when Manasseh or Massey, as he elected to be called, was created a baronet in 1805, there was an exceptional remainder to his nephew Ralph Franco.[56] Massey was first elected to parliament in 1802 for New Romney, and he subsequently represented Barnstaple, Grampound, and Westbury.[57]

In spite of his conversion to Christianity, local tradition in the Jewish community of Plymouth asserts that he asked for a rabbi on his deathbed.[58] After his death, his family is reputed to have given a scroll of the Torah which belonged to him to the Plymouth Congregation. A scroll of Esther was found in his belongings as late as 1970.[59]

Another Sephardi family which settled in Devon at the beginning of the nineteenth century and which became active in high society was that of Lousada, who made their seat at Peak House, Sidmouth. There they attracted other titled members of their family to come to stay for short periods. The *Western Luminary*[60] records that Sidmouth was a favoured spot for Sir Moris and Lady Ximenes, as well as for Mr and Mrs David Lousada.[61]

Plymouth also attracted a number of Sephardim. In 1808, a Solomon Sebag[62] was a member of the Meshivat Nefesh Society,[63] and in 1814 there was a Mrs Pereira, probably Jewish, a tea dealer in Little Church Lane.[64]

The influence of the Sephardim on the Congregations of the South West of England throughout their existence has been negligible, both in respect of numbers and financial contributions, as well as upon the development of Jewish life.[65]

To complete the picture of the composition of South-West Jewry in the latter part of the eighteenth century it is necessary to add that there were also a few East-European Jews. The earliest of them was employed by the Plymouth Congregation and indeed most of those known to us had some synagogal function. Moses Isaac, beadle, trusty, and probably also *mohel*[66] to the Plymouth Congregation was born in 1728 in Mezeritz, Poland, the home town of Rabbi Israel Baal Shem Tov, the founder of modern Hasidism.[67] He landed at Harwich in 1748, and was acting as beadle in Plymouth in 1778,[68] but it is not known when he actually arrived in Plymouth. Another very early East-European immigrant was Joseph Cohen of Brod, Poland, who arrived in England in 1775 and subsequently made his way to Plymouth.[69] Two brothers, Mordecai Levy and Joseph Levy were born in Lissa, Poland in 1770 and 1771, respectively.[70] They both came to England in 1789, Joseph arriving in Plymouth in 1795 when he was appointed *shochet*, beadle and teacher.[71] In 1799, David Jacob Coppel arrived in Plymouth from Belleye (Biala, Podlaska),[72] and a year later Moses David Angel, a fellow citizen and possibly a relative, followed him.[73] The 26-year-old ill-fated Isaiah Falk Valentine, murdered in Fowey in 1811,[74] came from Breslau (Wrocklaw).[75] These East-European Jews had an importance beyond their numbers. They provided, in the main, the religious leadership. They also provided the base on which further immigrants from Russia and Poland in the latter part of the eighteenth century could anchor themselves.

Having described the overall composition of the South-West Jewries until the end of the eighteenth century, let us now attempt to assess the total number of Jews in Devon and Cornwall at various points in time. It is possible to do this for 1765, which was two or three years after the Plymouth and Exeter Synagogues had been built, and just before the Falmouth Synagogue was built, on the basis of the seating capacity of these three synagogues. The Plymouth synagogue, when it was originally built, had 142 seats,[76] that of Exeter about 100, and that of Falmouth about 50,[77] giving a total seating capacity of 292. It is likely that in a period of rapid expansion the synagogues were built with provision for future immigrants, and indeed when the Plymouth Synagogue was built in 1762 it had 89 seats for males but only 52 male members.[78] It seems therefore that some 40 per cent of the seating was installed in anticipation of future expansion. As seats were provided at that period only for adults (children were expected to stand or squeeze in somewhere)[79] it follows that there were some 175 or perhaps 200 adult Jews in the South West. Allow that there were some 250 children, since it was after all a young immigrant society, then there must have been, in all, some 450 Jewish souls in Plymouth, Exeter and Falmouth. Perhaps another 50 should be added for Penzance and other towns where there were only a few families, so that there were in all some 500 Jewish souls in Devon and Cornwall, which represented a little over six per cent of all the Jews in England in the 1760s.

After the rapid increase in the early 1760s there appears to have been a drop in the overall number of Jewish residents in Plymouth, and probably in the rest of the South West, as the Plymouth Congregation had only 40 members in 1785,[80] a drop of 10 from 1760. The check to immigration, if there was one, was only temporary because by 1815 the membership of the Plymouth Congregation had risen to 97,[81] whilst Exeter in 1820 had 53 members,[82] and Falmouth and Penzance each had at least twelve or possibly more members. In all, the four communities had at least 174 members and perhaps as many as 200. One must take into account, moreover, that there were poor families whose heads could not afford to be contributing members of the Congregation a factor which would increase the numbers, and offset those who were bachelors or had no offspring. It may therefore be hazarded on the basis that each member represented four individuals that there were between seven hundred and one thousand Jewish souls in Devon and Cornwall at the end of the Napoleonic Wars.

In the period 1825 to 1845, the composition of South-West Jewry remained much as it was in the eighteenth century, in so far as the overwhelming bulk of the membership of the Congregations remained Ashkenasi, together with a fair sprinkling of Sephardim in the two counties. By 1840, however, there was this difference—the Ashkenasim were mostly British-born. According to the 1841 census,[83] of 126 Jews in Exeter only 20 were foreign-born and the rest were born in England. Although at first sight it seems that the community was now overwhelmingly British-born, it is not likely that this was the impression gathered by the Gentile onlooker, because of the 106 Jews who were English-born, only 41 were over 20 years of age. In other words, of the adult Jews the British-born slightly outnumbered the foreign-born.[84] The Jewish community in Plymouth was even more English, there being only 18 foreign-born Jews to 192 British-born.[85] These figures compare closely to those given by Lipman for the proportion of foreign-born Jews to British-born in London in the mid-nineteenth century.[86] These British-born Jews were for the most part born in Devon and Cornwall. In most cases there is no evidence to show whether or not their parents ever resided in one or other of the towns where the four Congregations were established. But in the latter part of the nineteenth century they provided a reservoir which replenished to some extent the losses due to Jewish emigration from the South West.

Besides the changing aspect of the South-West Congregations in as much as they were becoming predominantly British-born, there was also a change in the foreign element, as in the quarter century after the ending of the Napoleonic wars there was an increasing immigration from East Europe. Plymouth attracted, amongst others, a *Zvi ben Judah Lyons* who was a member of the Congregation in 1829 and who came from Warsaw.[87] The Polish-born Shemoel Hirsch landed in England about 1821,[88] and in the same year *Joseph ben Samuel* of Brisk, the scribe *Michael ben Abraham* of Vilna, *Nathan ben Reuben* from Hungary, and Lazarus Solomon of Lublin, all passed through Plymouth.[89] The last named Lazarus Solomon settled in Plymouth. Some years later he sent back to Poland twenty-five pounds left by a *Ze'ev ben Judah* from Shatwinitz who died of cholera at a Plymouth inn in 1832;[90] and in the same month died *Jacob ben Uri Shragai* from Lontshotz.[91] Mathias Cohen of Sonnhaus, Poland, died in Plymouth at only 'half his days' in 1833.[92] Revd S. Hoffnung, later *chazan* in Exeter, arrived in England from Poland in 1836.[93] Perhaps the first recognizably East European name which is met in Plymouth is Mandovsky, who offered comparatively large sums to the Plymouth Congregation in 1822.[94] There was a more extensive movement of Sephardim in Devon towards the end of the first quarter of the nineteenth century than has perhaps hitherto been recognized. The chance survival of the Plymouth Congregation's cash book for strangers' offerings which was kept by its beadle, H. Issacher, for the years 1821–23 reveals quite a procession of Sephardi men who worshipped at the synagogue, and who were called to the Torah and made offerings during 1822 and 1823. They were *Moses ben Solomon Delavayo, Moses ben Hayyim Portuguese,* a Turk called *Jacob ben Joseph Portuguese, Jacob ben Shalom Mogadore,*[95] and an unnamed person from Madagascar.[96] Some of these six Sephardim who passed through Plymouth during an eighteen-month period may have been *en route* to the West Indies, whereas others were probably trading in the area.[97]

Among the Sephardim who were resident in Plymouth about this period were Abraham Franco who was a resident in Plymouth at least since 1821,[98] until his death in 1832,[99] and who may have been a poor relation of Manasseh Massey Lopes whose sister was married to an Abraham Franco; Joseph and Sara Montefiore (née Mocatta) who had a son at Stoke, Devonport, on 29 July 1828,[100] possibly the mother came down to Devon for the benefit of the climate during her pregnancy; and *Isaac ben Joseph Bosca,* who died in Plymouth 1833, whose surname may indicate that he was a Sephardi.[101]

Then there were a few Sephardim resident in Plymouth who apparently had no connection at all with the synagogue there. Aaron Lara, for example, a relation of Lopes, died in Plymouth in 1813, leaving an estate of £1,000.[102] In 1828, a Jose Bento Said, a Portuguese Jew from Angra in the Azores, fell ill in Plymouth. In the preface to his *Descripsam das 3 cidades unidas, Plymouth, Stonhause, e Devonport*[103] he says that he was helped during his illness by friends. And finally, a Lydstone Pereira, possibly a Sephardi Jew, is listed in the Plymouth Directory of 1830.[104]

Exeter, too, had a number of Sephardi visitors some of whom had connections with the Congregation. In 1815, there was a Turkey Rhubarb

Jew from Mogadore in Morocco who together with three fellow countrymen had a shop in Exeter from about 1815 for some five years;[105] in 1827, the Gutteres family supported Anglo-Jewish charitable institutions from nearby Sidmouth;[106] in 1828, a Mrs Lopes bought *kosher* meat in Exeter;[107] in 1839, Hannah, relict of Moses Ancona, was buried there[108] and so was an Aaron Amzalek in 1838, the Congregation writing to the Sephardi Congregation at Bevis Marks, London, asking for the funeral expenses and the cost of the gound to be refunded.[109] A number of Sephardim also made donations to the Exeter Congregation, presumably because they worshipped there at some time. There was David Lindo, uncle of Lord Beaconsfield, who died in 1852 and left the Exeter Congregation a legacy of £10;[110] and a Miss Guedela who donated £2 to the Congregation in 1855.[111]

In spite of the growing immigration from Eastern Europe the total number of Jews in Devon and Cornwall declined between 1820 and 1845. From the 1841 census the names have been extracted of 119 Jews in 27 households in Exeter, and 193 Jews in 42 households in Plymouth. There were some 100 Jews in Falmouth and Penzance at this period,[112] so in all there were some 450 Jews in the South West of England or about half the number that resided there just after the Napoleonic Wars. These figures are confirmed by the answers to a questionnaire which was circulated by Nathan Adler shortly after his induction as Chief Rabbi of the British Empire in 1845.[113] The information in the questionnaire is summarized in Table 5, the 1841 census figures being quoted in brackets.

A comparison of the 1841 census and 1845 questionnaire figures indicates approximate agreement for numbers of individuals in the case of Plymouth, but a considerable disparity in that of Exeter. This may be

*Table 5: Number of Jews in the South West in 1845 (1841 census figures in brackets)*

| | *Members* | *Individuals* |
|---|---|---|
| Exeter | 32 (27 households) | about 175 (119) |
| Plymouth | 52 (42 households) | 205 (193) |
| Falmouth | 12 | 50 |
| Penzance | 11 | 51 i.e. 20 males, 20 females, 11 children |
| Total | | 481 |

(Source: Chief Rabbinate Archives, 104.)

explained by assuming the figure of 'about 175' Jews in Exeter reflected the situation some years earlier, but that in fact, there were about 145 Jews in Exeter in 1845.

Some of the foreign-born Jews who settled in Devon maintained a close interest in their towns of origin. Moses Mordecai, who was settled in Exeter by 1788 and had been an established goldsmith and bookplate designer in Portsmouth before that date, bequeathed in his will dated 7 December 1808, ten guineas to the synagogue at Maintz so that his soul might be commemorated there on Festivals, as well as numerous legacies to his family who lived in that locality.[114] Michael Solomon Alexander took the opportunity whilst journeying through West Prussia to visit his native town of Schönlanke to see his surviving family, even though he anticipated considerable hostility from the rest of the Jewish community on account of his apostasy.[115] Where a man's birthplace is recorded on his tombstone it often implies that he had not long left his native place, as most tombstones of foreign-born Jews do not record their birthplace. David Jacob Coppel who is described as 'from Bialin in Poland' on his tombstone in 1805, had landed at Gravesend only six years earlier.[116] A Mr Woolf who died in the cholera epidemic of 1832 had no relatives in England[117] and is described as 'from Shetvinitz in Poland'.[118] However, if a person had been in England for many years and his native city or land was nevertheless commemorated, it probably indicates that there was a strong sentimental tie. Thus Jacob Philip Cohen who died in 1832 is described as 'from Lontschutz' and yet he was well established in Plymouth by 1819.[119] Similarly, when a Jew emigrated from England his town of origin might be mentioned on his tombstone. Joseph Marks, for example, was born in Portsea in 1805, moved to Exeter about 1833 where he lived and worked for the next twenty years. He emigrated to Australia in 1853 and died in Melbourne in 1872. His tombstone records that he was 'formerly of Exeter', and one can imagine his widow and children visiting his grave, and conjuring up visions of their native Exeter which they were unlikely ever to see again.[120]

After 1850, the Jewish communities of Devon and Cornwall changed both in respect of their distribution as well as in their composition. From this period the two Cornish communities dwindled away until they all but vanished. At this stage the Cornish Congregations were formally disbanded and their synagogues sold at the end of the nineteenth century. The gradual decline of the Penzance Congregation is apparent from its membership, which is summarized in Table 6.

In 1881, a letter to the *Jewish Chronicle* signed by 'An Occasional Visitor to Bath' complained that he had found the Bath synagogue closed on a Friday night and Saturday morning. He pointed out that 'in Penzance

*Table 6: Seatholders in the Penzance Hebrew Congregation, 1841–1874*

| | |
|---|---|
| 1841 | 12 seatholders |
| 1845 | 11 seatholders |
| 1849 | 8 seatholders |
| 1854 | 5 seatholders |
| 1874 | 4 seatholders |

(Sources: Roth MSS 271, p. 91; Chief Rabbinate Archives, MSS 104; PenHC Minute Bk 1843, pp. 23 recto, 37 verso; A. Myers, *Jewish Directory for 1874* (1874).)

with only three members the synagogue was open every *Shabbat* and frequently made *minyan* from visitors like himself'.[121]

In Falmouth the number of Jewish families similarly declined rapidly from 1842 to 1874, as Table 7 indicates.

From the middle of the nineteenth century, the Exeter Congregation also began to decline steeply in numbers. The names of 128 Jews have been noted in the 1851 census returns for Exeter, and only 73 in the following decennial returns in 1861. By 1875 there were only 40,[122] and about 1880 the Congregation disbanded.[123] The number of Jews in Plymouth also declined after the Crimean War, as the 1851 census reveals the presence there of 278 Jews whilst the 1861 census shows a drop to 233. This level gradually increased. In the 1871 census there are 268 names which are almost certainly Jewish, with a further 27 who are possibly Jewish. By the end of the 1880s, there was sufficient Jewish life to attract some of the new wave of immigration which came from Eastern Europe. The numbers may be expressed in tabular form as in Table 8.

It is difficult to pinpoint the causes for the decline of the Jewish communities in the South West of England, and more particularly in Cornwall. Some emigration either to more prosperous centres in England

*Table 7: The Size of the Falmouth Hebrew Congregation, 1842–1874*

| | |
|---|---|
| 1842 | 14 families |
| 1845 | 50 individuals, men, women and children |
| 1865 | 9 males aged over 13 years |
| 1874 | 3 families |

(Sources: *JC*, 14 March 1842; Chief Rabbinate Archives, MSS 104; Roth, MSS 204; A. Myers, *Jewish Directory for 1874* (1874).)

*Table 8: The Number of Jews in Plymouth and Exeter according to the Censuses of 1841, 1851, 1861, 1871 and 1881*

| | *Number of Jews in* | |
|---|---|---|
| *Census dates* | *Exeter* | *Plymouth* |
| 1841 | 119 | 193 |
| 1851 | 128 | 278 |
| 1861 | 73 | 233 |
| 1871 | not searched | 268 |
| 1881 | not searched | 280 |

or abroad was occasioned by a falling off in local commercial opportunities. This was largely caused by the spread of the railway system to the South West which enabled local shopkeepers to obtain goods more easily from the large industrial centres and deprived the Jewish tradesmen of their advantage in national connections in the business world. Better transport also enabled the inhabitants of villages and farmsteads, who had previously found it difficult to get to main trading centres and had relied on travelling Jewish salesmen for their supplies, to come into the market towns to do their own shopping. In Falmouth, Andrew Jacobs, the secretary of the Congregation, recorded in the 1851 census: 'Since the giving up of the foreign packet here, the Jewish population has decreased with the inhabitants generally'.[124]

Whilst the members of the South-West Congregations were experiencing economic difficulties, and perhaps to some extent because of them, they also suffered from internal dissension, possibly occasioned in part by business rivalry and for the rest by a clash of personalities.[125] This, too, may have had the effect of driving away those who wanted to live in a more pleasant atmosphere.

There is no evidence whatsoever that local hostility was driving away the Jews, nor are there any grounds to suggest that most Jews in the South West lost their identity by merging into the general background. Some few men may have married Gentile women; if the latter were not converted their children were not Jewish and were lost to the Jewish community. Even a well placed contemporary observer, Isaac Aryeh Rubenstein, the *shochet* and communal factotum in Penzance in 1886, was at a loss to explain the exodus:

> Many Jews flourished here in abundant plenty and some acted in a representative capacity in local and national government equally with Christians. Moreover those who live here today live on the fat of the

land and enjoy unhindered and uninterrupted peace. In spite of this, our brethren have forsaken this place . . . Why, I know not. It is a riddle without interpretation . . . they leave a blessed land . . . without any compelling reason.[126]

The trend noted above during the period 1820–45 of increasing East-European immigration continued in the third quarter of the nineteenth century. Numerically, the number of German-born Jews steadily dropped after 1851 whilst the number of Polish- and Russian-born Jews rose until the end of the century. This trend is illustrated by the figures in Table 9 which are extracted from the 1851, 1861, 1871 and 1881 censuses. Table 9 shows that the general increase in the foreign-born element in the new Jewish communities of the new industrial towns due to an influx of East-European Jews was reflected in the South-West Congregations.[127]

The first known large scale influx of East-European Jews to the South West of England was composed of prisoners of war brought back from the Crimea. A number of these were imprisoned at Plymouth and Dartmoor and some even brought their wives and families with them.[128] The Plymouth Congregation helped to look after these prisoners and some of them probably settled in Plymouth. Once again they probably wrote glowing letters to their families and friends back in Russia and Poland extolling the wonderful opportunities in England and freedom from oppression and persecution which was characteristic of the British way of life. A good example of this type of panegyric, though of a somewhat later period, is the letter already quoted in part, written by

*Table 9: Number of East-European Jews in Plymouth and Exeter, 1851, 1861, 1871 and 1881*

| | *1851* | | *1861* | | *1871* | *1881* |
|---|---|---|---|---|---|---|
| *Place of birth* | *Plym.* | *Ex.* | *Plym.* | *Ex.* | *Plym.* | *Plym.* |
| Poland | 20 | 9 | 27 | 7 | 41* | 44 |
| Russia | 1 | – | 2 | – | 8 | 2 |
| Hungary | – | – | 1 | – | 5 | – |
| Rumania | – | – | – | – | – | 7 |
| Total | 21 | 9 | 30 | 7 | 54 | 53 |
| Germany | 14 | 9 | 9 | 4 | 25 | 14 |
| Prussia | – | – | – | – | – | 8 |

* Including 5 from Russian Poland and 1 from Prussian Poland.

Isaac Aryeh Rubinstein from Penzance in 1886, and which was printed in the widely-read Russian-Jewish journal, *HaMelitz*:

> In the provinces there are many great cities, and many large factories of different types. One can easily learn a craft or trade which provides a livelihood. The hearts of the English manufacturers are favourably inclined to us, and they make no racial or religious distinctions . . .
>
> Our brethren in Russia and other lands complain that the State closes the way before them, not allowing them to settle in the country—would it not be better for them in the lands of liberty and freedom where there is no Pale of Settlement, where there is none who says to them: 'until here and no further'.[129]

Notwithstanding such glowing reports only a trickle from the flood of East-European immigrants made its way to the provinces. In spite of Rubinstein's warning that in the two or three centres where Jews settled in England in large numbers—'Each grabs his neighbour's food, for seven eyes are cast upon one loaf'—the new immigrants preferred to cluster in ghettos of their own making rather than spread to Cornwall. The Russo-Jewish Committee explained why Jews would not leave the slums of the East End of London and go to the provinces:

(1) No desire to be amongst strangers.
(2) Local prejudices against foreigners.
(3) Some refugees would not learn English.
(4) Lack of opportunity for Jewish education.[130]

There may well have been some force in the second reason—local prejudice against foreigners. The 1871 census returns for Plymouth pointedly identify some of the heads of family as Jews. Thus, the occupation of Leah Morris, born in Germany and not naturalized, is given as 'Jewess', and she has four English-born children. Harriet Bellem's occupation is listed as 'supported by the Jewish community' and she was born in Dartmouth in 1810. Ninety-year-old Rachel Karbman, born in Prussia and still a foreign citizen, is recorded as being on 'Jewish relief'. Isaac Neuman, born in Souack, Poland, and a British Subject is called a 'Pedlar Jew'. And although the door-to-door enumerator recorded the occupation of Samuel Levy as a 'Second-hand clothes dealer and outfitter', the Superintendent Registrar crossed this out and substituted 'JEW'.

The four reasons adduced by the Russo-Jewish Committee may help to account for the reluctance of new immigrants to move to the South West but they do not go far to explain why the local Jews left.

Those East-European immigrants who did go to Plymouth were probably attracted by family or friends. From the Plymouth returns of the

1871 census records we learn the overseas origin of 66 Jewish men and women. The place of origin of a further eight Jewish men in Plymouth who applied for naturalization between 1879 and 1898 has also been noted.[131] In the census returns, sometimes only the country of origin is noted, more often the district is listed, and, occasionally, the town. Of the 74 men, eight came from the district of Suwalk in Russian-Poland, and five of these, including the Roseman and Fredman families, came from the same town, Saki (variously spelled Schaki and Shakie).[132] Tabulating the countries of origin of Plymouth Jews from the 1871 and 1881 census returns, the following picture emerges:

| | *1871* | *1881* |
|---|---|---|
| Austria | — | 1 |
| Germany | 21 | 14 |
| Poland (including Russian-Poland | 40 | 44 |
| Prussia | 13 | 8 |
| Rumania | — | 7 |
| Russia | — | 2 |

From the above figures it will readily be seen that a substantial proportion of the active, adult Jewish population of Plymouth came from Poland and Russian-Poland. Together with the Rumanian and Russian Jews they spoke a common language—Yiddish. The strong influence of family connections on immigration is clearly shown in the case of the Fredman family. The 1871 census reveals three Fredmans known from other sources to be brothers or cousins, Lavine (or Levin as he is called in the 1881 census), Samuel Wolf (he was later called just Wolf), and Jacob David. A fourth Fredman, Levy, was almost certainly also a brother of these three.

By noting where and when their children were born it is possible to determine that Jacob was the first to come to England. He was in Birmingham in 1862 where his first daughter, Leah, was born, and then moved to Devonport where his second daughter, Phoebe, was born in 1863. Levin and Levy arrived in England between 1866/7, when they each had a child born in Saki, and 1870, when Goldie, the last of his children to be born in Saki, was born, and 1871, the year of the census.

Reading between the lines, it appears that once Jacob had established himself as a clothes dealer with a shop in busy Queen Street, Devonport, he called over his brother Levy when there was a job for him as beadle of the Plymouth synagogue, and then called over his family, who started at the bottom of the ladder, as so many other immigrants did at the end of the nineteenth century, as hawkers of sponges and leathers. Sponges were then widely used to rub down horses and leather was needed for harness.

Once an immigrant became established, if he was kind-hearted he would bring struggling immigrants from London who had even a remote tie with him by marriage or common provenance and help them to find work and accommodation and to settle in Plymouth or Devonport. Thus, 'the father of Joe Greenburgh, a most philanthropic gentleman, brought my uncle, Samuel Wolfson, to Devonport before the First World War. My uncle and aunt, being childless, in turn brought me over, and "adopted" me, as my father and step-mother were having a hard time in Lithuania.'[133] This *landsmanshaft*[134] possibly accounts for the survival of Plymouth Jewry well into the twentieth century when many other communities of similar size and nature disintegrated. There is a similar instance of such a cohesive community in Sunderland where the East European emigration came almost entirely from one town, Krottingen.[135] This provided a continuity of tradition which enabled the Sunderland Jewish community to maintain its orthodox character until well into the twentieth century. Possibly something similar happened in Plymouth where two large families, the Rosemans and the Fredmans, together with their *landsmen* dominated the community for the best part of a century.

A large proportion of the nearly three million Jewish emigrants who left Eastern Europe, passed through Germany; 70,000 settling there between 1870 and 1910.[136] The acculturated and assimilated German Jewish community looked on these *Ostjuden* with loathing born of embarrassment and fear. The fear was that the *Ostjuden* would reinforce German anti-Semitic stereotypes and the embarrassment was lest native-born Jews should be identified with the 'dirty, hirsute, kaftan-clad and shabbily-dressed beggars'.[137] The establishment in London reacted in a similar way,[138] but there is no evidence to suggest that the English-born Jews in Plymouth displayed any hostility towards the East European immigrants. Rather, the new immigrants in the latter part of the nineteenth century were probably greeted as a welcome reinforcement to a declining community.[139] Putting it another way, neither in Plymouth nor in Devonport were there sufficient Jewish men to establish a viable rival Congregation.[140] Moreover, the new immigrants trickled in slowly and were more easily absorbed and assimilated by the existing Jewish community. The same welcome was extended in the twentieth century to the Jews who settled in and around Plymouth during the Second World War. From the 1880s, the Plymouth Jewish community was the only viable and active one in the South West. In 1883 it numbered about 230 souls.[141] It continued to grow, if slowly, from 300 in 1906,[142] until there were about 400 in 1935.[143] After the Second World War there was a decline in numbers, an accurate count in 1965 indicated 202 Jewish souls. In 1970, in the whole of the South West of England there were about 350 Jewish souls,[144] but that number has further declined, so much so that by the beginning of the 1990s only intermittent services are held in Exeter and Torquay and it has become impossible to ensure a regular *minyan* for Friday night and Saturday morning services in Plymouth.

## *Part 2*

## *The Towns of Minor Jewish Settlement in the South West of England*

Besides the principal centres of Plymouth, Exeter, Falmouth and Penzance, where Jewish life was organized on traditional lines, traces of individual Jewish families have been noted in another 15 Devon and 12 Cornish towns in the period 1750–1900. In these towns there is rarely evidence of more than two or three Jewish families resident at any one time, more often there was only one, but it should be borne in mind that there may have been several or even many more.[145] In spite of their small numbers these isolated and scattered families represent an important factor in the development of provincial Jewish life, as they provided a kind of cross-country hostelry, always ready to welcome a Jewish hawker (or commercial traveller as they preferred to style themselves in the latter part of the nineteenth century), or simply to help a poor Jew on his way to the next Jewish community where he would find food, shelter, the offer of a job, and financial help.[146]

The Jews who were settled outside the four Devon and Cornish communities generally maintained close links with the nearest of them. Frequently we owe our knowledge of their settlement in places like Tavistock, Dartmouth and Totnes to the fortunate usage of synagogal officials who, in Congregational records, appended a man's town of residence to his name. This was often done to distinguish two men of the same name, and also when the man was long resident in a town and was perhaps the only Jew there,[147] hence *Libche Truro*,[148] *Alexander Truro*,[149] *Jacob Jukel Tavistock*,[150] *Izak Totnes*,[151] and *Nathan Dartmouth*.[152] On the other hand it appears that Jews sometimes settled in towns where there was no organized Jewish life because they wished to cut their ties with their faith, frequently after marriage to a Gentile. This perhaps explains the presence of 'Silas Dursley, died Thursday sen-night a Jew, in the 109th year of his life, at Chudley, Devon' as early as 1729,[153] though his name does not seem to indicate a Jewish origin.[154] Another Jew who settled in Devon at an early date was one Samuel Hart, brother of the Moses Hart who was one of the pillars of the London Jewry.[155] Samuel's will was drawn up in Yarnscombe on 1 August 1744 and from its provisions it is apparent that he had married out of his faith and left it.[156] It is likely that there were other such Jews, but unless their name was typically Jewish, once they severed their connection with the synagogue it is extremely difficult to find any trace of them.

Occasionally a Jew achieved national fame, and his birthplace consequently became public knowledge. Such a one was Elias Parish-Alvars, famous English harpist and composer, who was born in Teignmouth in 1808.[157] The search for traces of Parish-Alvars led to the discovery of a Joseph Parish, 10 Wellington Row, Teignmouth, who was a music seller there in 1827.[158] Without any other indication it is difficult to identify Joseph Parish as a Jew, but the similarity of name and musical occupation suggests that he was related to Parish-Alvars. Another Jew, Isaac Gompertz, member of a famous family of that name, published a poem called *Devon* in Teignmouth in 1825. The book contains an epitaph to his brother Barent who was buried in the Exeter Congregation's cemetery. The *Gentleman's Magazine* mentions the presence of Jews in Barnstaple from 1765–1805 and in Dartmouth in 1764.[159] A record at the Royal Institution of Cornwall, Truro, shows that Moses Jacob was well established in Redruth in 1767.[160] The census of 1851 shows, perhaps best of all, the widespread birthplaces of the members of the South-West communities in the first half of the nineteenth century. There were the Ezekiel and Levi families in Newton Abbot from 1780 to 1800; Joanna, wife of Henry Moses, was born in Hayle in 1812; Betsy, wife of Abraham Abrahams, was born in Callington in 1799; Moss J. Jacob and Amelia wife of Henry Joseph of Penzance were born in Camborne in 1813 and 1812; Esther, daughter of Agnes and Jacob Moses, was born in Yealmpton in 1829; Martha, wife of Henry Woolf, was born in Bodmin in 1829; and Edel, wife of Joseph Joseph was born in Liskeard in 1771. Towards the second half of the nineteenth century English-born Jews in Plymouth came from a narrower spread of birthplaces. From the 1871 census returns for Plymouth we learn of only two other families in Cornwall. There was the Joseph family in Redruth in 1842, and the family of Mrs Henrietta Jacobs in Truro in 1845. In each of these instances it may be assumed that there was at least the one Jewish family resident in the town at the time of birth. The Aliens List of Plymouth reveals Jews settled in Cornwall from the first half of the eighteenth century. Levy Emanuel was a silversmith in Truro from 1748–63; Isaac Van Oven was a spectacle maker there from 1771–85, Barnett Levy was a silversmith in St Austell from 1758–75; Moses Israel made his home in Tavistock from 1797; from 1764, Moses Mordecai was a silversmith in Dartmouth, where he was joined by Nathan Joseph in 1784; Mordecai Jacobs plied his trade as an umbrella maker in Cornwall, probably travelling from place to place without any settled abode, from 1753–73.[161] Local directories also reveal the presence of Jews at Truro,[162] St Austell,[163] Liskeard,[164] Newton Abbot,[165] Bideford,[166] Tiverton,[167] Dawlish,[168] and Torquay,[169] particularly in the early part of the nineteenth century.

It may be that Jews settled in all the main centres of trade where it was possible for yet another all-round trader, particularly one specializing in jewellery and the watch trade, to make a living. As soon as the man ceased to trade, either because of old age or death, his widow and children might well move back to a neighbouring organized community. This process may be illustrated by the Bellem family of Dartmouth. The progenitor of the family was *Matathias Hyman ben Elijah Bellem*, who was a member of the Plymouth Congregeation before 1786, at which time he was in Dartmouth.[170] He had a son Aaron who was born about 1780,[171] probably in either Dartmouth or Plymouth.[172] Aaron married a Plymouth girl, Rachel, born in 1783,[173] and she joined him in Dartmouth where their children Harriet, Hannah, Jacob, Abraham, Esther and another were born.[174] Aaron died on Wednesday, 23 October 1833 in Dartmouth, and was brought to Plymouth for burial on the following day.[175] By 1841, his widow and all her children were back in Plymouth. She died there in 1853, leaving her deaf and dumb daughter Harriet behind her to be supported by the Plymouth Jewish community until she too died in 1890.[176] After the death of their mother the other children left Plymouth for unknown destinations.

Table 10 summarizes the minor Jewish settlement in Devon and Cornwall in the eighteenth and nineteenth centuries (see Map 4).

In the twentieth century, individual Jewish families, generally with a weak attachment to their religion, settled for short periods throughout Devon and Cornwall. Occasionally, particularly in the second half of the twentieth century, such a family would join the Congregation at Plymouth or Torquay. During the Second World War a number of Jewish children were evacuated to the South West. Mrs Doreen Cohen (née Barnet) was one of a group from the West Ham Secondary School, London, who were evacuated to Helston in Cornwall. She recalled:

> Passover was a memorable time. Together with the Jewish parents we prepared everything ourselves. The provisions came from London but arrived very late, so for the first few days we only ate *matzah*. The minister, Rabbi Salomon, kept trying to reassure us that food was on its way. I'll never forget slaving over the frying pan preparing the *Seder* meal. I had to fry enough fish for 50 children and parents.

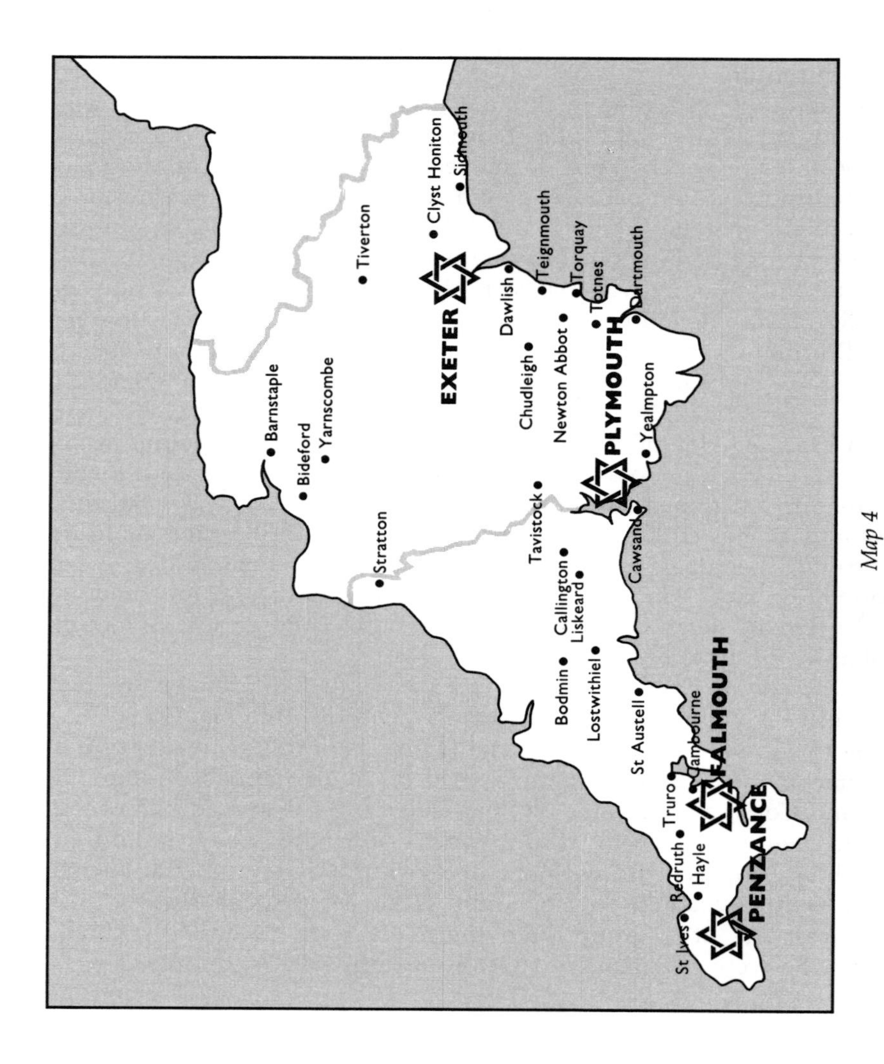


*Map 4*

Major and minor settlement of Jews in Devon and Cornwall in the eighteenth and nineteenth centuries.

*Table 10: Minor Jewish Settlement in Devon and Cornwall in the Eighteenth and Nineteenth Centuries*

| *First and last known dates of Jewish residence* | *Number of Jewish families* | *Devon* | *Cornwall* |
|---|---|---|---|
| 1729 | 1 | Chudleigh | |
| 1744 | 1 | Yarnscombe | |
| 1748–1844 | 14 | | Truro |
| 1758–1820 | 5 | | St Austell |
| 1764–1850 | 7 | Dartmouth | |
| 1765–1844 | 6 | Barnstaple | |
| 1767–1849 | 6 | | Redruth |
| 1771–1823 | 2 | | Liskeard |
| 1786–1830 | 7 | Newton Abbot | |
| 1786–1839 | 3 | Totnes | |
| 1788–1837 | 5 | Tavistock | |
| 1794–1825 | 2 | Bideford | |
| 1799 | 1 | | Callington |
| 1808–74 | 8 | Teignmouth | |
| 1812–35 | 3 | | Hayle |
| 1813 | 1 | Clyst Honiton | |
| 1813 | 1 | | Cambourne |
| 1815 | 1 | | Cawsand |
| 1818–41 | 2 | Tiverton | |
| 1819–82 | 5 | Dawlish | |
| 1821 | 1 | | Lostwithiel |
| 1821 | 1 | | Stratton |
| 1823–57 | 6 | Torquay | |
| 1827–54 | 3 | Sidmouth | |
| 1829 | 1 | Yealmpton | |
| 1831 | 1 | | Bodmin |
| 1831 | 1 | | Saltash |
| 1856 | 2 | | St Ives |

Habonim, a Jewish youth movement, set up South-Devonshire hostels, in Exmouth, Dawlish and Teignmouth, for the new concentrations of evacuated Jewish children.[177]

Some Jewish families from Plymouth moved inland to escape the bombing. One family moved to Horrabridge, and Mr J.B. Goodman could be seen walking the ten miles or so, there and back, every Saturday morning, under the hot summer sun or through the deep Dartmoor snow drifts of winter, to attend the Sabbath service in the Plymouth synagogue.

## *Part 3*

## *The Area of Town Occupied by Jews*

Various factors appear to have determined the part of town where Jews settled. Foremost, probably, in the early years of settlement, at least, was the economic factor. In the first stages of settlement Jews seem to have lived near the docks and traded mostly with visiting seamen and the local poor.[178] As they became better established and more prosperous they gradually moved away from the old part of town towards the newer and better-class trading areas. When, with the passage of time and increasing prosperity, it was no longer customary to live over the shop, they tended to leave their businesses in the commercial centre and moved out to the more fashionable suburbs.[179]

This pattern of movement is well illustrated by the Jewish community of Plymouth. Of the 24 addresses which have been noted of Plymouth Jews from 1750 until 1795, ten of them were in Southside Street and one on Southside Quay.[180] The Plymouth Town Rental Books of the eighteenth century list those who utilized the Plymouth Conduit System. Again it may be observed that Jews who were prominent in the affairs of the Plymouth Congregation and amongst its generous benefactors lived near the dock area, as Table 11 indicates.

From 1798 until 1814, Jewish settlement increased substantially. During this period at least 29 Jews lived and traded in and near the docks, 11

*Table 11: Plymouth Jews Utilizing the Conduit System 1751–1800*

| *Name of User* | *Name of street in which premises were located* | *Year in which use of the conduit began* |
|---|---|---|
| Joseph Jacob Sherrenbeck | Southside Street | 1751 |
| Solomon Abrahams | St Andrews Street | 1755 |
| Jacob Myer Sherrenbeck | Southside Street | 1759 |
| Heart Abraham | Trevail Street | 1762 |
| Lyon Homberg | New Quay | 1767 |
| Abraham Symons | Southside Street | 1767 |
| Mr Mordecai | Southside Street | 1773 |
| Solomon Abraham | St Andrews Street | 1776 |
| Abraham Alexander | Old Town Street | 1776 |
| Mr Nathan | Old Town Street | 1777 |
| Mr Hart | Southside Street | 1777 |
| Mr Mayer | Southside Street | 1777 |

(Source: Plymouth Town Rental Books at the Plymouth Library.)

were in the market area, whilst there were 32 in the centre of the town. Once again, the entries in the Plymouth Town Rental Books for 1801 indicate a concentration of Jews in Southside Street and nearby, as Table 12 illustrates.

At the beginning of the nineteenth century there was a westerly movement towards Plymouth Dock; four Jewish traders opened businesses in Stonehouse and 27 settled in Plymouth Dock,[181] then thriving under the stimulus of prize-money, high wages, and pilferage from the Dockyard estimated at half a million pounds worth a year.[182]

Another factor influencing place of residence of Jews in towns, apart from proximity to employment or place of business, was the need to be within walking distance of the synagogue. Observant Jews refrain from riding on Sabbath and Festivals. Until the breakdown of Sabbath observance in the mid-nineteenth century the great majority of Jews lived in close proximity to the synagogue. A further factor which led to Jews inhabiting one section of the town, was that when they first came to a town they tended to live with relatives,[183] and later, naturally enough, took premises in the near vicinity.[184]

As members of the community became more settled the proportion of Jews living and trading in the more fashionable streets rose. Thus *White's Directory of Devonshire*, 1850, lists 33 Jews in Plymouth, and of these the business of 28 is mentioned. The Jews moved away from the quayside and dock areas into the more fashionable shopping streets. The few without trade or profession tended to live in the expensive residential

*Table 12: Jews Utilizing the Conduit System in 1801*

| *Name* | *Address* | *Annual Rental* |
|---|---|---|
| Abraham Aaron | Southside Street | 8s. |
| Sander Alexander | Southside Street | 8s. |
| David Cohen | High Street | 8s. |
| Benjamin Hart | Southside Quay | 16s. |
| Henry Hart | Southside Street | 8s. |
| Samuel Hart | High Street | 8s. |
| Solomon Isaac | Southside Street | 12s. |
| Jacob Jacobs | Southside Street | 8s. |
| Sarah Jacobs | Southside Street | 8s. |
| Joseph Joseph | Southside Quay | 8s. |
| Nathan Joseph | Great George Street | 12s. |
| Rosey Joseph | Bilbury Street | 16s. |
| Sarah Moss | Southside Street | 8s. |

(Source: Plymouth Town Rental Books at the Plymouth Library.)

districts. Stonehouse had not yet become a virtual extension of Plymouth and White records only three Jews there. Devonport's Jewish population dwindled after the Napoleonic Wars; in 1814 Rowe listed 23 Jews in business there, whilst White in 1850 noted only eight.

This pattern was maintained until the end of the nineteenth century, except that during this period Stonehouse became a busy commercial centre and more Jews moved there. Thus in Eyre's *Plymouth Directory*, 1896, 44 businesses owned by Jews are listed, 26 of them are in Plymouth itself, 10 in Devonport, and 8 in Stonehouse, and most of them are in the busy trading areas. Most of the heads of Jewish households in the Three Towns have been listed by this directory as there were only 46 members in the Congregation in 1896.

Immigrants, whether Jewish, Asian, Chinese or Greek-Cypriot, tend to use their homes as shops. There are sound psychological as well as economic reasons for this. The family stays together, double rent is avoided, and all members of the family can help in the long hours the shop is open.[185] After the first World War, with growing prosperity, it became unfashionable for people to live above their shops and there was a general movement in Plymouth, as elsewhere, to the suburbs. The Jews of Plymouth followed the general trend. Most of them disregarded the Jewish law which forbade them to ride on the Sabbath, so they no longer needed to live within walking distance of the synagogue in the centre of the town. By 1970, all but a few of the 90 or so Jewish families in Plymouth resided in the fashionable Mannamead or Hartley areas.

## *Part 4*

### *The Movement of Jews Within and Without the South West*

Sometimes an immigrant would settle in one of the South-West Jewish communities and remain there until his death. Perhaps just as often his stay was limited to a number of years and when new opportunities arose he moved on elsewhere. Even if the immigrant himself remained, then almost invariably some of or all his children, and especially daughters, would move either to other towns in Great Britain or to other countries. The remainder of this chapter is an attempt to answer the historian L.P. Gartner's question, 'Whither did these Jews go?'[186]

Throughout the period dealt with in this work the main attraction was no doubt London, containing as it did the largest Anglo-Jewish community in the country. Thus Gumpert Michael Emdin, silversmith of Stoke Dammerel, earlier than 1761,[187] after his bankruptcy in 1767 moved to

St Martins-in-the-Fields, London, where he died in 1775.[188] Men with a skill which had no sufficient local outlet had to go to London. In 1819, Solomon Hart RA, who later became librarian of the Royal Academy, moved to London from Plymouth with his father, the latter being in some financial trouble.[189] From the same city left More (Morry) Mordecai, perhaps the same as Moses Mordecai of the Aliens List,[190] 'He left from here *Rosh Hodesh Ellul*, [5]580 [Thursday, 10 August 1820], for the holy congregation of London', as did *Moses Hayyim ben Abraham Ralph* who 'journeyed from here for London' in 1816.[191] From Exeter, Jacob Solomon and his wife, née Philips, sailed to London in March 1830,[192] where Myers Solomon died in 1874.[193] From Dawlish, Leon Solomon moved to London about 1860.[194] Two of the founders of what was to become a leading London synagogue, the Hampstead Synagogue, came from Plymouth and Falmouth. They were Frank I. Lyons, who was born in Stonehouse in 1846, and Alexander Jacob, son of Moss J. Jacob, born in Falmouth in 1841. In 1859, he joined the gold-rush to British Columbia; he returned to England *c.* 1861, stayed a short time in Birmingham and then settled in Hampstead.[195]

Others went to the neighbouring towns with which the South-West Jewish communities had always maintained close ties, and of these Portsmouth was the most important.[196] Those who had lived there before settling in Plymouth include Michael Barnett Levy who was there from 1766 until 1776; Barnet Levy, 1765 until 1775; Eleazer Emdin, from 1794 until 1798;[197] Phineas Levy who was born there in 1784 and moved to Plymouth Dock sometime before 1812;[198] *Moses ben Joseph Gosport* who married Zirrele, daughter of Benjamin Jonas of Plymouth, and who settled in Plymouth before 1813;[199] and Barrow Moss (1782–1817) who was a bachelor in Plymouth in 1805,[200] and was married to Sarah (Sally) daughter of Solomon Isaac[201] by 1810.[202] Exeter, too, had a representative of Portsmouth Jewry in the person of Joseph Marks who was born in Portsea and moved to Exeter after his marriage to Julia Solomon of Exeter in the early 1830s.[203] It is natural that Plymouth Jews with Portsmouth connections returned there for personal reasons. Caroline, orphan daughter of Barrow Moss, married George Jackson of Liverpool at the Portsmouth Synagogue in 1845,[204] and Abraham Ezekiel after 50 years in Exeter moved to Portsmouth some three years before his death in 1799.[205]

Birmingham attracted Jews from Devon in the latter half of the nineteenth century. The Misses Silverstone of Exeter, Honiton Lace manufacturers, moved to 29 Paradise Street, Birmingham, in 1864.[206] Hyman Hyman[207] moved there to 55 Vyse Street, between 1852 and 1865, when his daughter Lizzie married the Revd George J. Emanuel BA.[208] In 1843, Ellen Ezekiel married Abraham Mosely, then living in Bristol, son of Moss Mosely, a member of a numerous and well-known Birmingham

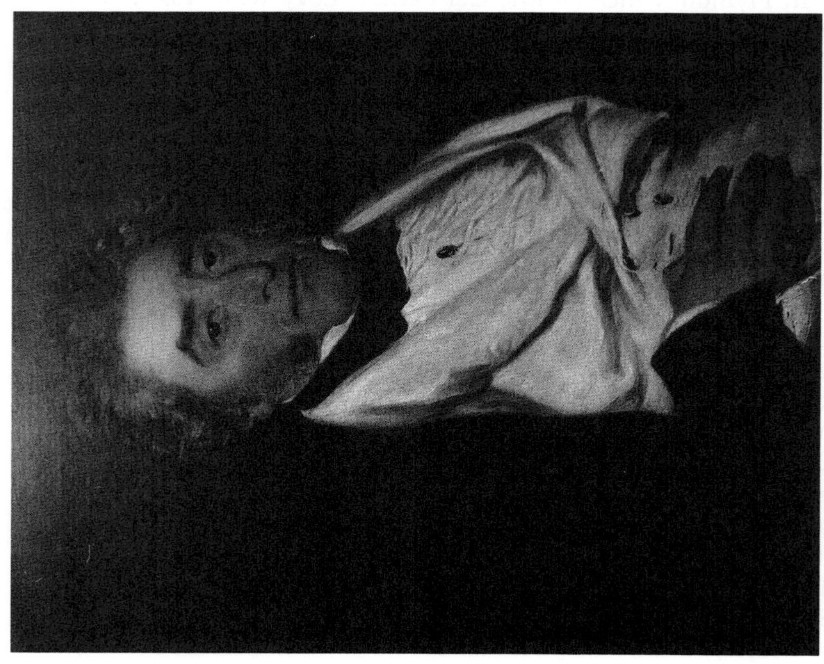

*Illustrations 1 and 2*

Jacob (1782–1857) and Sarah (1789–1875) Solomon of Exeter.

family.[209] As early as 1821, Jewish travellers from Birmingham were visiting Plymouth.[210]

Most of the other communities in Great Britain in the nineteenth century had a representative from one of the South-West Congregations. Lyon Joseph, formerly of Falmouth, died at Bath in 1825.[211] Jacob Abraham, optician and precision instrument maker of Fore Street Hill, Exeter, set himself up in Bath in 1800.[212] Samuel (Mark) Lazarus was installed as Master of the Royal Sussex Lodge in Bath in 1827. His daughter married a Plymouth Jew called Lyon. When she was widowed, she returned to Bath with her adult daughter Phoebe who married Solomon Wolfe, the *chazan* of Bath from 1816 to 1866.[213] *Naftali ben Judah,* born in Plymouth 1788, is described as 'of Chatham' in 1821.[214] At Liverpool, Barnet Joseph, son of the Lyon Joseph who retired to Bath, founded the Hope Place Synagogue,[215] whilst Abraham Hoffnung, son of an Exeter minister, was largely responsible for the building of the Princes Road Synagogue between 1870 and 1874.[216] At Merthyr Tydvil in 1873 was Mrs Isaacs, daughter of the Revd B.A. Simmons of Penzance.[217] Manchester's connection with Plymouth goes back at least to 1821 when an *Isaac ben Joseph* was in Plymouth for the Days of Awe, perhaps at the same time as two other travellers who made offerings to the Congregation after the Festival of Tabernacles.[218] To Newport went Catherine, another daughter of Revd Simmons, who married Hyman Feinburg there in 1850,[219] and yet another daughter of his married the Revd Harris Isaacs there.[220] Revd Stadthagen's daughter, Phoebe, married Abraham Isaac, pawnbroker of 36 Commercial Street, Newport, in 1858, and settled there.[221] A Nathan Jacob and his wife Miriam of Dartmouth settled in Sunderland at the beginning of the nineteenth century.[222] Jewish brides from Penzance went to Pontypridd (1865), Dowlais (1869), Aberdare (1870) and Newcastle (1876)[223] and from Plymouth twelve went to London,[224] six to Birmingham,[225] three to Bristol,[226] two to Manchester,[227] and one each to Cambridge and Cardiff.[228]

About the end of the first quarter of the nineteenth century there was a large emigration of London Jews overseas. In a plaintive aside in the London Beth Din minutes, Rabbi Hirschell records in 1833: 'in our generation America, Asia, and Africa have become like the environs of London'.[229]

There are three main reasons for mass migrations of people—to escape persecution, to find a haven from tyranny, and the desire to better one's economic condition. Religious persecution and political tyranny may be discounted as factors prompting Jewish emigration from England in the nineteenth century. Economic opportunities no doubt provided the main stimulus, and they were seized in the first instance by young men of middle-class background. Personal factors such as poor health, frustrated love and family incompatibility may also have been factors. Jews from the South West followed the same trend as their brethren from London and other parts of England. In the early part of the nineteenth century some returned to Germany[230] where one man, Jacob Hart, obtained diplomatic distinction as the British Consul in Saxony.[231]

Others went to America.[232] One group of former Exeter Jews founded the Jewish community of Cincinnati, Ohio.[233] In October 1816, Joseph Jonas[234] of Exeter arrived in New York from Plymouth and went on to Philadelphia. He wanted to continue to Ohio but was told, 'In the wilds of America and entirely amongst the Gentiles you will forget your religion and your God'. On 2 January 1817, he set off for Pittsburgh. The Ohio was frozen, so he waited until it thawed and on 8 March 1817 he arrived in Cincinnati, the only Jew amongst some six thousand inhabitants. He soon established a prosperous business, but for two years he kept the High Festivals in solitude.[235] His coming created quite a stir. An old Quakeress came to see him and asked, 'Art thou a Jew? Thou art one of God's chosen people! Wilt thou let me examine thee?' She turned him around and finally said, with a tinge of disappointment in her voice, 'Well, thou art no different to other people'.[236] He kept his shop closed on Saturday, the Jewish Sabbath, and when a farmer who had left his watch with Jonas for repair came back and found the shop closed, he thought at first that Jonas had absconded. The neighbours explained the situation and he went back home to explain to his wife why he could not obtain the watch. She went to see Jonas and asked him, 'Are you a Jew?' 'Yes', he replied; she raised her eyes, 'How I thank you, O God, for allowing me to see one of the sons of Abraham before I die'.[237]

In December 1818, his friend David Israel Johnson[238] of Portsmouth came and stayed awhile. He was followed in June by Lewis Cohen of London, Barnet Levi of Liverpool and Jonas Levy of Exeter,[239] and a few days later by his brother Abraham,[240] his sister and her husband, Morris and Sarah Moses,[241] as well as Philip Symonds, together with his wife and child, all originally of Portsmouth. In 1820, three German Jews arrived. They were followed in 1821 by another Portsmouth Jew, in 1822 by Phineas Moses[242] and Samuel Jonas[243] brother of Joseph Jonas, and in 1823 by two more Jews from Portsmouth and one from Barbados. They formed themselves into a Jewish community; Joseph Jonas was the first President, and Jonas Levy of Exeter was the first *shochet*.[244] In 1826, they performed the last rites for the two daughters of the Revd Gershom Seixas of New York who were married to Abraham and Joseph Jonas.[245]

Two years later, they lost another member when Samuel Joseph, originally of Plymouth and later of Philadelphia, died.[246] Joseph Jonas continued to be a pillar of the community which had grown to 2,000 souls by 1844, and at his death in 1869 the Board of Trustees resolved:

> That we recognise in Joseph Jonas, the Israelite, indeed, and free from guile, whose course through a long life has been such as all good men may study, and whose peaceful end at the age of 77 all may envy. *Resolved*, that as the founder of our holy congregation, and as the first Jew that trod this city, we owe him a debt of gratitude which we can only pay by acknowledgement . . .[247]

His brother, Abraham Jonas, had an even more distinguished career. In 1815, he and two of his brothers, Baruch and Jonah, were members of the Plymouth Congregation. Abraham, however, was not apparently an active one, there being no cash entries on his ledger sheet.[248] When Abraham arrived in Cincinnati he went into partnership with his brother-in-law Morris Moses and they opened up as auctioneers. Abraham stayed on only a few years, before moving to Williamstown, Grant County, Kentucky. From 1834 he became a close friend of Abraham Lincoln and an influential leader of the Republican party. Jonas was admitted to the Bar in Quincy, Illinois, about 1839 and Lincoln did most of his work at that period in Jonas' office. In one instance Lincoln directed his Secretary of War to dispose of the case of a man arrested for disloyalty 'at the discretion of Abraham Jonas, whom I know to be loyal and sensible'. A pen portrait exists of him:

> Abraham Jonas was tall, of medium weight, rather inclined to leanness than flesh, with black eyes and hair and complexion between dark and fair. His features were very strong, with a serious face, which broke into a very pleasant expression when amused. He was a very intellectual man and full of humour and wit; and benevolence was well marked in his countenance.[249]

The Samuel Joseph referred to above was the older[250] son of Abraham Joseph I and was born in Plymouth in 1759. He became a Vestry Member of the Plymouth Congregation in 1803.[251] He was married to Rebecca Myers before 1805, and their only child Jane was born in Plymouth in November 1806.[252] He was an active member of the Meshivat Nefesh Society from 1795 until 1817 when his account is marked 'gone to America'. In 1819, the Samuel Joseph's were living in Philadelphia and about 1825 they moved to Cincinnati, a centre of wine growing, where Samuel set up as a beer and cordial purveyor.[253]

While it is rare to have the early beginnings of a community so well documented within the lifetime of the founders, in the case of the Cincinnati community there is a twofold significance for the Jews of the South West. It firstly illustrates the influence of a few men from Devon and Cornwall who set the impress of their religious faith on the development of the community. Secondly, it demonstrates the processes, *mutatis mutandis*, by which the South-Western communities themselves were formed, relatives and friends attracting other relatives and friends in an ever-widening circle.

Other Jews who went to America from Devon include Baruch (Barrow) Jonas, another member of the Exeter Jonas family, who in 1832 remitted from Buenos Aires a small debt which he owed to the Exeter Congregation.[254] There was also a *Samuel ben Alexander Aryeh*[255] who paid five shillings in part payment of his debt of one pound to the Plymouth Congregation on 25 July 1819—'and travelled from here to the country of America on the holy Sabbath of the pericope Deuteronomy, 5579 [= Saturday, 31 July 1819] by ship'.[256] Henry Hyman, later active in American synagogal affairs, went to Virginia from Plymouth about 1830 and moved to New York during the American Civil War.[257]

The preliminaries of emigration and the part played in it by chance are well illustrated by the story of Israel Solomon, a British Jew who had earlier lived in Falmouth. His own memoirs provide a graphic picture:

> My mother died in the early part of 1832, at Bristol, to which city we came from Falmouth, and in Bristol I carried on a retail silver and jewelry trade, combined with pawnbroking. My brother Barnet was there apprenticed to the cabinet and upholstery business. After the death of my mother, we broke up our residence and business in Bristol, with the intention of emigrating to Australia, but by the advice of our cousin Benedict Joseph, we determined to go to New York and in that year, 1832, our business transactions in England were almost completed, so that my brother and the late Benedict Joseph went down to Liverpool to secure three berths on a clipper ship sailing to New York, leaving me in London to close up all business left unfinished. Upon their arrival in Liverpool, the government had issued an order that all passenger ships must have a doctor on board, and on this account the price of passage would be increased five pounds for each passenger. To save the ten pounds that it would have cost them had they waited for me, they started for America without me. When I arrived at Liverpool with the intention of following them, my cousin Barnet Joseph advised me to go to Paris and become agent or commissioner for purchasing French manufactured articles to send to England. His arguments being strengthened by an acquaintance of mine, one Behrends, I followed his advice, and on the saving of ten pounds passage money all my future depended, until I abandoned England forever in June, 1881.
>
> My brother Barnet arrived in New York after a nine weeks' trip. The cholera was then raging, which cause prevented him from getting any position in his trade, and after travelling over a portion of the United

> States he returned to New York and was induced to open a cigar store in which he continued for about one year, when, through the advice of a friend, he renewed his own trade in the year 1834, occupying a store on Broadway, between Grand and Canal streets. The location was considered then far uptown. He succeeded in business, and in the year 1835, he married Julia, a daughter of John I. Hart, of New York. My brother's family consisted of four sons and five daughters, all of whom married, and are now living happily in New York, excepting, however, the youngest, who died in 1879, leaving a daughter to emulate her virtues. My brother remained in active business for nearly fifty years, retiring in 1878, since which time the firm he established has been continued by B.L. Solomon's Sons.[258]

From the 1870s, there was a slow but steady emigration to America by Plymouth Jews. Members of the Pearl family went out, and some came back. And then others of the family followed in their footsteps. According to the 1881 census, Abraham and Henrietta Cohen were born in Germany. Their first two sons, Joseph and Pinchus were born in the USA in 1877 and 1878 respectively. Their third son, Isidore, was born in Plymouth in 1880. Also in the 1881 census are six Jewish visitors staying at the Duke of Cornwall Hotel. There is no way of telling from the census returns whether or not they were related, but they were all born in London except ten-year-old Sydney Abrahams who was born in New York. He was in the hotel with an Alfred Abrahams who was probably related to him. In the twentieth century, Solomon (Spencer) Orgel went to America to try his luck in 1929. His bad fortune was to land in the week of the Wall Street crash. He could not find a job and became desperate. His good fortune led him to a factory where the owner, as a child, had heard his parents speaking of the time when their family was *en route* to America and had been given shelter at a time of desperate need by 'an Orgel family in Plymouth. When you told me your name and where you came from, do you think that I could turn you away without a job?'

To the West Indies went Angel Emanuel of Plymouth, a nephew of Abraham Joseph I, as the Plymouth Meshivat Nefesh account book recorded:

> May his name remain a monument of his virtues, this dear worthy and valuable member went to the West Indies and unfortunately died of the fever then prevalent there, 24 January 1797, 26 Teves 5557.[259]

Abraham Joseph I's daughter Esther married Mozely Isaac Elkin of Barbados in 1821, and died only two years later.[260] A nephew (or niece) of Henry Ezekiel wrote to him from the West Indies in Exeter in 1832,

> the whole town is under Arms—I refer you to the 'Times' 20th–21st–and this day 23rd—the revolting of the Negroes in the Parishes and Neighbourhood where I had spent 19 of my best years and so well know the kindness I experienced from the West Indians . . .[261]

In 1843, Issacher H. Hyman,[262] son of Hyman Issacher, the Plymouth Congregation's beadle, went to Jamaica, and only 10 weeks later, he died of yellow fever.

It was not until the third and fourth decades of the nineteenth century that Jews began to settle in any numbers in South Africa. The three pioneer Jewish congregations were at Cape Town, Port Elizabeth, and Kimberley (Griqualand West), and the three pioneer ministers were Joel Rabinowitz, Samuel Rapaport, and Meyer Mendelssohn.[263]

Mendelssohn had ministered to the Exeter Congregation from 1854 until 1867.[264] He was born in Prussia in 1832 and in 1858 came to Exeter where he married Rebecca, daughter of Israel Silverstone,[265] shopkeeper, of 107 Fore Street, Exeter. In Exeter, his firstborn son, Sidney, was born in 1861. From Exeter, Revd Mendelssohn moved to Bristol where he remained until he received a 'call' from Kimberley in 1878, probably through his brother-in-law, the Revd Berthold Albu, who was appointed the first minister of the Griqualand West Congregation in 1876.[266] Albu[267] himself had been appointed *chazan/shochet* in Exeter on 3 July 1853 and acted in that capacity until April 1854 when he resigned. He stayed on for a few months to marry Bella, another daughter of Israel Silverstone. Revd Mendelssohn held the post as minister to the Kimberley Congregation until 1884, and even in his retirement helped out in times of need, until his death in 1889.[268]

The Plymothians who went to South Africa included a branch of the Emdon family of Plymouth;[269] some of the descendants of Henry and Brayna Morris towards the latter end of the nineteenth century;[270] some of the relatives of Abraham Robins, whose occupation was a 'South African shipper',[271] and the Sulski family.[272]

Besides America, and to a less extent, South Africa, the other great goal for emigrants was Australia, where Jewish communities began to appear about the end of the first quarter of the nineteenth century. Though there may have been individual Jews leaving for Australia from Devon and Cornwall at that early date, the main outflow from the two counties dates from the mid-nineteenth century.[273]

The careers of some of these mid-nineteenth century emigrants may be followed in some detail. There was Henry Joseph (1832–1888), second son of Abraham Joseph II, who set sail for Australia about 1853. He is probably the Mr Joseph referred to in the unpublished diary of John James Bond concerning the voyage of the *Lady Flora* from Gravesend (17 April 1853) to Melbourne (14 August 1853):

> *April 20.* We awoke to hear the seas roaring. Mr Joseph who has eight in his cabin, all ill, came to ours begging to be allowed to say his prayers there as it was not possible in his own.[274]

He made his way to Ballarat, then the centre of the gold rush. He may have started as a prospector, but he was later a gold assayer at Ballarat and Bendigo. About 1867, he married Rebecca, daughter of Samuel Lyons, an auctioneer, and the couple moved to Gympie in Queensland.[275]

Another son of Abraham Joseph II who emigrated to Australia was Solomon Joseph, born on 15 June 1834.[276] His father wanted him to be a rabbi but, as his grandson Wilfred Jessop suggests, he either rebelled against his father's will in this matter or in the choice of a prospective bride, so 'he was to all intents and purposes exiled by his father with £300 in his pocket'.[277] On his 25th birthday he started a sorrowful journey from Plymouth to his New World, and from that day until the day before he disembarked at Melbourne on 5 September 1859 he kept a diary.[278] In Australia he met Caroline Cohen, daughter of Abraham Cohen who had in 1832, in conjunction with George Nicholls, a well known Sydney solicitor, founded *The Australian* newspaper, and they married in 1867. They first lived in Dunedin, New Zealand, where Solomon was employed for a short time by Julius Vogel, later Prime Minister of New Zealand. In 1868, he returned to Australia and after an unsuccessful business career, he founded and edited a weekly communal paper *The Australian Israelite*. In 1882, he acquired a tenuous control of the *Tamworth News* which he edited until his death in 1900.[279] The descendants of these two brothers are today scattered across Australia.

In September 1853, *The Jewish Chronicle* printed a long letter from some of the 350 passengers of the S.S. *Great Britain* which had earlier sailed to Australia.[280] Sixteen of them were Jewish, three each from Birmingham and Liverpool, one each from Bristol and Hamburg, and no less than four from Plymouth. The Plymothians were Simeon Cohen, who was one of the signatories to the letter,[281] Isaac Isaac,[282] Mark Levy,[283] and Abraham Marks.[284]

Isaac Stone was another Jew from Plymouth who went to Australia. He was born in Poland in 1828, came to Plymouth in 1846 as a Hebrew teacher, married Anna Mordecai in 1857, and shortly after went to Australia for the sake of his health.[285] Simcha Stone, a relative, was born in Poland about 1837 and came to Plymouth when he was 18 or 19 years of age. In Plymouth, Simcha Stone sold clothes to sailors, probably not on his own behalf. He suffered from asthma, and about 1883 sailed to Australia with his wife and ten children. One of these, Joseph, had already gone out there in 1876 when he was only sixteen years old, but returned to help with the emigration. The father became a photographer in Melbourne and died there aged 54. Joseph, his son, was a sponge merchant in the same city.[286] Finally, it may be mentioned that Rose Marks, born in Plymouth in 1832 and daughter of Charles Marks, married J. Solomon Henry of Adelaide in 1855 in England.[287]

The mid-nineteenth century Exeter Congregation also had its representatives in Australia. Let us examine one family in detail because its story exemplifies so many of the emigrants. Joseph Marks, an uncle of the Rose Marks just mentioned, was born in Portsea about 1806. He joined the Exeter Congregation in 1832 and within a year married Julia, the Exeter-born daughter of Isaac Solomon who was born in Prussia, and Rosetta Solomon who was born in Rochester, Kent. In 1834, he became a vestry member of the Congregation. In 1838, he was the tenant of a shop in Fore Street Hill, Exeter, with the rateable value of £33 per annum; and in 1844 he is described as a clothier in King Street, Exeter, where he was instrumental in catching a notorious 'fence'. In 1843 and again in 1849 his name appeared on the Voters' Register at 113 Fore Street Hill. In 1853, the Exeter Congregation paid a farewell tribute to him for 'having sat among us for 20 years and filled the office of President and Treasurer'.[288] The Marks family, twelve of them apparently, left Bristol in the *Cotfield* on 1 August 1853 and arrived in Adelaide on 30 November 1853, the arrival being noted in the *South Australian Register*.[289]

Joseph and Julia Marks had thirteen children. Of these, Solomon born 1837, died a few months later; Samuel died and was buried in Exeter in 1870;[290] Isabella (b. *c.* 1845), Alexander (b. *c.* 1849), and Miriam (b. *c.* 1852), were alive in Australia in 1884 but were lost sight of thereafter; Isaac (b. *c.* 1834), Sarah (b. *c.* 1836), Charles (b. *c.* 1839), Josiah (b. *c.* 1841), Ellen (b. *c.* 1842), Rosetta (b. *c.* 1843), Henry (b. *c.* 1847) and Catherine (Kate) (b. *c.* 1850) all died in Australia. Joseph and Julia first settled in South Australia and in Victoria in 1856. There he became a wholesale grocer in Elizabeth Street, where an adjoining clothing shop was apparently run by his wife. Joseph died in Melbourne in 1872. He does not seem to have left very much behind him in the way of worldly goods—furniture valued at £50 and some parcels of land at £20. Letters of Administration were only taken out in 1889, five years after the death of Julia, by which time the land had probably greatly appreciated in value.

As to their children: Isaac was an accountant, married late in life and died in Melbourne in 1904; Josiah left an estate of some £4,000 when he died in 1902. He, too, was an accountant, the founder of a well-known Melbourne Building Society, an active member of the Melbourne Hebrew

Congregation, and a leading Freemason; Alexander had a jeweller's shop in Ballarat; Henry had a successful furniture shop in Melbourne. Their grandchildren included: Julia Marks, a literary figure who wrote poems, novels and songs; Henry and John Harris Marks who dominated the wholesale jewellery trade in Melbourne in the 1920s. One great-grandson, Samuel Clement Leslie (né Lazarus), was a Rhodes scholar who later accompanied Australian Prime Minister Bruce to the 1926 Imperial Conference, and was a UK Ministerial adviser during World War II. Another, Ken Marks, is a judge of the Supreme Court of Victoria, whilst yet another is a County Court judge. Today, the descendants of Joseph and Julia Marks are scattered throughout the length and breadth of Australia, and number in their midst doctors, lawyers and businessmen, both Jewish and non-Jewish.

From Penzance, three of the children of the Revd B.A. Simmons went to Australia—his son Abraham Barnett Simmons, who officiated as lay reader from the 1870s at St Kilda's and Ballarat, and who died in Ballarat in 1908;[291] Amelia, who married Isaac Davidson of London, and Arthur.[292] Benjamin Aaron Selig, a watchmaker, who was born in Penzance in 1812 and was president of the tiny congregation there in 1852, sailed for Australia in 1854. He later moved to New Zealand and he became minister of the Wellington Congregation until his death in 1872.[293]

Emigration offered opportunities for commercial and social success to sons of poor men. One son of the Revd Samuel Hoffnung,[294] Sigmond (1830–1904), or Sidney as he became known, emigrated to Australia from Exeter in 1852, and founded what was to become one of the country's largest commercial organizations.[295] He was educated in Liverpool. Lack of capital forced him to leave home. He became a salesman for a West Country firm and became friendly with one of his Jewish customers, Henry Nathan of Plymouth, who lent him £500 to buy goods and take them to Australia.[296] Another son, Abraham, who travelled extensively on his own behalf as well as on behalf of his brother's firm, cultivated extensive trade links with the Hawaiian Islands. He was appointed Hawaiian Commissioner of Immigration, accredited to the Government of Portugal, and in 1879, at the request of the Hawaiian Government, he organized an emigration from Madeira and the Azores to the Hawaiian Island of about fifteen thousand agricultural labourers. For his services the King of Portugal personally decorated Abraham Hoffnung with the title of Chevalier of the Order of Christ of Portugal, one of the oldest orders in existence.[297] In 1881, he was appointed the King of Hawaii's Chargé d'Affaires at the Court of St James, and he was the 'first Jew who had penetrated the charmed circle of the "Corps Diplomatique" at the Court of the Sovereign of the British Empire'.[298]

*Illustration 3*
Betsy Levy of Barnstaple (1759–1836) who married Isaac Jacobs of Totnes.

*Illustration 4*
Henry Ezekiel of Exeter (1755–1831).

*Illustration 5*
Moses Solomon (born London 1775, died Plymouth 1838).

*Illustration 6*
Solomon Alexander Hart R.A. (1806–1881).

Throughout the ages, the Land of Israel has attracted elderly Jews from the diaspora who go there to spend their last years and be buried in its holy soil.[299] There is no record of any Jew from the South West of England going to Palestine until Abraham Greenbaum, who was the beadle and collector of the Plymouth Congregation some time before 1883, went there about 1900 after his retirement. Old habits die hard, and in a letter dated 3 September 1913 he enclosed a receipt for donations he had solicited in Plymouth on behalf of 'a *Yeshivah,* Ohel Torah in Gibath Shaool, near Jerusalem—where people recite [Psalms] day and night, and men learn and *davven* and pray to the Almighty that all should have health and prosperity'. He died in Jerusalem just before the First World War.[300] The family of Tobias Shepherd, who was in Plymouth about 1876,[301] went out to Israel in the early part of the twentieth century. His daughter, Lily Tobias, became a well-known Anglo-Palestinian writer. In the mid-1960s, the families of Jack and Evelyn Smith, Jack and Eve Cohen, Jack and Shirley Richman, and Harold and Betty Richman all made *aliyah* (emigrated to Israel). Their going halved the number of children in the *cheder,* made it more difficult to assemble a *minyan,* and seriously reduced the viability of the Congregation. They were followed in the 1980s by Maurice and Ruth Overs and the late Reginald and Francis Lewis. In Israel they all made, and those still here, continue to make, a notable contribution to the nascent State.

Such then were the outward peregrinations of some of the Jews of Devon and Cornwall. The emigrants beyond the seas could have provided only a small proportion of the Jewish communities in the New World and British colonies. But from what has been noted above they appear to have provided notable religious and lay leadership, possibly disproportionately to their numbers. Moreover, the accounts of their wanderings that have survived often throw light on the new communities which they formed. This in turn illuminates the steps taken in the formation of Jewish communities in England in the eighteenth century, of which, as has been noted particularly in the South West, little evidence has survived.

CHAPTER THREE

# The Demographic Structure of the Jewish Communities of South-West England 1750–1900

In this chapter aspects of the demographic structure of the Jewish communities of Devon and Cornwall will be investigated. A statistical examination will be made of both the immigrants, as well as of the longer-settled Jewish inhabitants, in respect of their composition by sex, age and marital status. An assessment of health standards will be attempted. Special reference will be made to the great cholera epidemic of 1832 and its effect on the Jews in Devon and Cornwall.

Unfortunately, it will rarely be possible to describe the demographic structure of the four communities over the whole period, as no systematic records have survived (even supposing that they were kept). But enough has been preserved to be able to draw a number of conclusions and to compare and contrast statistical data of the Jews in Devon and Cornwall at certain times with trends in the general population of Great Britain and with other Jewish communities in England and other countries.

## *Part 1*

## *The Immigrants*

Lloyd Gartner has pointed out that younger people tended to dominate the age strata of Jewish immigrants to England in the late nineteenth century, as they generally do in most migrant societies, but that convincingly satisfactory demonstration of this tendency is hard to come by.[1]

There is, however, clear evidence that immigrant Jews—at least the males, as there is little or no information available about females—in Plymouth at various times did conform to the supposed general tendency.[2]

All the male Jewish immigrants who were in Plymouth between 1798 and 1803 were registered on an Aliens List.[3] Their ages on landing in England are given, and during the second half of the eighteenth century it appears that they were predominantly under thirty years of age when they arrived. Table 13 clearly indicates that the majority of the men were in their twenties. In addition, two of the five oldest men had emigrated from their home towns when at most in their early thirties, and possibly much younger.[4]

A similar pattern is apparent in the early nineteenth century. Myer Stadthagen arrived in England as a bachelor in 1828 when he was 24 years old,[5] and Henry William Morris from Prussia was married in Exeter in 1830 aged 23.[6] From the 1851 census it may be seen that of sixteen Jewish aliens in Plymouth, ten arrived before they were 30 years old, four before 40 and two by their early forties. In each of these cases the date of arrival has been computed by noting the birth of the immigrant's first English-born child, and assuming that he arrived at least one year before this event.

In the latter part of the nineteenth century, a similar pattern of young immigrants seems to have been repeated in the Russo-Polish influx. Unfortunately, it is not now possible to identify all the Jewish aliens who were in the South West at this period. But reference has already been made to ten Jewish men in Plymouth who filed applications for naturalization between 1879 and 1897.[7] From the information which they gave,

*Table 13: Age of Jewish Male Immigrants in Plymouth in the Period 1798–1803, on their Entry into England*

| *Age Group* | *Number* |
|---|---|
| Under 12 years | – |
| 13–15 years | 3 |
| 16 years | 6 |
| 17–19 years | 9 |
| 20–30 years | 35 |
| 31–40 years | 3 |
| 48–51 years | 2 |
| over 52 years | – |
| Total | 58 |

(Source: Lipman, 'Aliens List'.)

as well as from other sources,[8] the following picture of their age at the time of their arrival in England appears:

| *No of immigrants* | *Age* |
|---|---|
| 1 | 14 |
| 1 | 16 |
| 2 | 20 |
| 1 | 22 |
| 1 | 23 |
| 2 | 25 |
| 1 | 32 |
| 1 | 38 |

Similarly, the Naturalisation Certificate of Meyer Mendelssohn, minister in Exeter from 1854 to 1867, shows that he landed in England in 1850 when he was eighteen years old.[9]

There is some slight evidence relating to Plymouth from 1841 until 1881 which suggests that more Jewish male immigrants than female arrived in England at that period. The figures extracted from the census returns are shown in Table 14. Examining these figures it is clear that, in Plymouth at least, foreign-born Jewish men far outnumbered foreign-born Jewish women. It is tempting to explain the increase from one foreign-born Jewish woman in 1841 to seven in 1851 on the basis that the immigrant

*Table 14: The Number of Foreign-born Men and Women in the Jewish Community in Plymouth, according to the Decennial Census, 1841–1881*

| *Year* | *Number of foreign-born Jewish males in Plymouth* | *Number of foreign-born Jewish females in Plymouth* | *Total number in Jewish community in Plymouth* |
|---|---|---|---|
| 1841 | 18 | 1 | 193 |
| 1851 | 31 | 7 | 278 |
| 1861 | 32 | 8* | 233 |
| 1871 | 46 | 24† | 268 |
| 1881 | 34 | 27‡ | 280 |

* Plus two boys and two girls under 16 years old who came with their parents.
† Plus three boys and six girls under 16 years old who came with their parents.
‡ Plus six girls and eight boys under 16 years old who came with their parents.

(Source: Decennial census returns, 1841–81.)

*Table 15: The Number of Jewish Immigrants in Plymouth in 1851, 1861, 1871 and 1881 who Arrived as Bachelors*

| | *1851* | *1861* | *1871* | *1881* |
|---|---|---|---|---|
| Total number of adult foreign-born Jewish men in Plymouth | 31 | 32 | 43 | 33 |
| Number of foreign-born Jewish men with foreign-born wives | 7 | 5 | 18 | 23 |
| Number of foreign-born Jewish men with English-born wives | 12 | 12 | 13 | 7 |
| Number of foreign-born Jewish bachelors | 7 | 9 | 5 | 2 |
| Number of foreign-born Jewish men whose marital status or wives' birthplace are not stated in the census | 5 | 6 | 7 | 1 |

(Source: Census returns, 1851, 1861, 1871 and 1881.)

males noted in 1841 had brought over their wives by 1851. But this is not so, as only two of the immigrant males in Plymouth in 1841 were also noted there in 1851, and both of these had English-born wives.[10]

There is reason to believe that most immigrant Jewish males on their arrival in this country in the mid-nineteenth century were unmarried, though the picture seems to change as the century wore on. Table 15 extracted from the decennial census returns of 1851–81, bears out this proposition. From these figures it is apparent that at least 19 of the 31 immigrants in 1851[11] and 21 of the 32 immigrants in 1861 arrived as bachelors. These figures approximate to those for emigrants from England in 1861, which the Registrar-General put at rather less than three unmarried adults to every one married emigrant.[12] The proportion is reversed by the next census, because in 1871 and 1881, the respective figures are 19 out of 70 and 9 out of 61.

## *Part 2*

## *Births, Marriages and Deaths*

At no period has it been possible to make any accurate statistical survey of births, marriages, and deaths, in the Jewish community, whether of the South West or elsewhere in England, because of a lack of detailed precise information. Before 1837, communal records relating to these matters, when kept, were done so haphazardly, and after that date, when the state compelled the registration of such data, synagogues kept only a copy of marriages performed under their auspices but not of all births and deaths. However, such surviving evidence as there is, does permit some observations to be made.

In the South West, records which relate to the births of Jewish boys in Plymouth have survived in the form of Joseph Joseph's Circumcision Register (1784–1834)[13] and entries in the Plymouth Hebrew Congregation's Book of Records (1829–67).[14] Evidence on Jewish male births in Penzance is to be found in the Circumcision Register of B.A. Simmons, in which he recorded the circumcisions he performed there from 1821 to 1847.[15]

In Plymouth, Joseph Joseph performed 74 circumcisions between 1784 (when he was about eighteen years old) and 1816, and then 2 more in 1834. Fifteen of the babies were his own sons, grandson, and nephews, 4 were the grandchildren of Abraham Ralph of Barnstaple, a relative, 7 were Samuel Hart's children, another relative, and 10 others he himself designated as relatives: that is, just less than half were his own sons and sons of his relatives. In addition to these Joseph performed two circumcisions in Exeter in 1786 and 1816, either because the parents were close friends or perhaps Exeter was temporarily without a *mohel*. He also performed a circumcision in Tavistock and one in Totnes, but these were for members of the Plymouth Congregation (see Table 16).

The question arises, did Joseph circumcise all the Jewish boys born in Plymouth in the period 1784–1816 or was there another practitioner also at work? If he was the sole practitioner in Plymouth at this period then the average number of Jewish births per annum in that town can be esti-

*Table 16: The Number of Circumcisions Performed by Joseph Joseph of Plymouth, 1784–1834*

| *Period* | *Number of circumcisions performed* |
|---|---|
| 1784–89 | 17 |
| 1790–99 | 16 |
| 1800–09 | 24 |
| 1810–16 | 17 |
| 1834 | 2 |
| Total | 76 |

mated as twice 2.3, i.e. 4.6.[16] In Portsmouth, over a similar period from 1762 until 1807, Reb Leib Aleph performed 113 circumcisions, an average of about 2.5 per year.[17] In Liverpool, Samuel Yates performed 76 circumcisions in the first quarter of the nineteenth century over a period of 24 years, a little over 3.1 per year.[18] It appears that other *mohelim* lived in Plymouth and may have been active in the period 1784–1816, the beadle Reb Hayyim Issacher being one of them[19] and Lipe Levy (died 1836) another.[20]

Reference has already been made to a register of the births kept in the Plymouth Congregation's Book of Records. The first birth noted was in June 1829 and the last in October 1837. Altogether 43 births were recorded, 25 boys and 18 girls.

The 43 children born over eight years represent a birth rate in the Jewish community of Plymouth of just under 5.4 per year. However, by no means all the Jewish births which took place in Plymouth in that period were recorded, as is evident from the 1851 census returns. Twelve girls and four boys of Jewish parentage were noted in the census who were born in Plymouth between 1829 and 1837 but whose births were not recorded in the Plymouth Congregation's register. When these 16 are added to the 43 in the register, a birth rate of 7.3 per year is obtained. This is almost double the birth rate derived from Joseph's Circumcision Register and, unless there was a marked increase in the birth rate in the later period, further indicates that Joseph was not the sole *mohel* in practice in Plymouth between 1784 and 1816.

A slight record of Jewish births in Cornwall is contained in the circumcision register of B.A. Simmons who was the *mohel* for Cornwall. He records sixteen circumcisions which he performed between 1821 and 1847. There were two in Falmouth, in 1821 and 1823; five in Truro, in 1821, 1823 (two), 1825 and 1826; and the rest in Penzance, which include four of his sons and a grandson. This very paucity of circumcisions in Cornwall is an indication that the communities there were hardly self-generating, though again a caveat must be entered—there may have been other *mohelim* at work.

In his circumcision register, Joseph Joseph of Plymouth happily recorded the date of both birth and circumcision. This throws light not only on the health of the children but also on the number of caesarean births in the period covered by the register, 1784–1816. This is because Jewish law demands that the child should be circumcised on the eighth day after birth unless the child be ill, in which case it is forbidden to proceed until he is better. The usual illnesses which cause a postponement are jaundice or a yellow body, discharge from the eyes, difficulty in feeding, or underweight.[21] Modern *mohelim* anticipate postponement in about 10 per cent of cases, it is therefore noteworthy that all of the 74 circumcisions performed by Joseph Joseph was 'in its time'.[22] Jewish law permits circumcisions to be performed on the eighth day when that is the Sabbath or a Festival only where the birth is a natural one, but if the child is born by a caesarean delivery then the circumcision must be postponed at least to the Sunday, and even then it only takes place if the child is otherwise well. Joseph performed 14 circumcisions on a Saturday[23] and one on a Festival,[24] indicating that there were no caesarean births amongst these 15.[25]

It would be useful to compare the annual birth rate per thousand in the Jewish community of Plymouth at various times with that of all England. Unfortunately, it is rarely, if ever, possible to assess accurately the number of Jews present in Plymouth at one time. It is even more difficult to do so over a period long enough to have a sufficiently large number of births—and there are no accurate statistics of those available—to make any meaningful comparisons possible.

Turning from births to marriages much more can be said. Firstly, the decennial census returns throw light on the marital status of Jews in the South West of England at that time. Secondly, certain marriage statistics may be extracted from the almost complete set of marriage registers recording all marriages which took place under the aegis of the four Jewish communities in Devon and Cornwall from 1837 until 1912.[26]

Utilizing the 1851, 1861, 1871 and 1881 census returns it is possible to obtain a detailed picture of the marital structure within the Plymouth and Exeter Congregations in the mid-nineteenth century. Table 17 summarizes the figures.

We may now examine the marriage registers of the South-West Congregations. Since 1837, each Anglo-Jewish Congregation has kept two identical marriage registers. When they are filled up, one is retained by the Congregation and the other returned to the Registrar-General. The Plymouth Congregation has preserved a complete set of marriage registers from 1837 until the present day. These registers record 149 entries in the first volume which spans the period 16 August 1837 to 16 October 1912.[27] The data which can be extracted from these entries will be discussed in detail after briefly describing the entries in the marriage registers of the other three South-West Congregations.

The Exeter Congregation's registers have now disappeared, but the Registrar-General has his own copy of the registers and they were made available for inspection, with the proviso that the particulars of no individual entry be disclosed and no reference by name to any actual person be made or individuals in the matter be approached.[28] There were 30 weddings celebrated under the auspices of the Exeter Congregation between 1838 and 1872, and one more each in the years 1902 and 1907. The first 20 and last 2 entries in the surviving Exeter register give only

*Table 17: The Marital Status of Jews in Plymouth and Exeter in the Mid-nineteenth Century*

| | Plymouth | | | | Exeter | |
|---|---|---|---|---|---|---|
| *In town on the night of the census* | *1851* | *1861* | *1871* | *1881* | *1851* | *1861* |
| Husbands and wives | 62 | 56 | 90 | 86 | 26 | 18 |
| Married women (husbands apparently away temporarily) | 8 | 1 | 3 | 6 | 6 | 2 |
| Married men (wives apparently away temporarily) | 3 | 2 | 5 | 2 | 3 | 1 |
| Total number of married Jews noted in census | 73 | 59 | 98 | 94 | 35 | 21 |
| Add number of husbands and wives apparently away temporarily | 11 | 3 | 8 | 8 | 9 | 3 |
| | 84 | 62 | 106 | 102 | 44 | 24 |
| Widowers | 3 | 4 | 6 | 2 | 2 | 2 |
| Widows | 10 | 9 | 6 | 5 | 3 | 1 |
| Non-Jewish spouse | 1 | – | – | – | 1 | – |
| Marital status not mentioned | 1 | 1 | – | – | – | – |
| Total number of Jews noted in the census | 278 | 233 | 268 | 280 | 128 | 73 |

(Source: Decennial census returns, 1851–81.)

the names of the bride and groom and the place of marriage, i.e. there are only ten entries in which full information is given about ages, occupations, marital status, and father's occupation.[29]

The Penzance Congregation's copy of its marriage register was deposited with the Board of Deputies, London. According to it there were 17 marriages between 1838 and 1892. The Falmouth Congregation's marriage register has survived only at the General Register Office, and contains only 3 entries. Because there are so few entries in the other three Congregations, any meaningful use of the Jewish marriage registers of the South West for statistical purposes is here necessarily restricted to those of the Plymouth Hebrew Congregation. It should be noted that Jews may contract marriage in England according to Jewish usages lawfully only if both parties are Jewish.[30]

According to Joseph Jacobs, European Jews in the nineteenth century married less than non-Jews of the same country in proportion to their numbers. In Prussia, for example, between 1820 and 1876 there were 75 Jewish marriages per thousand of population compared with 88 amongst Christians. For Russia, the corresponding figures are: for the period 1852–9, 82 Jewish marriages per thousand compared with 95 Christian, and in 1867 there were 87 Jewish marriages as against 100 Christian marriages per thousand.[31] For comparison, the frequency of Jewish marriages in Plymouth for the period 1837 until 1906 may be determined by an analysis of the marriage registers. In this seventy-year period, 135 marriages were celebrated under the aegis of the Plymouth Congregation,[32] which, with an average Jewish population of 266 in any one year, gives an annual rate of 8.0 marriages per thousand.[33]

This figure being barely one-tenth of the marriage rate given by Jacobs is so low that it calls for comment. There is evidence from at least two other Anglo-Jewish communities which suggests that the marriage rate in the Plymouth Congregation was by no means exceptionally low for England. G.H. Whitehill quotes an annual marriage rate of 8.5 per thousand Sephardi Jews in London for the years 1851–1860.[34] The other Jewish community which was not greatly dissimilar to the Plymouth Congregation in the South West of England was the Sunderland Congregation in the North East of England. In the decade 1847–56, 15 marriages were celebrated in the Sunderland Congregation[35] at a time when there were about 150 Jews resident in the town.[36] This would give a marriage rate of 10 per annum per thousand. At the turn of the century from 1897 to 1906 there were 62 marriages under the auspices of the Sunderland Hebrew Congregation when the Jewish population was about 1,100.[37] This gives a marriage rate of about 5.6 per thousand.

Tables 18 and 19 set out in detail the figures on which the annual rate per thousand of marriages celebrated under the auspices of the Plymouth Congregation from 1837–1906 has been calculated, together with the comparative rates for Sunderland Jewry and the London Sephardim.

It would appear that either Jacobs or his printers mistakenly multiplied the marriage rate by ten. This conclusion is confirmed by the marriage rate for the Jewish population of Britain in the period 1901–20 of 9.0 per thousand (dropping to 4.3 per thousand in the period 1971–75) and for the general population of England which varied from 7.5 to 8.6 per thousand between 1901 and 1975.[38]

Age at marriage affects the physical, mental and social traits of a people. Jacobs and other nineteenth-century statisticians defined 'normal' marriages as those effected by brides between the ages of 18 and 40 years and bridegrooms under the age of 40. 'Abnormal' marriages, then, are those where the bride is under 18 or over 40, or bridegrooms over 40. All

*Table 18: Frequency of Jewish Marriages in Plymouth, 1837–1906*

| Period | Number of marriages celebrated | Estimated Jewish population at any one time | Rate per 1000 per annum |
|---|---|---|---|
| 1837–46 | 18 | 200 | 9.0 |
| 1847–56 | 18 | 250 | 7.2 |
| 1857–66 | 15 | 240 | 6.2 |
| 1867–76 | 23 | 230 | 10.0 |
| 1877–86 | 18 | 230 | 7.8 |
| 1887–96 | 17 | 250 | 6.8 |
| 1897–1906 | 26 | 275 | 9.4 |
| | 135 | Average | 8.0 |

(Source: Plymouth Hebrew Congregation Marriage Registers.)

*Table 19: Comparison of the Annual Marriage Rate per 1000 of Jews in Plymouth, Sunderland, London (Sephardim) 1837–1906*

| Period | Plymouth | Sunderland | London (Sephardim) |
|---|---|---|---|
| 1837–46 | 9.0 | 8.0 | – |
| 1847–56 | 7.2 | 10.0 | – |
| 1851–60 | – | – | 8.5 |
| 1897–1906 | 7.4 | 5.6 | – |

(Sources: For Jewish marriages in Plymouth, PHC Marriage Registers; for Jewish population of Plymouth see above, p. 39; for Sunderland and London rates see above, p. 76.)

the brides in the marriages celebrated under the auspices of the Plymouth Congregation in the period 1837–1906 were 'normal', even the two widows remarrying whilst still young.[39] As to 'abnormal' bridegrooms, only one, aged 43, occurs in the 111 entries where the age is given. Again, even the widowers remarried whilst still comparatively young, between 37 and 44.[40] This situation may be contrasted with central Europe where in 1873, 12 per cent of Jewish marriages were 'abnormal' against 35 per cent for Catholics and 33 per cent for Protestants.[41]

The average age of Jewish grooms on marriage in Plymouth from 1850 until 1906 was comparatively young, grooms being 25.7 years of age and brides 22.4. This period may be conveniently broken down into two, the first from 1850–78 when the marriages were predominantly from the

English-born community descended from the Central European immigrants, and the second from 1882 until 1906 when the marriages were mainly from the new Russo-Polish immigration. In the first period the average age of grooms was 27.2 years, and of the brides 22.9. In the second period grooms married nearly three years earlier and brides one year earlier, as Table 20 shows.

These figures compare closely with those obtained from the marriage registers of the Sunderland Hebrew Congregation in the second half of the nineteenth century.[42] In Sunderland from 1852 until 1893 there were 130 marriages where the ages were noted, the average age of grooms being 25.3 years and of brides 22.0 years. The Sunderland Jewish community's registers show a tendency for the average age to increase slightly where the corresponding period in Plymouth shows a decrease, perhaps because the Russo-Polish immigration arrived in the North East earlier than in the South West. The comparison of the average age on marriage of Jewish grooms and brides in Plymouth and Sunderland may be expressed in tabular form, as in Table 21.

By far the largest number of brides and grooms in the marriages solemnized under the aegis of the Plymouth Congregation, and likewise under the Sunderland Hebrew Congregation, in the second half of the nineteenth century fall into the 20–30 year-old age group. Very few Jewish grooms in Plymouth were under 20 years old. Indeed, from 1850

*Table 20: Age at Marriage of Jews in Plymouth, 1850–1906*

| Period | Number of marriages | Average age of grooms in years | Number of Grooms | | |
|---|---|---|---|---|---|
| | | | Aged 19 or below | Aged 20–30 | Aged 31 and over |
| 1850–1906 | 111 | 25.7 | 1 | 96 | 14 |
| 1850–78 | 54 | 27.2 | – | 42 | 12 |
| 1882–1906 | 57 | 24.3 | 1 | 54 | 2 |

| Period | Number of marriages | Average age of brides in years | Number of Brides | | |
|---|---|---|---|---|---|
| | | | Aged 19 or below | Aged 20–30 | Aged 31 and over |
| 1850–1906 | 111 | 22.4 | 14 | 96 | 1 |
| 1850–78 | 54 | 22.9 | 6 | 47 | 1 |
| 1882–1906 | 57 | 21.9 | 8 | 49 | – |

(Source: PHC Marriage Registers.)

*Table 21: Comparison of Age at Marriage of Jews in Plymouth and Sunderland, 1850–1906*

| | | *Average age in years* | |
|---|---|---|---|
| *Town* | *Period* | *Grooms* | *Brides* |
| Plymouth | 1850–78 | 27.2 | 22.9 |
| Sunderland | 1852–74 | 25.0 | 21.8 |
| Plymouth | 1882–1906 | 24.3 | 21.9 |
| Sunderland | 1874–93 | 25.4 | 22.2 |
| Plymouth | 1850–1906 | 25.7 | 22.4 |
| Sunderland | 1852–93 | 25.3 | 22.0 |

(Source: Plymouth and Sunderland Hebrew Congregations' Marriage Registers.)

(when exact ages were first entered in the marriage registers) until 1878 there were no grooms below 20 years old, and from 1882 until 1906 only 1.7 per cent of the grooms were under 20 years old. In a society where the groom was expected to provide a home and keep his wife it is not surprising to find very few young men marrying under the age of 20. Conditions in the latter part of the nineteenth century in Russia were very different. Nearly half the grooms there were under 20 years old,[43] but that was because it was thought that married Jewish men might be exempted from army conscription—virtually a twenty-five year hard labour sentence from which few returned sound in mind and body.[44] Few Jewish women in Plymouth married when they were over 30 years of age. As Table 22 shows, only 1.8 per cent of Jewish brides in Plymouth in the 1850–1878 period were over 31 years old, and in the 1882–1906 period there were none at all in this age group.

Table 22 shows the percentage of Jewish brides and grooms in Plymouth in the three age categories, with comparative figures of Jews in Russia and Germany and another Jewish community in England.

Before leaving the question of age it may be remarked that 70 per cent of Jewish husbands in Plymouth during the period 1850–1906 were older than their wives. In Budapest, at the same period, 73 per cent of Jewish husbands were older than their wives, compared to 64 per cent of the general population.[45] The average difference of age between Jewish husbands and wives in Plymouth in the nineteenth century was 3.9 years, in Budapest it was 8.7 for Jews, compared with 6.7 for others.[46]

In the 135 Jewish marriages which took place in Plymouth between 1837 and 1906 six widowers remarried but only two widows. This was in

*Table 22: Percentage of Marriages of Jews in England, Russia and Germany according to Age Groups, 1850–1906*

| | | Under 20 years old | | Between 20–30 years old | | Over 31 years old | |
|---|---|---|---|---|---|---|---|
| Place | Period | % of grooms | % of brides | % of grooms | % of brides | % of grooms | % of brides |
| Moscow | 1868–72 | 6.2 | 49.3 | 76.6 | 48.5 | 17.2 | 2.2 |
| Russia | 1867 | 47.6 | 63.2 | 37.9 | 29.4 | 14.5 | 7.4 |
| Posen | 1867–73 | 0.7 | 17.8 | 65.7 | 69.1 | 33.6 | 13.1 |
| Plymouth | 1850–78 | 0.0 | 11.1 | 77.7 | 87.0 | 22.2 | 1.8 |
| | 1882–1906 | 1.7 | 14.0 | 94.7 | 85.9 | 3.5 | 0.0 |
| Sunderland | 1852–93 | 2.3 | 20.8 | 86.1 | 76.9 | 11.5 | 2.3 |

(Sources: The figures for Moscow, Russia and Posen are taken from Jacobs, *Jewish Statistics*, p. 50; for Plymouth and Sunderland from the Marriage Registers of the respective Congregations).

line with the national trend in England where in 1861, for example, twice as many widowers married spinsters as bachelors married widows.[47] The two widows represent 1.5 per cent of the brides. This figure may be contrasted with that of widows remarrying in the general population of England in the period 1851–1861, which was 9 per cent of all the brides.[48] The six widowers represent just under 5.2 per cent of the Jewish grooms and this figure may be contrasted with that of widowers in the general population of England in the period 1851–61, which was 14 per cent of the grooms.[49] This markedly lower remarriage rate of Jewish widows and widowers in Plymouth may be due to the smallness of the numbers or other unknown factors.

From 1837 until 1906 it was seemingly the custom in the South-West Congregations for brides to be married in their home town. Of the 135 Jewish marriages performed in Plymouth between 1837 and 1906, the bride was a Plymothian in every instance except two,[50] and the father of one of these two was Abraham Emdon, a member of a long established Plymouth family, and the other was from Birmingham, probably a Sephardi, and there was no Sephardi synagogue in Birmingham,[51] whereas at least 35 of the 135 grooms came from various other towns.[52]

There is one aspect of the address given by Jewish brides and grooms in Plymouth during the period under discussion which calls for special comment. It is that a surprisingly high number of the grooms who were listed as living in Plymouth (31 out or 100) had the same address as their brides. It is highly unlikely that all these couples were cohabiting before marriage. The possible explanations include:

(a) Some of the grooms may have come from overseas shortly before the wedding, and therefore had no 'usual place of residence' in England. In such cases it would be natural for the groom to give his bride's address.

(b) The party giving notice of the marriage to the Superintendent Registrar may have entered the same address for bride and groom merely as a matter of convenience, either to save the trouble of giving notice in two separate districts, or to save fees which were lower when both parties lived in the same registration district. In either event, such a proceeding was illegal and could, in theory, have led to a prosecution.

(c) There may well have been cases where bride and groom lived in the same street but in different houses, or even the same house but in different parts.[53]

There appears to have been only one certain case of marriage between an Ashkenasi and a Sephardi in the South West of England, and in view of the small numbers of the latter this is hardly surprising. Moreover, the Sephardi community in London positively discouraged such mixed marriages, at least until the middle of the nineteenth century.[54] One such mixed marriage took place in Plymouth in 1823,[55] when a *Jacob ben Shalom* of Mogador, a North African Sephardi, married Deborah, daughter of Benjamin Levy. The only other marriages known, of which evidence has come to light where possibly one party was Sephardi, were those of the brother and sister Marcoso of Birmingham, who married Hannah Samuels and Solomon Wolf II of Plymouth in that town in 1872.[56]

In order to estimate the number of Jews in a community at any particular time it is often necessary to estimate how many individuals were living at home in the average Jewish family, because the size of a community is often given in terms of family units. It is possible to assess accurately the size of the Jewish family in the South West in 1851 by noting the size of Jewish families enumerated by the census of that year in Exeter, Falmouth, Penzance and Plymouth. For the purpose of estimating the size of family only near relatives such as grandfather, father, son, husband, brother, nephew and their female counterparts, as well as in-laws have been counted. Those described as 'visitor', 'lodger', or 'servant' have not been counted.

From Table 23 it appears that the average number of related individuals in the Jewish family in the South West was a little over 4.1. This compares closely with the size of the average family in the whole of England in 1851 which was 4.0.[57] V.D. Lipman, when attempting to establish the number of individuals in the mid-nineteenth century Jewish community of London from a source which gave only the number of families,

*Table 23: The Number of Related Individuals in Jewish Households in the South West of England, 1851*

| *Family size in persons* | *Number of families in* Exeter | *Falmouth* | *Penzance* | *Plymouth* | *Number of individuals* |
|---|---|---|---|---|---|
| 1 | 6 | 2 | 1 | 12 | 21 |
| 2 | 4 | 1 | 1 | 10 | 32 |
| 3 | 6 | – | 3 | 9 | 54 |
| 4 | 1 | – | – | 7 | 32 |
| 5 | 4 | 2 | 1 | 5 | 60 |
| 6 | 3 | – | – | 6 | 54 |
| 7 | 1 | 2 | – | 9 | 84 |
| 8 | 1 | – | – | 4 | 40 |
| 9 | 1 | – | – | – | 9 |
| 10 | – | – | – | 2 | 20 |
| 11 | – | – | 1 | – | 11 |
| 12 | 1 | – | – | – | 12 |
| 13 | 1 | – | – | – | 13 |
| Totals | 29 | 7 | 7 | 64 | 442 |

Total number (29 + 7 + 7 + 64) = 107 family units representing 442 individuals.

(Source: 1851 census.)

assumed a multiplier of five.[58] By the South West experience it would appear that Lipman's estimate was a little on the high side.

Marital fidelity and sexual morals of the Jews in England and the countries from which they originated have been accorded much adulation,[59] though it is as difficult to prove the general proposition as it is easy to find exceptions. Apparently, sexual misconduct was sufficiently frequent in the Plymouth Congregation at the end of the eighteenth century to warrant a special rule: 'One who lives with a non-Jewess . . . may not come to any holy matter' (i.e. he may not be called to the Reading of the Law).[60] Of course, such unions need not necessarily have been immoral as, in the absence of civil marriages, there was no easy way of regularizing them. Abraham Daniel of Bath or Abraham Franco of Plymouth, for example, could only have married their common law wives in church, and though such a church wedding was technically possible for an unbaptized Jew, they were probably reluctant to take advantage of the facility. For in the eyes of the Jewish community such a step would be tantamount to apostasy, whereas both Daniel and Franco retained close feelings of attachment to the Jewish faith: Daniel was mindful of the Plymouth synagogue in his will, and Franco wanted his wife to be accepted by the Jewish community.[61]

The rule quoted above takes official cognizance of Jews living in sin with non-Jewish mistresses, but as there was no similar rule promulgating any punishment for a Jew living with a Jewess without being married to her, it may be assumed that such conduct was virtually unknown in Plymouth in the eighteenth and nineteenth centuries. The virtue of some Jewish women in the South West, however, was not above suspicion as at least one Jew in Exeter apparently suspected his wife's affections. In his will dated 1808, Moses Mordecai left the bulk of his fortune to his wife, Hindla, with the proviso,

> I solemnly prohibit my wife to marry Mr. Bendix,[62] nor shall she take him as a lodger and in case of her doing it nevertheless, then she shall have no more than her *ketubah* [marriage settlement] £20, or to arrest her £1,000.[63]

The minutes of the London *Beth Din* for 1812 record, 'a letter which the head of the *Beth Din* wrote to Plymouth to that man who wishes to live with his prostitute wife'.[64] It is not certain in this case whether the woman was a prostitute in the generally accepted sense of the term, or whether she was one according to Jewish law which defines a prostitute as a woman who has had intercourse with any man to whom she is forbidden. Thus an unmarried woman could have paid intercourse with any number of men, provided that they were not blood relatives, and yet not be classed in Jewish law as a prostitute, whereas a married woman having intercourse but once with someone other than her husband and without a fee is termed a prostitute. The same minute dated 13 *Tammuz* 5572 [= 23 June 1812] refers to a '*Moses ben Judah Mannheim*[65] who has two children from his wife's daughter [i.e. his step-daughter], one, Isaac, born during his wife's lifetime, and the other, ten months after her death, the two children are regarded as biblical bastards,[66] the name of the second is Eleazar Lazer'. There were four men in Plymouth who were called *Moses ben Judah*. They were: Moses Lazarus, son of Lippe Lazarus, who had a son Isaac in 1832;[67] Wolf of Poland who died in the 1832 cholera epidemic;[68] Moses who unsuccessfully tendered for the baking of *matzah* in 1834;[69] and Pike from Exeter who visited Plymouth in 1821.[70] None of them fit well with the man referred to in the *Beth Din* minute, and he may not even have been a Plymothian, in spite of the proximity of the two cases in the one minute.

In the nineteenth century, where matters could be put right the Chief Rabbi in London and the Plymouth Congregation did all they could to facilitate proper marriage. For example, Dr Adler wrote to the president of the Plymouth Congregation on 8 September 1851 that Salmon Eliezer[71] of Plymouth 'eight months ago formed a clandestine union with a Jewish girl in Hull, and now wants to get married'. Adler askes the president to institute enquiries about Eliezer's status. This was followed by another letter saying that everything was satisfactory and the wedding could take place, the fees being waived.[72] Adler again wrote to the president in 1861, this time asking him to investigate the morals of a Mr Marks, and later saying that there was a strong presumption that Marks was guilty and therefore could not read the prayers until the case was heard.[73] Towards the end of 1867 or 1868,[74] Adler wrote to Revd Mendelssohn in Exeter that he had met 'the Zamoisky girl' who had left her mother to be with her father,[75] without her mother's consent. Adler said that the girl could go back to Exeter and people need not think she had been immoral, 'so please avoid calumny as Mrs. Adler is convinced the girl has not done anything wrong'. At the same time Adler wrote to Mrs Zamoisky to say that she should do a mother's duty and take the girl back, as 'she has not yet fallen but there would be great danger if she stayed any longer'.[76]

A few months later, Adler wrote to the president of the Plymouth Congregation saying that he would not send a marriage authorization until he was satisfied that (unnamed) parties had not been previously engaged,[77] 'and if she is pregnant she must confirm on oath before you and the *shochet* that Mr. — is the natural father of the child'.[78]

But if the morals and reputation of some were questionable it is also true to say that the Jewish community of Plymouth took pains to protect the reputation of respectable girls. Two instances of a similar nature confirm this. The first was when an accident befell Esther daughter of Judah on the sixth day of Passover, 1794:

> She was standing on the table to watch the sight of the army to amuse herself like the other girls, as she was at that time between eight and nine years of age. She fell backwards from the table on to a chair which was standing there. Her mother looked at her at once and immediately to see whether she had any wound or bruise. Then she saw blood on her dress there. Then her mother sent to three women—elders, by name Beilah wife of Moses Yorkshire, also the woman Deichah wife of Moses, beadle of the holy congregation, may its Rock and Redeemer preserve it, and also Reichla, wife of Naftali, to rule whether it was blood of virginity or not. They thought that they were in doubt if it was blood of virginity or not. Therefore let this be for a memorial so that there should not be found in her, God forbid, an evil name, for there are men of Belial who say that she whored in the house of her mother and father.[79]

A similar accident befell another girl, Hannah, the ten-year-old daughter of *Manasseh ben Zvi*,[80] who fell from a chair and left blood. She was examined by a doctor who was in doubt whether it was blood of virginity or not. Hayyim Issacher, beadle of the Plymouth Congregation in 1821, recorded the incident in his account book 'so that the *kosher* daughters of Israel shall not be put to shame'.[81]

One indication of the prevalence of pre-marital sex is the number of children born out of wedlock. In all the records of Jewish births in the South West of England from 1784–1867, the name of the father is given. It may be assumed, therefore, that the father was invariably the husband of the mother, and that there were no recorded illegitimate births in this period. This state of affairs may be contrasted with that obtaining in the Jewish community in Newcastle in the six year period 1881–7 when Elias Pearlson recorded 4 male births from unmarried mothers in 94 circumcisions.[82]

It is possible that to some extent sexual morals amongst the Jews of the South West, and perhaps in the rest of England as well, were influenced by, or perhaps just reflected, a discreditable trade which was carried on by some Jews. Southey in his *Letters from England*, (1809) writes, 'but when the Jews meet with a likely chapman they provide prints of a most obscene and mischievous kind . . .'[83] Their participation in this trade is confirmed from another source. At the end of 1814, there were some 2,350 American prisoners of war in Dartmoor Prison with whom 'the Jew traders did a roaring trade in watches, seals, trinkets, and bad books'.[84]

Only one case of apparent criminal immorality certainly involving a Jew in the South West has come to light. Joseph Jacob Sherrenbeck, probably the founder of the Plymouth Congregation, and certainly its leading light until his death about 1780, was fined twenty pounds and imprisoned for two years and had to find sureties for his good behaviour for seven years in 1734 at the Taunton Assizes when he was found guilty of criminal conversation with the wife of one Lazarus Chadwick.[85] Although the offence was described as one of 'criminal' conversation it was almost certainly a case of adultery and not of rape, which would have carried a far heavier penalty.[86] In 1805, a Solomon Hymes was acquitted on a charge of rape at the Lent Assizes for Cornwall and Launceston.[87]

There are no records available to indicate the prevalence of divorce in the nineteenth century either for the Anglo-Jewish community in general or the South-West Jewish community in particular.

Turning to statistics of deaths, unfortunately there is no evidence to make any realistic estimate of the life spans of Jews or Jewish mortality rates in the South West of England at any period. Congregational records which have survived relate to comparatively short periods of time. Tombstones, which in the nineteenth century invariably recorded age at death, were not generally erected for the very young nor for the poor (whose lives were likely to have been shorter than the rich) and may therefore be supposed to give an exaggerated picture of longevity. Nor, apart from the cholera epidemic of 1832, which will shortly be discussed in detail, and a little information in the decennial censuses on the number of deaf and dumb, or blind, or lunatic Jews in Devon, is there any evidence about the type or prevalence of diseases in the Jewish communities of the South West as distinct from the rest of the population of England.

Cholera struck Sunderland in 1831 and rapidly spread throughout England. The seven[88] Jews in Plymouth who died from 'the plague' between 1 August and 2 September, 1832, are listed in Table 24.

The Hebrew entry in the Plymouth Congregation's Book of Records[89] relating to the death of the Franco's reads, in translation:

> *Abraham ben Nathan Portuguese* went to his rest with his Gentile wife with whom he lived without marriage and from whom he has three sons and two daughters. They died and were buried both of them in one day and in one grave by decree of the plague. Monday 8 *Ellul* (5)592.

*Table 24: Plymouth Jews who Died in the Cholera Epidemic, 1832*

| *Name* | *Date of death* |
|---|---|
| Moses ben Judah Woolf | 1 August[a] |
| Shina Assenheim (a child) | 3 August[b] |
| Abraham Lazarus (a child) | 4 August |
| Hannah Woolfson | 14 August[c] |
| Jacob Philip Cohen | 17 August[d] |
| Mrs Fanny Lyons | 28 August[e] |
| Abraham Franco | 2 September |
| Mrs Abraham Franco | 2 September |

(a) He was a recent immigrant from Poland and died in an inn, the Ring of Bells. He left £25 with Lazarus Solomon of Plymouth to be sent back to his family, which was done through the Chief Rabbi in London.

(b) Daughter of Isaac and Hannah Assenheim (PHC Bk of Records, p. 55b).

(c) She was the wife of Jacob Woolfson of London who settled in Plymouth about 1819 (PHC Min. Bk II, 245; PHC A/c. 1821). He subsequently married Rose bas Alexander of Portsmouth and they had a son in October 1834 (PHC Bk of Records, p. 12).

(d) He came to Plymouth from Lantshotz in Poland some time before 1819 and was a bachelor (PHC Min. Bk II, p. 159). He seems to be identical with Meyer Jacob Cohen, dealer in hardware, whose will was proved at Totnes on 22 August 1832, effects under £100 (Devon Record Office, Wills, C794).

(e) In 1822 she was a straw hat maker in Pike Street (*Ply. Direct. 1822*). She was married to Solomon Lyons who married again a year or two later.

(Source: PHC Bk of Records, pp. 55, 56.)

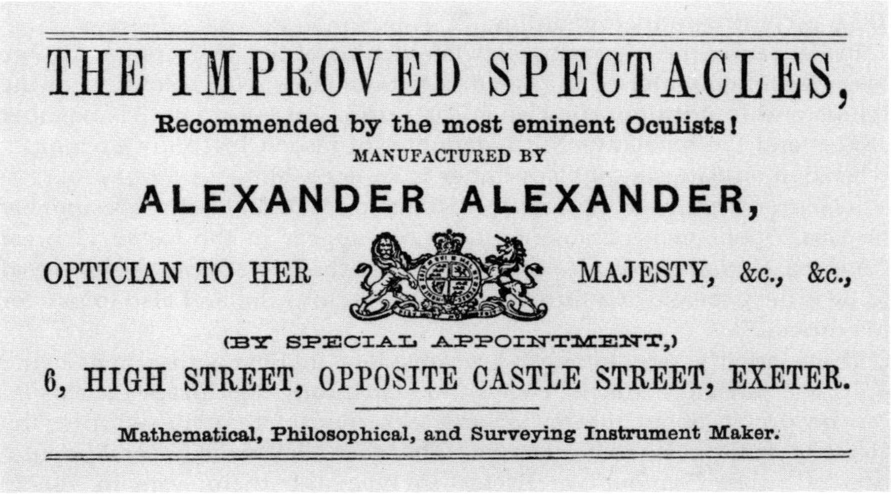

*Illustration 7*
Advertisement of Alexander Alexander, 1857.

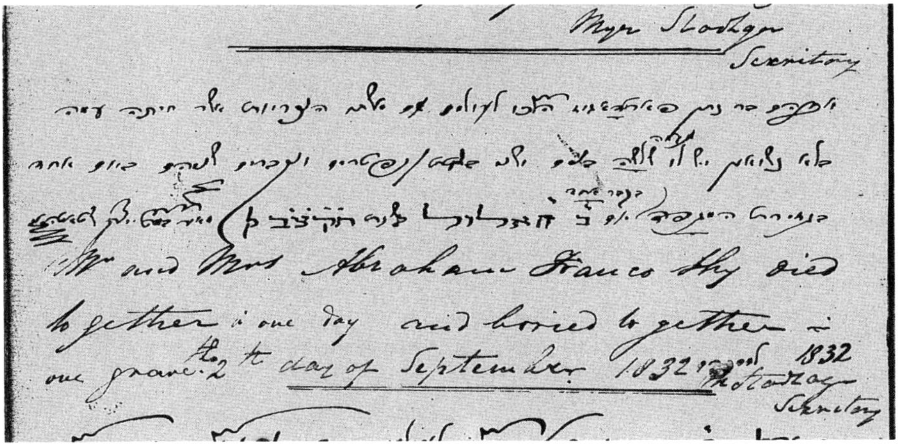

Mr and Mrs Abraham Franco they died together in one day and buried together in one grave 2 th day of September 1832

*Illustration 8*
Mr & Mrs Abraham Franco's death notice, 1832, in the Plymouth Hebrew Congregation's Book of Records.

The English entry reads: 'Mr and Mrs Abraham Franco they died together in one day and buried together in one grave, the 2nd day of September 1832, according to their counting'.[90]

In Penzance, there were two Jewish victims of the 1832 epidemic. One was Pessia, daughter of the Revd B.A. Simmons, who according to the tombstone inscription, died and was buried on Saturday, 7 November 1832,[91] and the other was Ruth, daughter of Joseph Joseph of Plymouth, who died on Saturday, 10 November 1832, aged 20.[92]

In Exeter there was only one Jewish victim, Nathan Harris. Presumably he was a poor man because he does not appear in the Exeter Hebrew Congregation's records as a contributor and the Exeter Congregation had to bear the expense of a nurse (and brandy!) for him, and also to pay for his funeral.[93]

The mortality rate amongst Jews resident in Plymouth—there being only one Jewish victim in Exeter, no statistically significant conclusion can be drawn about the Jewish mortality rate in that city—during the 1832 outbreak seems to have been higher than that of the general population of either Plymouth or Exeter. In Plymouth there were in all 779 deaths in a general population of about 76,000;[94] in Exeter there were 440 deaths in some 28,000 inhabitants. Expressing these figures as deaths per 1,000 the result is:

| | *Deaths per 1,000 of Jewish population* | *Deaths per 1,000 of general population* |
|---|---|---|
| Plymouth | 28 | 10 |
| Exeter | 9 | 15.7 |

The high rate in Plymouth is truly surprising. The report of the General Board of Health on cholera presented to Parliament in 1853[96] pointed out that Jews suffered much less in proportion to their numbers, their deaths being only 0.6 per 1,000 compared with 6.0 and 9.0 per 1,000 of the general population in Whitechapel and Shoreditch, which in any case had a large number of Jewish residents. The reasons for the Jews' comparative immunity were listed by the report:

1 Even the poorer Jews never crowd more than one family to a room.
2 Jews do not abuse alcohol.
3 Jews are particular about food.
4 Jews refresh their bodies better on their Sabbath rest.
5 Poor Jews do not enter workhouses, primary sources of infection, and being fewer are more easily relieved and hence no extreme destitution.

6 Jews annually cleanse their houses at Passover, the poorer ones limewhiting their rooms.

The factors which promoted the spread of cholera applied to Jew and non-Jew alike. The poor were debilitated by long hours of work, and low wages, no sanitation in their houses and but one necessary or convenience per street and even this not always utilized, excrement often being thrown into the street or river. Drinking water was scooped out of the river by bucket, fetched from the public pumps, or delivered through the conduit system, though the latter mode was sometimes available only for two or three hours a day.[97] It has already been pointed out that about one in ten of Jewish families in Plymouth in the early nineteenth century took advantage of the conduit system.[98]

A possible explanation of the comparatively high incidence of cholera amongst the Jews in Plymouth in 1832 may lie in the fact that the men tended to travel in the course of earning their livelihood and frequently stayed at inns where the standards of cleanliness were low. At least one of the five or six adult victims was a recent immigrant and lived at an inn,[99] and two others, the Franco's, were impoverished.

The cholera epidemic of 1832 was closely observed in Exeter by Thomas Shapter and recorded by him.[100] An Order of Council of 20 July 1832 directed local Boards of Health to purchase suitable land for burial grounds and then peremptorily ordered that all cholera victims should be interred in those special grounds.[101] Until 17 August 1832, the dead were buried in their usual cemeteries and up to that date 209 were buried in the usual cemeteries and 21 in the Jews' and dissenters' burial grounds.[102] Shapter does not further break down the figure so that the exact number of Jews might be ascertained, but as there are no tombstones of that date and no reference to burials in the Congregational accounts (other than Nathan Harris who died on 22 August)[103] it may be assumed that there were no Jewish deaths in Exeter whilst the usual cemeteries were in use.

The first Jewish death took place on 22 August, a day of prayer for the city, when 'the parochial churches, the Roman Catholic chapel, the dissenting places of worship and the Jews' Synagogue were alike open and crowded'.[104] The death of Nathan Harris raised a problem, because after 17 August all cholera victims had to be buried in the special cemeteries at Pester Lane or Bury Meadow, an exception could only be made if the regulation could be 'safely dispensed with in any particular case . . . which . . . shall be entered in the minutes . . . and transmitted . . . to the Privy Council'.[105] To enforce this regulation would have interfered with the feelings and religious rites of the Jews who must have their own burying grounds. The Jews made representations to the Health Board asking for exemption and the mayor wrote to the Lords of the Council on 18 August 1832 to say that the basis of the Jews' application for exemption was,

> that it was their universal practice based on religious scruple never to open a grave in which a body has once been interred and that therefore no danger of infection could arise.

The Lords replied that they had no power to interfere, whereupon the Jews themselves suggested that another and separate burial ground should be provided. Shapter writes,

> Whilst this proposition was being entertained one of the Jewish persuasion died from cholera. It was however thought better not to interfere in any way with their religious feelings . . . the Board of Health, as the death had not been reported to them, assuming an official blindness. On 25 August a small portion of Bury Meadow was fenced off for them, but as no death from cholera subsequently occurred amongst these people, it was never used for this purpose.[106]

Before a separate part was fenced off, however, the Bishop of Exeter was asked his opinion and he wrote back,

> I have nothing whatever to do with the Jews' burying ground, and no control over it . . . I have no right, and certainly no wish to exercise my power or form any opinion.[107]

It appears from the foregoing account that the authorities in Exeter leaned over backwards to avoid giving offence to the local Jewish community. Their tolerance was probably motivated, in part, by a desire to avoid trouble of any kind, as there had already been civil riots in Exeter against the use of Bury Meadow and because coffins were carried underhand instead of upon the shoulders of the pall bearers.[108] Disorders accompanied cholera throughout Europe, the poor imagining that it was spread by the rich prompted by Malthusian doctrines.[109] The Jews do not appear to have been blamed for the spread of the disease, unlike after the Black Death when Jews who survived the plague were often massacred by fear-crazed mobs.[110]

The aftermath of the cholera brought its own problems with a large number of orphans. Longmate quotes one case which he says was typical of the generosity of working class families to one another in this time of crisis, which occurred at Bilston, Staffs. where three young children aged 12, 8 and 3 were seen wandering about the streets not knowing where to go, frightened and hungry, with a six-months-old baby at home. Four neighbours came in and took a child each.[111] Similarly, in the Jewish community of Plymouth some kind soul or souls must have taken pity on the orphaned children of the Franco couple referred to above, as at least three of them were formally converted to Judaism just two months after the death of their parents.[112]

Finally, on the subject of disease, it may be added that according to the census returns there was in 1851 one deaf and dumb woman, Harriet Bellam,[113] amongst the 278 Jews living in Plymouth, and in 1861 two deaf and dumb children out of the three belonging to Aaron and Sarah Aschfield.[114] In 1881, two tailors, Abraham Jacobs, 75, and Abraham Morris, 57, together with Morris' wife Harriet, 57, were all listed as blind. The 1881 census also lists a young woman, 35, the mother of four children, as a lunatic. The numbers are too small to be statistically significant.

# CHAPTER FOUR

# The Occupations of Jews in the South West of England

Throughout Europe in the medieval period, Jews were in most cases rigorously excluded from ordinary trade and crafts. Money lending, old clothes dealing, and peddling were the only callings generally permitted to Jews in Europe down to the period of the French Revolution.[1] In England, from the resettlement in 1656, policy was particularly liberal and apart from restrictions on retail trade imposed on Jews within the boundaries of the City of London until 1831, there were few limitations on the trades or callings open to them.[2]

In the second half of the eighteenth century, well-to-do Jews in London, where the bulk of the Jewish population was to be found, engaged in wholesale commerce, brokerage, stock-jobbing, and trade in precious stones. Then came a middle class of shopkeepers, silversmiths, and watchmakers. Lower down the social scale were the artisans—pencil-makers, tailors, hatters, and a sprinkling of artists. New immigrants who had difficulty in establishing themselves and were less fortunately placed than the resident Jews, earned their keep by the two activities which required little capital or training—trading in old clothes and peddling.[3] Jews living in the South West of England engaged themselves in a similar occupational pattern, as will now be described in detail.

Two literary sources clearly indicate that the first Jews to settle in Devon in the early 1730s and in Cornwall about 1760 were, in the main, pedlars. The first was the polemic published by Joseph Ottolenghe in which he mentions Jewish pedlars staying at an Exeter inn about 1733 in sufficient numbers to provide him with at least a partial living as a *kosher* slaughterer.[4] The second source is a family history compiled by Israel Solomon in which he recorded a family tradition that about 1760 his ancestor Alexander Moses of Falmouth subsidized young Jews with money and goods thus enabling them to peddle around the neighbourhood of Falmouth, provided that they returned at the end of the week in time to form a quorum for the Sabbath services.[5]

From both these accounts it is evident that besides the Jewish hawkers in Exeter in the 1730s and Falmouth in 1760 there was in each of these towns at least one Jew with sufficient capital to open a shop. Indeed, Cecil Roth has suggested that the first Jew to settle in many a country town or village throughout England was a comparatively well-to-do merchant,[6] and it is indeed reasonable to assume that the presence of such a man attracted Jewish pedlars to the neighbourhood as they could anticipate credit and help from their well-off co-religionist.[7]

The earliest written record of Jews settled in Plymouth is of a land transaction carried out by Sarah, wife of Joseph Jacob Sherrenbeck, in 1744, indicating the presence of at least one comparatively wealthy family.[8] His co-founders of the Plymouth Congregation numbered at least one shopkeeper among them, a Joseph Cohen who became bankrupt in 1749.[9]

There are slight pointers from time to time in the second half of the eighteenth century which indicate the presence of a growing class of Jewish shopkeepers in the South West. The bankruptcies, for example, of Sampson Cohen of Dartmouth in 1764, Gompart Michael Emdin and Isaiah Samuel of Plymouth in 1765 and 1768,[10] suggest that there were a number of Jewish shopkeepers. In Exeter, from about 1740, one of the founders of the Congregation was Abraham Ezekiel, variously described as a silversmith,[11] engraver in general, optician, goldsmith and printseller;[12] and 'for fifty years and upwards a respectable tradesman of Exeter'.[13] By 1796, five Jews had shops in the fashionable shopping area of Exeter sufficiently well established as to warrant inclusion in the *Exeter Pocket Journal*.[14] There is also some evidence that in Cornwall by the 1780s, sales of watches and clocks by Jews had made serious inroads into sales by their Gentile competitors. This is the inference to be drawn from an advertisement by the watchmakers and goldsmiths of Cornwall in 1783 that 'for very substantial reasons' they had resolved not to repair watches bought of Jews.[15] Apart from this isolated affair no other evidence of any organised opposition by Gentile traders against Jewish competitors in the South West, nor any widely based criticism of sharp practices or unfair trading methods has been noted.[16]

On the contrary, there is reason to assume that in the eighteenth century there were friendly relations between the Jews and their Gentile neighbours in the South West, as A. Arnold has noted at least four Jewish lads who were apprenticed between 1762 and 1769 to masters not of their faith.[17] They were:

(a) Moses Moses,[18] apprenticed in 1762 to Jason Holt of Plymouth, Jeweller, for £80 10*s*.[19] He might be the Moses Moses who married Sarah Jonas of Plymouth Dock in 1815 and later emigrated to Ohio, though it is more likely that he was the Moses Moses who was apprenticed to Benjamin Levi of Portsea as a pen engraver for £20.[20]
(b) Francis Lyon (1752–1837), apprenticed in 1767 to John Lakeman of Exeter, clock and watchmaker, for £48.[21]
(c) Henry Solomon, apprenticed in 1753 to Jn. Lievre of Stoke Damerell, carpenter, for £9.[22]
(d) Solomon Solomon, apprenticed in 1769 to John Sampson of Penzance, watchmaker, for £69.[23]

Another Jewish apprentice in the South West at this period was Judah Lyon who was apprenticed in 1772 for £42 to Solomon Nathan of Plymouth, jeweller and engraver, the only recognized Jewish master in the South West whose apprentice was duly registered.[24] Other apprentices in the eighteenth century who may have been Jewish were Benjamin Gubby,[25] apprenticed in 1738 to a barber in Tiverton for six guineas; two apprentices called Nathan Harris, one apprenticed in 1737 to a clockmaker in Crediton,[26] and the other to a Liskeard gunsmith in 1762;[27] Abraham Abrahams, apprenticed in 1764 to a Plymouth watchmaker,[28] and Isaac Abrahams, apprenticed in 1768 to a Plymouth tailor;[29] Reuben Phillips apprenticed in 1761 to an Exeter druggist for £80;[30] and the only girl, Rebecca Phillips, who was apprenticed to a Crediton mantelier (i.e. a mantle maker) for ten guineas.[31] It is unlikely that any of the other ten Devon and Cornish apprentices listed by Arnold were Jewish, and the only evidence of apprentice Jews in the South West in the eighteenth century which has come to light is that which has been published by Arnold.

A further indication of the amicable relationship which apparently prevailed is the employment by Jewish shopkeepers and tradesmen of non-Jews as pupils and assistants. After the death of E.A. Ezekiel of Exeter in 1806, for example, 'James Rickard, Engraver, pupil to the late Mr Ezekiel of this city . . . informs the public . . . that he intends carrying on the above business . . .'[32] and 'C. Frost, senior pupil of the late E. A. Ezekiel . . . solicits the patronage . . .'.[33] It looks as though Ezekiel himself was the pupil of one J. Woodman, because he advertised in somewhat similar terms on the death of Woodman in 1784.[34] About 1767, Abraham Joseph of Plymouth engaged a journeyman[35] watchmaker called James Dawson for one year for £18, with 'heat, drink, washing, and lodging, and shaving twice a week'.[36] From Joseph he went on to Moses Jacob of Redruth on a two year contract 'at the wages of sixteen shillings for every five days in each week, excepting Jewish festivals and holidays'.[37] In Exeter, Morris Jacobs advertised in 1828 that he had a vacancy for a tailor's apprentice, and in 1835 James Levandor, a dentist, also advertised for an apprentice.[38]

The names of two Jewish artisans who were active in Plymouth at this period have come to light. A list of the members of the Plymouth Congregation who donated to a War Levy in 1779 includes the names of *Abraham ben Solomon* and *Solomon Ze'ev ben Meir KZ* who are respectively described as a carpenter and an embroiderer in gold.[39]

In the eighteenth century, too, and especially its latter part, some of the more assimilated Jews in the South West engaged in artistic and learned occupations. There was, for example, an Isaac Polack, described in 1760 on his marriage in Cornwall to Mary Stoughton, widow, as 'a Jewish Priest',[40] who was an interpreter and translator of commercial and legal documents. He lived at Penryn and advertised in 1776 as follows:

> TRANSLATION OF LANGUAGES.
>
> ISAAC POLACK, of Penryn, in Cornwall most respectfully acquaints those Mercantile Gentlemen who have connections in foreign countries, such as France, Germany, Holland, etc. that he writes and translates into English (and vice versa) letters, invoices, bills of lading, and other incidental circumstances of commercial intercourse, stiled in either the FRENCH, HIGH GERMAN, or LOW DUTCH Languages, with the utmost propriety, accuracy, and expedition.
>
> Also protests in the foregoing Languages carefully copied and translated for Attornies, Notaries, and Tabellions; and the greatest attention will be paid, so as to merit their kind favours. The said Isaac Polack begs leave to assure the Gentlemen above addressed, (or any other employer) that SECRECY will be the principal object attached to; and their respective commands from any part of this county assiduously accomplished at a reasonable charge.
>
> PETITIONS, MEMORIAL, etc. drawn for disabled seamen, sailors, widows, orphans, or other persons, to any public office in this kingdom, and to foreign Courts, or to their Ambassadors, Residentaries, and Ministers of State, at very moderate fees.
>
> Dated Penryn June 27, 1776.[41]

Among the artistically inclined and successful of the Jews in the South West in the eighteenth century were the three Daniell brothers, one of whom, Abraham, was largely based in Plymouth, as well as Samuel Hart, also of Plymouth, and E.A. Ezekiel of Exeter, all of whom, together with other Jewish artists of the South West, will be discussed in detail in a later chapter.

It is only at the very end of the eighteenth century that it is possible to draw an overall picture of the occupations followed by Jews in Plymouth. This can be done in the first place from the pages of the *Universal British Directory* published in 1798 which list 12 men and one woman who, to judge by their names at least, were Jewish, as well as a further 5 men with Jewish-sounding names who have not otherwise been identified. The woman, a Sarah Abrahams, is listed simply as a shopkeeper. Abraham Daniell, the miniaturist, is listed. So are Henry Hart, Aaron Aaron, Solomon Isaac and Joseph Joseph who are all listed as silversmiths. Benjamin Levi, optician, and Samuel Cohin together with Emanuel Hart, watchmakers, were probably skilled workmen, as no doubt in his own fashion was Levi Levi the umbrella maker. Surprisingly, perhaps, only one Jew, Abraham Emanuel, is listed in Plymouth-Dock, and he was a jeweller and silversmith. Jews are listed in the *Directory* in two more South West towns. In Barnstaple were Abraham Ralph, described as a silversmith and dealer in wearing apparel and his son, Leape Ralph who was listed as a pawnbroker and engraver. Father and son probably had very similar establishments. In Falmouth, the two Jews who were listed, Alexander Moses and Benjamin Woolf, could probably have described their businesses in much the same terms, though they were simply described as pawnbrokers.

The occupations listed in the *Universal British Directory* can be supplemented by a register of 58 aliens compiled in 1798 and revised in 1803 which has survived in the records of the Plymouth Congregation.[42] Naturally, the *Directory* does not list pedlars or the poorer type of trader; nonetheless, using the two sources it is possible to make some observations about the overall occupational structure.

Of the occupations given in the Aliens List, at the top of the economic tree were probably the specialist craftsmen and retail traders in the precious metal, jewellery, watchmaking, and optical trades, of which there were 24 representatives.[43] Then came 4 shopkeepers, a hardware dealer and a dealer in pens.[44] Among the skilled artisans were a boxmaker, a pen-cutter, three hatters, three umbrella makers and a tailor.[45] Next, and possibly lower on the social scale, were 13 men described as dealers and chapmen or as dealers in old clothes,[46] and then, at the bottom of the ladder, three hawkers or pedlars.[47] Finally, there were two synagogal officals and a tutor.[48]

It is now opportune to consider the way of life of the Jewish pedlars in England, with particular reference to the South West, in the eighteenth century and in the nineteenth, at least until the spread of the railway system.

How did the immigrant Jew without family or friends and who arrived with little or no capital get started? According to an account published early in the nineteenth century, he merely entered a synagogue, made his needs known to the worshippers who then made a quick collection for him and within a day or two of landing would direct him to the provinces with a peddling tray suspended from his neck filled with two or three pounds' worth of stock with which to start trading.[49]

New immigrants presumably dispensed with the Hawkers' and Pedlars' Licenses, annual cost £4 for the hawker and a further £4 for each of his horses, which were required from 1790 onwards, and kept well out of the way of local constables.[50] The acquisition of a licence was probably a great occasion in a pedlar's life and at least one Jew from Exeter preserved his licence for posterity.[51] The process of becoming a pedlar as well as the trials and tribulations of the life has been well described in a biography of one Samuel Harris who arrived in England when he was 14 years old.[52] Being fitted out with a pedlar's tray and a small stock of goods by the worshippers in a London synagogue, he spent a week trying his luck in London but without success, so off he went to Birmingham.[53] He peddled his goods whilst *en route* and when he arrived there three weeks later he had made fifteen shillings over and above his expenses.[54] He remained in the vicinity of Birmingham for a few months and did so well that the local Jews said to him, 'You'll drive your own carriage, yet'. But he went to Bristol, and there did so badly that he became as poor as ever. Moreover, he fell ill, was removed to hospital, and survived only through the charity of the Bristol Congregation.[55] Stocked out by that Congregation he made for Bridgwater, passed through Taunton and came to Exeter.

> Here I received seven shillings from the Jewish Poor Strangers Fund and a number of donations from Jews who reside in the town, and was able to lay in a nice stock of hardware and again do business, and as I was travelling in Devonshire and Cornwall, the Jews so kindly assisted, that I was soon raised higher than before my illness at Bristol.[56]

Apparently he travelled in the South west during 1821 and 1822, as his name figures in the Plymouth Congregation's accounts for an offering of 1*s*. 8*d*. he made when he was called to the Torah in November 1821.[57] In 1822, he returned to Birmingham 'in prosperity', went to Newcastle fair in 1823 and was there robbed by a fellow Jew of all he possessed, so that 'in the twinkling of an eye I was reduced from a gentleman to a beggar.[58]

Once again the local Jews fitted him out and within a few months when he arrived in Manchester he could boast that he possessed stock worth £25 and several pounds in cash.[59] Again illness laid him low, and after a few weeks of sickness was down to his last threepence.[60] The stuffing seems to have been knocked out of him by this time, and although the local Jews collected thirty shillings for him to start him off once again, he turned to a Methodist and asked if he could get a situation in a Christian house.[61] Eventually he was baptized, apprenticed to a watch-movement maker, then to a hairdresser, and finally became a house servant.[62] At this stage he disappears from sight.

It may here be added that besides vicissitudes of the type described by Shemoel Hirsch, there were dangers of a more violent description lurking for pedlars on the lonely roads. An eighteenth-century German writer travelling in England informed his readers that all foot travellers were liable to be treated with disdain.[63] Reference has already been made to a Jew driven to suicide after some outrage committed against him near Herland Cross in Cornwall in the mid-seventeenth century.[64] Then there was the murder of one 'Little Isaac' by a militia man, Edward Jackson, in a wood near Plymstock, Devon, in 1760. According to the report in the *Gentleman's Magazine*,[65] Jackson and Little Isaac met and had a pint of beer together. They then walked together for about two miles, and when they entered a wood and the Jew sat down to rest himself, Jackson struck him on the head with a cudgel. The murderer stole Isaac's watch and took some articles out of his pedlar's box. Unfortunately for him, he offered these articles for sale to a Plymouth Jew, Mr Sherrenbeck,[66] who asked him how he had come by the goods. Eventually Jackson confessed and was hung for the murder. Commemorating this sad event, the hill from Hooe Manor to Staddon Heights by Radford Woods is still known as Murder Hill, and the woods were renamed Jew's Woods.[67] At the Exeter Assizes in 1768, a W. Killard was capitally convicted for 'robbing and barbarously using a Jew, between Newton-Bushel[68] and Totnes.[69] In the Red Lion Inn, Upshay, 'say the old people, on the oral testimony of their grandsires [i.e. *c.* 1780–1810], a Jew pedlar was murdered, and on that account the inn was shunned and at last pulled down'.[70]

In spite of the dangers on the road some wives of pedlars accompanied their husbands, and the peddling Jewess was apparently not uncommon, at least in Devon. In the Tavistock Subscription Library there are two dolls which represent a late eighteenth-century Jewish pedlar couple, each with his and her own typical dress and pedlar's tray, whilst at a Grand Ball in Exeter in 1814 amongst the various 'characters' was 'a Jew pedlar and his wife'.[71] By the very nature of their occupation pedlars leave few permanent records of themselves; and in particular it is hard to consider to what degree they were materially successful. Very likely some made good whilst others did not. But it is significant that the three men listed as hawkers or pedlars in the list of Jewish aliens who were in Plymouth, 1798–1803,[72] do not appear in any other record of the Plymouth or other South-West Congregations.[73] This would seem to indicate that these three, at least, were unsuccessful or did not settle in the South

West.[74] Similarly, no record has come to light of Jacob Israel[75] who gave his address as Mary Arches, Exeter (which is perhaps the synagogue) other than his application for a Pedlar's Certificate in 1873.[76]

Despite the failures, some pedlars did very well for themselves in a comparatively short time, particularly before the railways spread over the face of the land. Mayhew instances the case of a young man, not a Jew, who accumulated five pounds, bought Birmingham jewellery, became a foot pedlar, then bought a wagon, and was worth £500 by the time he was interviewed;[77] whilst an ex-convict peddling in Australia in 1854, turned £7 into £140 within twelve months.[78] It may be supposed, for clear evidence is lacking, that some nineteenth-century Jewish pedlars in the South West also prospered. Samuel Joyful Lazarus appears on the books of the Exeter Congregation for the first time in 1818 when he paid £2 15*s.* 6*d.*[79] He took out a hawker's licence in 1843,[80] married in 1846,[81] and appears as a householder in Paragon Place on the Voter's Register for Members of Parliament in 1849. Aaron Aarons was granted a pedlar's certificate in 1873,[82] but when he married in 1867 he was described as a jeweller.[83] Another Jew in the South West who rose from the lowly state of a pedlar to become a wealthy man was Lyon Joseph of Falmouth (1774–1825), though he subsequently lost most of his fortune in a series of disastrous trading ventures.[84]

Turning now to the Jewish shopkeepers in the South West of England in the eighteenth and early part of the nineteenth centuries, it is possible to give some idea, at least, of their trading activities by reference to their stock in trade. Reference has already been made to the advertisement of Gompart Michael Emdin who styled himself in 1761 as a

> Goldsmith and Jeweller . . . where a great variety of all sorts of silver plate, jewels, trinkets, and all sorts of watches, and haberdashery, wares, may be had as cheap as in London.[85]

E.A. Ezekiel of Exeter published a trade card in 1796 by which he informed the public that besides being an engraver and optician he was a goldsmith and printseller, and sold spectacles, telescopes, quadrants, cutlery, plate, gold seals, watches, prints and materials for drawing.[86] In 1800, he was advertising his stock consisting of:

> spectacles mounted in silver, tortoiseshell and steel . . . reading glasses, Claude Lorraines, opera glasses, acromatic telescopes, magic lanterns, microscopes . . . wheel and pedement barometers and thermometers . . . for the hot-house or brewery.[87]

Some idea of the varied nature of the trade carried on by Gershon Levy of Guernsey and Exeter is given by the articles he bought in 1803 at a prize auction. They included 3 black feathers, 7 neck handkerchiefs, 25 caps for 27*s*.; silk thread and national sash for 19*s*.; 29 pieces Nankeen at 5*s*. per piece; 6 men's coats for £3 3*s*. 0*d*.; 8,408 lbs of coffee for £421 10*s*. 0*d*.[88]

The Jew's shops were apparently so well stocked that they became proverbial; the very term *Jew's shop* conjuring up visions of a sort of Aladdin's Cave. In a work published in the mid-nineteenth century, a miner, on breaking a crystalline mass of quartz and pyrites, declared in awe that 'he thought he was in heaven. It was so beautiful, he could compare it to nothing else than a Jew's shop.[89]

By the first quarter of the nineteenth century there was a well established body of Jewish shopkeepers in the main towns of the South West ranging from the genteel jewellery shops to the coarser old-clothes shops and suppliers to ships' crews. From the second half of the eighteenth century when the British Fleet of 228 warships was manned by some 35,000 seamen and marines, the Jewish jewellers, clothesmen and petty traders of the maritime centres realized their trading opportunities. As a Naval historian has pointed out, 'why should they peddle their wares about the country when a more or less captive customer was to be found in the men-of-war lying at Portsmouth, Plymouth, Chatham and Sheerness?'[90] For the most part, at least until after the mutinies of 1797, the lower ranks, particularly pressed men, were not allowed to go ashore when a ship docked, for fear lest they desert. So traders were allowed to go on board, and even encouraged to do so as, until 1813, they were exempted from the £4 licence fee for hawkers and pedlars. They brought with them the goods which the simple seamen longed for—'old watches and seals, watch chains, rings, fancy shoes, scarlet and blue silk handkerchiefs, clay pipes and fresh food of every description'. The seamen were paid six months in arrears, the lowest paid received sixteen shillings a month, whilst an able-seaman had twenty-four shillings. Seamen who remained on board were paid in cash, but until 1792 those who were paid off received a wage ticket which was only encashable at the Navy Pay Office in London. This was an enormous inconvenience. The traders therefore performed a dual function. They provided the seamen with cash by buying up the wage tickets at a discount, and at the same time they supplied the goods which the seamen wanted. There were complaints that some traders, Jews and non Jews alike, sold faulty goods at inflated prices and cashed the wage tickets at excessive discounts. An honest trader with good references was, no doubt, welcomed. The letter which Joseph Joseph (*c*. 1766–1846) brought with him amounted to a royal command, if the signature is correctly deciphered:

> I do hereby certify that Joseph Joseph of Plymouth has at different times supplied the Crews of His Majestys Ships when under my Command with Cloathing to my entire satisfaction, and I do hereby recommend him to the Admirals, Captains, and Officers of His Majestys Navy, to be permitted to transact any Business that may be done on board the respective Ships under their Command.
>
> St. James's Palace
> Dec$^{r.}$ 2$^{d.}$ 1812.
>
> William, Ad$^{l.}$ of the Fleet.[91]

Twenty years later, Joseph Joseph was still interested in supplying ships and obtained a recommendation from the Flag Officer at Devonport:

> These are to certify that I knew Mr Joseph Joseph several years since, and that he was then, as I believe him to be now, a respectable Tradesman, and can recommend him as deserving the patronage of the Officers of His Majestys Ships.
>
> W. Twogood.
> Devonport
> 4th May 1883.[92]

The Joseph family were not the only ones to enjoy illustrious patronage. Phineas Johnson of Exeter and Plymouth prevailed on the London Jewish magnate Abraham Goldsmid to secure an introduction from Lord Nelson in the following terms:

> The bearer Mr Phineas Johnson being recommended to me by Mr Abm Goldsmid as a very honest man for transacting business for seamen—You'll please allow him on board Ivy at Plymouth Dock. Lord Nelson will feel much obliged to any Captain who may be pleased to show attention to the recommendation of Lord Nelson's friend and neighbour Mr Goldsmid.
>
> Merton Sept. 12th 1805.[93]

Moreover, they were now reaping the fruits of a reputation for industry and honesty. As early as 1794, the local press paid tribute in an obituary of him to the business ethics of Abraham Joseph I:

> He was one of the people called Jews, but the actions of his whole life would have done honour to any persuasion. He amassed a considerable fortune by very fair and honest means.[94]

Conditions in the Royal Navy in the eighteenth and early nineteenth centuries were hard for the officers and harsh for the men. The ever present hope of prize money made a cruel service bearable. In Plymouth, between 1793 and 1801 nearly 1,000 captures were sold off.[95] After the battle of Trafalgar each seaman received £6 10*s*., not a great sum in itself, but it did represent eight month's wages. The capture of a West India merchantman could earn a boatswain enough to buy a cottage and some land, whilst the captain could receive enough to buy an estate and set himself up for life. But the mere capture of an enemy prize or the seizure of contraband in a neutral ship did not of itself put money into the pocket of the sailor. The prize had to be dealt with in port, lawyers appointed, evidence and witnesses secured for the hearing in the Prize Court. There might be appeals and further litigation. If the capture was 'condemned', then ship and cargo had to be sold at auction, expenses paid, and five per cent of the net sum available for distribution sent to Greenwich Hospital. It might be years before any money was distributed. In the meantime, ship's crews could be broken up, men killed in action or died of disease, or dismissed after loss of limb or injury. Officers had their Prize Agents who looked after their interests, becoming in effect their bankers. The seamen, however, were forced to turn for assistance to the tradesmen of the naval towns who were the link between themselves and the naval authorities. At first, the system was unofficial and based on mutual trust. Abraham Joseph I certainly earned that trust, as his obituary in the local press indicates:

> As an agent for seamen, his practice was well worthy of the imitation of every person in that business, as several orphans and indigent widows can testify.[96]

In the course of time, however, it became clear that many seamen were being cheated by unscrupulous slopsellers, and especially by publicans and brothel keepers. Legislation was enacted in 1809 which required all lower deck seamen to be registered with a Licensed Navy Agent who was to protect their interest. To obtain a licence, a Navy Agent had to post a bond with two sureties, under penalty of £200.[97] The first list of 174 Licensed Navy Agents for the whole country in 1809 includes some 66 Jews.[98] The rapid proliferation of Jewish naval agents is itself an indication of the esteem in which Jewish traders were held. Between 1807 and 1814, whilst the number of navy agents in England as a whole increased sevenfold, the number of Jewish agents increased thirtyfold, as Table 25 indicates.

In Devon and Cornwall a large number of Jewish navy agents applied for licences and were appointed between 1812 and 1814, and although the

*Table 25: Jewish and Gentile Navy Agents, 1791–1818*

| Year | Non-Jews in all England | Jews | | Total |
|---|---|---|---|---|
| | | Devon and Cornwall | Rest of England | |
| 1791 | 27 | None | None | 27 |
| 1797 | 28 | None | None | 28 |
| 1801 | 35 | None | 1 | 36 |
| 1807 | 62 | None | 4 | 66 |
| 1809 | 108 | 14 | 52 | 174 |
| 1814 | 322 | 37 | 82 | 441 |
| 1818 | 282 | 35 | 121 | 438 |

(Source: Annual Register; Steel's List; Green, 'Royal Navy', p. 133.)

number gradually declined together with the volume of business after the Napoleonic wars,[99] the very number indicates a considerable degree of mutual trust.[100]

Naval agents could do well from their agencies. According to Green, 'Just one London Prize Agent between 1806 and 1814 paid £4,225 to six Plymouth-based Jewish Navy Agents'.[101] Joseph Joseph of the Barbican, Plymouth, received £2,258 on behalf of seamen over a period of nine years from just one London Prize Agent, whilst Samuel Hart of 33 Market Street, Plymouth, received £826 from the same source.[102]

As the Jewish traders increased their standing in the Gentile world it is not surprising that the most able of them received patronage from high places. The first of them in the South West was Abraham Joseph I (1731–1794) whose trade card proudly proclaimed that he was a 'slopman'[103] to His Royal Highness Prince William Henry'.[104]

Not all the Jewish traders were so well recommended and respectable—some sailed close to the wind, particularly when incomes dropped after the Napoleonic wars and yet prize money had to be paid out. Jonas Jonas of James Street, Plymouth Dock, and a Naval Agent for just seven months, had his licence revoked when he was committed to Exeter Gaol on suspicion of forging an order in the name of Thomas Warren of HMS *Lightening*, thereby receiving £31 15*s*. 2*d*. from Greenwich Hospital.[105] One man seems to have escaped a serious charge largely because of his lawyer, as the gossipy beadle of the Plymouth Congregation recorded in his account book:

> 4 February 1817
>
> By our many sins Judah ben Hayyim Mannheim was forced to go in chains of iron on his feet and hands to stand his trial at March

> Assizes, may they come to us for good, for there is a state prosecution against him to condemn him because they found that he forged on the Greenwich Hospital Navy Agent.
>
> 17 and 18 March 1817
>
> Judah was tried and is acquitted and came out free. His attorney was Alley Isaac of London.[106]

Some licences were revoked. Asher Nathan of 30 George Street, Plymouth lost his licence in 1812 for not giving notice of change of abode in order to evade just claims of seamen; Samuel Hart lost his on 20 January 1817 for not paying out prize money of £2 6*s*. 6*d*. but it was subsequently reinstated; Sampson Levy's was revoked on 15 January 1819 for not accounting to the widow of Joseph Hodgman, late of HMS *Triton*, for not paying prize money, and for moving to Liverpool without informing the Navy Pay Office; Lyonell Nathan's was revoked in July 1819 for not accounting to John Johnson, late of HMS *Spider*, but it was restored the following month as he was judged to have acted more from unintentional neglect than from a dereliction of his duty.[107]

Some Jewish traders in England sold cheap and shoddy goods and there is no reason to suppose that the Jews in the South West did not also do so.[108] *The Universal Songster*, for example, records a couplet which was commonly sung at the time: 'So sure as I'm a smouch[109] and my name is Mordecai I cheated all the world'.[110] But the book also quotes ballads which declare: 'Lipey Solomons, the honest Jew pedlar'.[111] and 'I am a Jew tailor, but a very good man'.[112] A Jewish shopkeeper in Exeter seems to have taken advantage of, or was imposed upon by, an extremely eccentric customer. He was Solomon Levy, a 'jeweller, carrying on business in a large way', who, in 1827, claimed some £463 for goods sold and £200 for cash advanced on a dishonoured cheque.[113] Jewish traders were at risk from their own co-religionist servants. In 1773 Mordecai Abraham advertised

> Run-Away
>
> on Monday 13 December Moses Isaac, a Jew, servant to Mordecai Abraham of Middle Lane, Plymouth, and carried off two boxes of great value . . . Whoever will apprehend Moses Isaac so as he may be brought to justice shall receive two guineas reward.
>
> He is a palefaced man about 30 years of age, black curled hair, 5ft 8ins high and apt to be merry in company.
>
> He wore a blue plush coat, reddish waistcoat and buckskin breeches.[114]

Shopkeepers selling second-hand goods are particularly likely to be offered stolen goods and may easily succumb to buying them. Even in the medieval period Jews were forbidden to take as pledges bloodstained clothes (which might have been acquired as the outcome of violent robbery),[115] and it does seem as though in the late eighteenth and early nineteenth centuries some Jews in England were prominently engaged with stolen goods. For at this period a thieves' captain was known as 'the Aaron'; and an 'Ikey Mo' or just 'an Ikey' was the cant term for a receiver of stolen goods.[116] Some Jewish dealers at Plymouth took full advantage of the pilferage, or rather pillage, at Plymouth's Naval Dockyard from which goods worth an estimated half million a year in the 1770s were spirited out,[117] and copper, in particular, 'was sold to the Hebrews who infested the lanes near the yard'.[118]

At least one Exeter Jew was fortunate to escape the halter. He was Uriah Moses, who was tried at the Old Bailey on 10 January 1798 for cutting out the glass of a draper's shop to steal the stock. He was sentenced to death. The sentence was later commuted to transportation, and he arrived in Australia on the *Royal Admiral* in 1800.[119] On the other hand, some pawnbrokers, like William Woolf of Plymouth or Joseph Marks of Exeter, were commended by the local magistrates for reporting stolen goods.[120]

It is possible to indicate the business activites of three Jews living in the South West in the latter part of the eighteenth and early part of the nineteenth centuries. The first, Joseph Hart of Plymouth, came to England from Mannheim in 1770 aged 14.[121] By 1779, he was sufficiently well established in Plymouth to contribute one guinea to the War Levy.[122] In 1798, he is described as a silversmith, and when he died in lodgings on 16 June 1822, he left four bank notes of one hundred pounds each, silver and gold coins and 'much gold and silver' a total value of seven or eight hundred pounds sterling.[123] His landlord refused, at first, to hand over the fortune, and the Plymouth Congregation had to take steps to secure it for the benefit of his heirs.[124]

The second businessman, Abraham Joseph I (1731–94), also of Plymouth, had wide interests as a mercer and draper on Barbican Quay,[125] and subsequently became a wholesale slopdealer to the Navy.[126] In 1781, he also had a property on Southside Quay, he bought the adjoining property in 1783 and another nearby in 1794 as well as one on Great George Street.[127] At his death in 1794, he had two leasehold dwellings on the Barbican, two others elsewhere, three dwellings on the quay in East Stonehouse, and freehold dwellings in Castle Street, Stillman Street, and Pyke Street. By his will, dated 22 October 1794, he left £500 each to his wife and to three children (the other two having each already received £500), about £4,000 to be divided among the rest of his family, two Scrolls of the Law with their appurtenances and £260 to various charities.[128] The house in which he had lived, 6 George Street, was sold after his death and appears to have been a desirable residence. The advertisement of its sale reads:

> . . . consisting of two good parlours, a drawing room, several lodging rooms and all convenient offices, together with a garden adjoining. The rooms are furnished with neat grates, and all necessary fixtures are on the premises. And the whole is fit for the immediate reception of a genteel family.[129]

In 1806, his son Joseph Joseph was paying conventionary rents on three houses on Southside Quay, one in Catherine Lane, and cellars at the lower end of Castle Street.[130] In 1817, of the 162 properties belonging to the Corporation of Plymouth which were let, eight were let to Jews, and five of these to Joseph Joseph. He had his father's cellars in Castle Street, a plot of ground near Dung Quay, a dwelling house on Southside Quay, a shop with chambers and lofts in Whimple Street, a cellar and loft in Middle Lane together with a messuage in Lower Lane. For the leases of all these Joseph had paid considerations totalling £367.[131] Despite the extent of his interests, his income did not match his needs, and in July 1817 his licence as a Navy Agent was revoked because of his bankruptcy.[132] By 1818, he had run up a debt of £90 to the Plymouth Congregation which he was unable to pay, and for the security of which he had to deposit the silver ornaments to the Scrolls of the Law which his father had left him.[133] Presumably on account of his financial straits, Joseph left Plymouth for London in December 1817, going on to Gibraltar on 11 June 1819,[134] and returning to Plymouth about 1824.[135]

The third South West Jewish businessman whose affairs are comparatively well-known to us was Lyon Joseph[136] of Falmouth (1774–1825). His career illustrates well the change of fortune that could befall a businessman in those times. Lyon Joseph began life as a pedlar, but he made his fortune shipping goods to the ports of the Peninsula not occupied by Napoleon. After goods so dispatched were declared contraband, Lyon Joseph sent further consignments which were seized and confiscated. In this bad speculation he lost £20,000. He despatched one of his ships, the *Perseverance*, to Gibraltar, but the captain ran it to Lisbon, misappropriated the cargo and absconded. Lyon set up some relatives in trade in Gibraltar and lost £7,300. He sent three thousand pounds worth of commodities to a Jew in Cadiz who stole them and became a Roman Catholic convert. When Lyon arrived to claim his goods the apostate threatened to denounce him to the Inquisition. He lost fifteen hundred pounds worth of uninsured goods in a vessel shipwrecked in the Cove of Cork which were salved and then seized by the Government who claimed Droits of Admiralty. He eventually retired to Bath, a broken man.

As the nineteenth century progressed, besides the shopkeepers and hawkers already discussed, a number of craftsmen and artisans can be identified who lived and worked in Plymouth. For instance, in the period 1812–1838, scattered references can be found to a feather dresser,[138] two fringemakers,[139] a huckster,[140] a shoemaker,[141] three straw-bonnet makers,[142] two tailors,[143] two umbrella makers,[144] and a boxmaker.[145]

There were probably many other Jews who were skilled workers but they were not of sufficient importance to be listed in the directories of the period. For example, of the artisans just mentioned, one of the fringemakers, the shoemaker, the two tailors, and the boxmaker, were not listed in any local directory and have been noted from other sources, primarily synagogal account books.

In the mid-nineteenth century a clear view of the occupations held by Jews in Plymouth, Exeter, Falmouth and Penzance can be determined using the 1851 census. From the census returns the occupations of 176 adult Jews of both sexes resident in the South West on the night of the census were noted. These occupations may be classified under the headings of Annuitant, Commercial, Manual and Professional/Clerical. Tabulating the occupations in their respective classifications produces the following picture:

| *Annuitant* | | *Commercial* | | *Manual* | | *Professional/ Clerical* | |
|---|---|---|---|---|---|---|---|
| Annuitant | 12 | General dealer | 14 | Bookbinder | 1 | Artist | 2 |
| | | Hawker | 12 | Dressmaker | 9 | Broker | 2 |
| | | Horseposter | 1 | Dyer | 2 | Clerk | 1 |
| | | Jeweller | 24 | Glazier | 1 | Druggist | 1 |
| | | Lace manufacturer | 6 | Housekeeper | 2 | Musician | 3 |
| | | Pawnbroker | 8 | Milliner | 5 | Optician | 2 |
| | | Pen dealer | 1 | Shoemaker | 7 | Policeman | 1 |
| | | Property dealer | 2 | Straw bonnet maker | 2 | Synagogal officials | 4 |
| | | Shop assistant/ apprentice | 18 | Tailor | 4 | Teacher/ Governess | 6 |
| | | Shopkeeper | 6 | Tinplate worker | 1 | | |
| | | Stationer | 1 | Watchmaker | 3 | | |
| | | Tailor (master) | 1 | | | | |
| | | Tobacconist | 1 | | | | |
| | | Traveller | 9 | | | | |
| | | Waterproof manufacturer | 1 | | | | |
| Total | 12 | | 105 | | 37 | | 22 |

The classification given above must be used with some caution, as some of the occupations could fall under more than one category. Some of the dressmakers, dyers, milliners, shoemakers and watchmakers, for example, might well have been described as 'commercial' rather than as 'manual'. Bearing this in mind, it may be said that in 1851 some 59 per cent of the Jews in the South West were engaged in one form or another of commerce and that three out of every four of these were probably self-employed, some 22 per cent were apparently engaged in manual trades, about 13 per cent were in what appears to have been professional or clerical occupations,[146] and about 7 per cent were retired annuitants. Table 26 sets out the figures broken down into the four main communities of the South West.

As earlier, in the mid-nineteenth century, some of the Jewish shopkeepers maintained establishments of high repute inasmuch as they had the patronage of royalty. In the period 1837–1840, eight Jews in England traded by appointment to Queen Victoria and no less than 3 of these were established in Devon and Cornwall. They were Alexander Alexander,[147], optician at Exeter; Henry Harris,[148] jeweller at Truro; and Aaron Levy,[149] jeweller at Plymouth.[150]

It is possible to make some assessment of the annual income of Jewish households in the South West in 1851, by the number of servants which they kept. One contemporary correlation of Victorian income groups with the number of servants kept indicated that a widow or spinster with £100 per annum could have one (perhaps part-time) servant; a family with £150–£300 per annum, one whole-time cook general; a family with £500 per annum, a cook and a maid; with £750 per annum, a cook, a maid, and a boy; and with £1,000 per annum, a cook, two maids, and a manservant.[151]

Before making any comparisons with the general population of England or with the Jewry of London, two observations must be made.

*Table 26: Occupations of Jews in the South West of England, 1851*

| *Occupation* | *Exeter* | *Falmouth* | *Penzance* | *Plymouth* | *Total* |
|---|---|---|---|---|---|
| Annuitant | 3 | 0 | 1 | 8 | 12 |
| Commerce | 25 | 8 | 7 | 64 | 104 |
| Manual | 2 | 3 | 3 | 30 | 38 |
| Professional/ Clerical | 4 | 1 | 1 | 16 | 22 |
| Total | 34 | 12 | 12 | 118 | 176 |

(Source: 1851 census.)

None of the Jewish families in the South West in 1851 kept a man-servant, so families with three or four female servants (who may also in some cases have doubled up as shop assistants) may have had a lower income than that suggested by Marion Lochhead which was based on two females and a male. Secondly, a household may have had no living-in servant, and yet be very comfortably placed. The bachelor establishment of the Nathan brothers, Nathaniel, Jacob and Henry, did not keep a living-in servant in 1851, nor in 1861, yet Jacob Nathan at his death in 1868 left an estate of some £14,000,[152] which at five per cent—and his pawnbroking business and properties[153] may well have produced a better return than that—would have brought in some £700 per annum.[154]

Bearing these points in mind, it can be shown that the Jewish community in the South West in the mid-nineteenth century may well have been comprised of a small upper-middle class, i.e. families with three or more servants; a larger middle and lower-middle class, with one or two servants; and a lower class with no servants[155] forming about half of the community.

Utilizing the 1851 census returns which give the number of servants in a household and on the basis of these criteria, it is possible that of the 57 Jewish families in Plymouth in 1851, 28, or about half, earned less than £150 per annum as they kept no servant, 19 may have been in the £150–£300 per annum income bracket as they had one servant; seven kept two servants and may have had an annual income of some £500; two families kept three servants, possibly indicating an income of about £750 per annum; and the one family with four servants may have had an annual income in the region of £1,000 per annum.[156]

The 27 Jewish families in Exeter in 1851 had a very similar income pattern. Seventeen families, or rather more than half, kept no servant and possibly earned less than £150 per annum; seven kept one servant indicating that they may have been in the £150–£300 per annum income bracket; two families each with two servants may have earned about £500 per annum; and the one family with three servants was possibly in the £750 per annum income group.

In Falmouth, four of the six Jewish families had no servant in 1851, indicating that their annual income may have been below £150; the one family with one servant could have been in the £150–£300 per annum income range; whilst the other family with three servants may well have been comfortably placed in the £750 per annum income group. In Penzance there were also six Jewish families in 1851, and financially, they may have formed a more homogeneous group, as two had no servants, possibly implying that they earned less than £150 per annum; three kept one servant and may have been in the £150–£300 per annum income bracket; and one household had two servants, possibly indicating an annual income of £500.

The class structure of the Jewish community in Devon in the mid-nineteenth century, expressed in percentages of the total number of Jewish families in Exeter and Plymouth, and compared with the figures given by Lipman for the Jews of London on the same basis, allows us to construct Table 27.

The figures set out in Table 27 for the South West and London correspond quite closely with regard to the upper-middle class, and particularly for Exeter and London in the other two classes. But Plymouth seems to have had more equally balanced lower- and middle-class Jews than London. On the national scale it may be observed that in 1851 there were 1,038,791 servants in a total population in Great Britain of 20,816,351 persons which is approximately one servant to every 20 people.[157] In 1851, there were 278 Jews in Plymouth with 43 servants attending them, that is one servant to 6.4 Jews, whilst there were 129 Jews in Exeter with 14 servants attending them, that is, one servant to every 9 Jews.

Just as there were Jewish apprentices in the South West in the eighteenth century,[158] so there were in the nineteenth. The census returns of 1841, 1851 and 1861 reveal the presence in Plymouth of four Jewish apprentices. In 1841, there were Phillip Ezekiel, aged 15, apprentice silversmith in the establishment of Aaron Levi in Bedford Street; and Henry, 15-year-old son of Mark Jacob, apprentice shoemaker in his father's business in East Street.[159] In 1851, there was George Morris, a 15-year-old apprentice watchmaker, who was the son of William Morris, jeweller of 35 Cambridge Street;[160] whilst in 1861, there was Henry, apprentice clerk, the 14-year-old son of Abraham Rosenberg, a pawnbroker of 13 Union Street.[161]

*Table 27: Classification of Jews according to Social Class in Exeter, Plymouth and London in the Mid-nineteenth Century*

| | *Exeter* % | *Plymouth* % | *London* % |
|---|---|---|---|
| Upper-middle class (3 or more servants) | 3 | 5 | 5 |
| Remainder of middle class (2 or 1 servants) | 33 | 45 | 30 |
| Lower class (no servants) | 63 | 49 | 60–70 |

(Source: 1851 census; Lipman, *The Structure of London Jewry*, p. 258.)

As noted before, at some time between 1851 and 1861 the Jewish communities of the South West began to decline.[162] This decline in numbers, though halted in Plymouth by the 1870s, may also have been associated with a decline in overall wealth. This may be seen by the lower ratio of servants in the total population of Jews in both Exeter and Plymouth, particularly the former. According to the 1861 census, the 73 Jews in Exeter had five living-in servants, that is one servant to every 14.6 Jews, whilst in 1851 there was one to every 9.0 Jews. In Plymouth, the gradual drop is perhaps best expressed schematically:

| *Year of census* | *Number of Jews noted in Plymouth* | *Number of servants* | *Number of Jews per servant* |
|---|---|---|---|
| 1851 | 278 | 43 | 6.5 |
| 1861 | 237 | 29 | 8.2 |
| 1871 | 268 | 26 | 10.3 |
| 1881 | 280 | 24 | 11.7 |

The same result is apparent from the occupations of Jews listed in the census returns, as may be seen in Table 28.

The decline is concealed in the commercial group listed in Table 28 because the 1881 figures contain a much higher proportion of hawkers of one kind or another than the earlier returns.

Further evidence on occupations of Jews in nineteenth-century Plymouth can be gained from the Marriage Registers of the Plymouth Congregation. The occupational groupings of the 287 Jewish men, grooms and fathers of brides and grooms, admittedly not all from Plymouth, involved in 149 marriages celebrated there between 1837 and 1912, are set out in Table 29.

Before attempting to draw any conclusions from the figures in Table 29, it is first necessary to detail the occupations in the various groups. The

*Table 28: Occupations of Jews in Plymouth according to the Decennial Census, 1851–1881*

| | *1851* | *1861* | *1871* | *1881* |
|---|---|---|---|---|
| Annuitant | 8 | 8 | 4 | 1 |
| Commercial | 64 | 63 | 66 | 64 |
| Manual | 30 | 29 | 15 | 16 |
| Professional/Clerical | 16 | 11 | 7 | 8 |

*Table 29: Occupations of Male Jews according to the Marriage Registers of the Plymouth Hebrew Congregation, 1837–1912*

| *Occupational Group* | *1837–75* | *1876–1912* | *Total* |
|---|---|---|---|
| Gentleman | 15 | 2 | 17 |
| Commerce | 151 | 116 | 277 |
| Manual | 23 | 54 | 77 |
| Professional/Clerical | 13 | 13 | 26 |

first figure after each occupation in the following list indicates the number of men following that occupation in the period 1837–1875, whilst the second gives the same information for the period 1876–1912:

*Commerce*

Barber 0,1; Bookseller 0,1; Butcher 0,1; Cement Merchant 0,1; Coal merchant 0,1; Corn merchant 0,1; Distiller 0,1; Draper 0,1; Furniture dealer 0,18; General dealer 53,25; Glass merchant 1,0; Grocer 0,2; Hairdresser 1,1; Hawker 5,7; Ironmonger 1,0; Jeweller 19,5; Linen merchant 1,0; Manufacturers 0,3; Marble and madrepore dealer 1,0; Master tailor 0,1; Merchant tailor 1,0; Naval agent 1,0; Pawnbroker 16,6; Photographer 0,2; Picture dealer 0,1; Property dealer 0,1; Rag merchant 0,1; Salesman 7,14; Shopkeeper 18,19; Silversmith 5,0; Tobacconist 2,0; Traveller 16,1; Wholesaler 3,0; Wine merchant 0,1.

*Manual*

Baker 0,1; Boxmaker 1,0; Cabinet maker 0,1; Cap maker 0,1; Electroplater 0,1; Farrier 0,1; Frame maker 0,2; Glazier 3,0; Last maker 1,0; Painter 1,5; Printer 0,1; Shoemaker 3,2; Tailor 9,32; Upholsterer 0,2; Watchmaker 5,5.

*Professional/Clerical*

Chemist 0,1; Clerk 1,1; Dentist 1,1; Doctor 1,0; Musician 0,5; Optician 2,0; Surgeon 1,1; Synagogue officials 7,4.

Once again,[163] this classification must be used with some caution, as it is difficult to be sure that all those described under the heading of 'manual' are correctly labelled. The baker, cap maker and electroplater, for example, may have been engaged largely in commercial activities in enterprises where these trades were carried on by employees. Similarly, perhaps the barber, butcher, hairdresser and photographer might have been better placed under 'manual' rather than 'commercial'. A further point to note is that not all the men whose occupations are given in the Plymouth Congregation's Marriage Registers were Plymothians. Many of the grooms and their fathers were from out of town. Nonetheless, it may be assumed that generally speaking the bride's family and the groom's came from similar backgrounds. Bearing these two points in mind, it is possible to make some general observations. Taking the period 1837–1912 as a whole, it is evident that those engaged in commercial activity in one form or another far outnumber all other groups put together, and that the numbers in the various groups roughly parallel the situation shown above for the year 1851.[164] The ratio of the various groups to the total number of occupations for the period 1837–1912 and for the year 1851 only, is given in Table 30.

The main observation to be drawn from Table 30 is that as the nineteenth century progressed there was a marked decrease in the number of Jewish men engaged in commercial pursuits and a corresponding increase in the number who were engaged in manual trades. This is a further pointer to an apparent decline in the overall wealth of the Jewish community in Plymouth in the latter part of the nineteenth century. There were, moreover, growing numbers of Jews in Plymouth after 1860 who earned their livings by occupations traditionally associated with impoverished immigrant Jews in England from about 1860 to 1910. The Marriage Registers of the Plymouth Congregation record one glazier in 1854 and two more in 1860;[165] two hawkers in the period 1837–59 and then ten more from 1860–1901;[166] the first painter appears in 1860 and there are five more between 1894 and 1906;[167] the first tailor is mentioned in 1838, followed by another two in 1854 and 1856 and then by thirty-five more from 1862 until 1912.[168] By 1860, there was a network of over 10,000 route miles of railway covering the face of England,[169] and the pedlar had become largely redundant. Instead of the

*Table 30: Distribution of Male Jews by Occupational Class in Plymouth, 1837–1912, and in 1851, Expressed as a Percentage of Those Gainfully Employed*

| | *1837–75* % | *1876–1912* % | *1837–1912* % | *1851* % |
|---|---|---|---|---|
| Annuitant/Gentleman | 7.4 | 1.1 | 4.4 | 6.8 |
| Commerce | 74.8 | 62.7 | 71.6 | 54.2 |
| Manual | 11.4 | 29.2 | 17.3 | 25.4 |
| Professional/Clerical | 6.4 | 7.0 | 6.7 | 13.6 |

(Source: Plymouth Hebrew Congregation Marriage Registers.)

road to riches, peddling was identified with pauperdom. Lloyd Gartner quotes one Joel Rabinowitz, Hebrew writer by choice and pedlar by necessity:

> . . . The peddler also trudges about from town to town and from city to city staggering under his burden. He is parched in the summer and frozen in the winter, and his eyes wither in their sockets before he gets sight of a coin. The farmers have wearied of these peddlers who stand before their doors daily. Still worse is the lot of the peddler who is faithful to his religion and refuses to defile himself with forbidden foods; he is bound to sink under his load.[170]

The element of danger which accompanied the pedlar in the eighteenth century did not disappear in the nineteenth. For example, one Tobias Tobias, an immigrant from Germany, was brutally assaulted in May 1853 near Exeter.[171] The poor fellow was so maltreated that he became deranged, his left eye and hearing on the left side were destroyed.[172] His pockets were rifled but his miserable stock of 8–10 pounds of red-wax beads on string in a cheap case was untouched.[173] When he recovered, a collection in the Anglo-Jewish community as a whole raised some £22 which was judged more than ample to put him back on his feet.[174]

As in the course of the nineteenth century new industries and avenues of trade emerged, some Jews in Plymouth were quick to seize the new opportunities. The new art of photography, for instance, attracted Jewish practitioners. As early as 1870, before the use of dry plates, Abraham Titleboam of Devonport was in business as a marine photographer,[175] as was Wolfram Ullman.[176] According to the 1871 census, the Exeter-born Joseph Jacobs of 23 Union Street, Plymouth, was an electro-gilder and tobacconist. The combination of occupations leads one to suppose that his tobacconist shop acted as a depot and he as an agent for an electro-plating workshop. On the other hand, 17-year-old Alfred Brock is listed in the 1881 census as an 'electro-plater',[177] and at his marriage in 1887 he still describes himself as such.[178] In 1896 he claimed that his were the oldest electroplating works in the West of England.[179] In the early part of the twentieth century, Tobias Brand of Frankfort Street, and later of Mutley Plain, put on sale the first refrigerators to be seen in Plymouth,[180] whilst, at the same period, the firm of J. Sanger changed naturally from being wholesale incandescent and gas fitting merchants to wholesalers of electrical goods.[181]

No description of the occupations of Jews in the South West would be complete without referring to the various kinds of work done there by Jewish women. The record is understandably incomplete, as it is in general Anglo-Jewish history, both because Congregational activity is

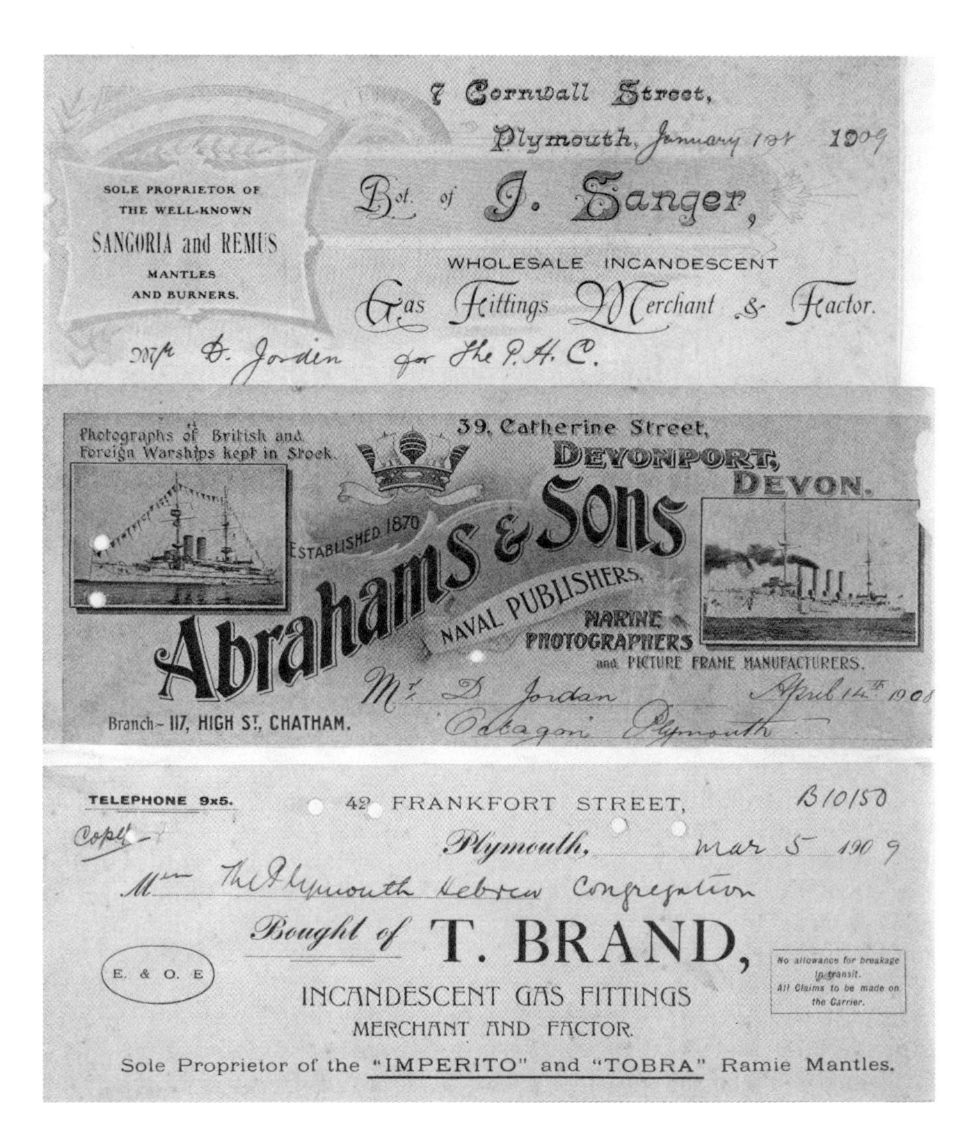

7 Cornwall Street,
Plymouth, January 1st 1909

SOLE PROPRIETOR OF THE WELL-KNOWN SANGORIA and REMUS MANTLES AND BURNERS.

Bot. of J. Sanger,
WHOLESALE INCANDESCENT
Gas Fittings Merchant & Factor.

Mr D. Jordan for The P.H.C.

Photographs of British and Foreign Warships kept in Stock.

39, Catherine Street,
DEVONPORT,
DEVON.

ESTABLISHED 1870
Abrahams & Sons
NAVAL PUBLISHERS,
MARINE PHOTOGRAPHERS
and PICTURE FRAME MANUFACTURERS.

Mr D Jordan April 14th 1908
Octagon Plymouth

Branch– 117, HIGH ST., CHATHAM.

TELEPHONE 9x5.

42 FRANKFORT STREET, B10150
Copy
Plymouth, Mar 5 1909
Mr The Plymouth Hebrew Congregation
Bought of T. BRAND,
E. & O. E.
INCANDESCENT GAS FITTINGS
MERCHANT AND FACTOR.
No allowance for breakage in transit. All Claims to be made on the Carrier.
Sole Proprietor of the "IMPERITO" and "TOBRA" Ramie Mantles.

*Illustration 9*

Letter-headings of Jewish-owned businesses in Plymouth, 1908–1909.

andro-centric, and also because women's activities are relatively under-represented in the documentary record.[182] Jewish women have been seen pre-eminently as homemakers, and although some attention has been given to their economic activity before marriage or as widows, by and large it has been assumed that they became financially dependent on their husbands in a framework of conjugal domesticity.[183] The debate and sharp distinction between 'home' and 'work' and a woman's place in them was not resolved until the latter part of the nineteenth century. But the stark alternatives often masked the economic contribution of working- and lower-middle-class women, partly because their activity was not quantifiable and, even when it was, because it was under-recorded, even in the census figures.[184]

To judge from later experience at the close of the nineteenth century in Manchester, it would be surprising indeed if none of the South-West Jewesses took in lodgers. They almost certainly ran the household economy, taking the wages from working children, and even their husbands, returning to the wage earners only pocket money. In this way, they continued the traditional role of the family as an economic entity. Many women helped their husbands, and reference has already been made to Jewesses who accompanied their pedlar husbands. It is surely significant that the very first record of the Plymouth Congregation is the use of land as a cemetery belonging to Sarah Sherrenbeck.[185] In the 1851 census a number of Jewesses style themselves as assistants to their husbands,[186] and it is therefore only natural that widows continued their husband's businesses.[187] Daughters also helped in their fathers' shops, and where they remained unmarried continued the business after the death of their father.[188] By 1881, however, perhaps in response to a change in social climate, only one wife, Charlotte Morris, is described as having an occupation—she was a dressmaker—and it is significant that her husband was apparently away on the census night. Otherwise, in 1881, only widows, and unmarried girls and spinsters are credited with a paid ocupation.

The census returns of 1851 provide clear evidence[189] that some 28 per cent of Jewesses in the South West were gainfully employed. In Falmouth 2 out of 6 Jewesses had occupations, in Penzance there were 3 out of 11, in Exeter 9 out of 44, and in Plymouth 32 out of 99. Table 31 gives the broad categories of occupations of all South-West Jewesses over the age of 14, listed in the 1851 census.

The occupations extracted from the four decennial censuses from 1851 for the Jewish women of Plymouth are detailed in Table 32. From Table 32 it appears that most Jewish women who were gainfully employed in the South West in the period 1851–81 worked with their hands in trades that were usually associated with their sex in the nineteenth century. The

*Table 31: Occupations of Jewesses in the South West of England, 1851*

| | |
|---|---|
| Annuitant, Independent, Lady | 12 |
| Commerce | 13 |
| Manual | 28 |
| Professional and clerical | 5 |
| | 58 |
| No occupation given | 97 |
| Total | 155 |

(Source: 1851 census.)

housekeepers all appear to have been running their own family domestic households. As the century wore on, the numbers of women listed as gainfully employed noticeably decreased, particularly in the manual trades. Table 33 expresses this as a percentage and compares it with the relative number of Jewish men in the corresponding groups in Plymouth in 1851.

It is perhaps worth noting that in 1851 of the 12 Jewesses listed as annuitants or of independent means there were 8 widows ranging in age from 45 to 89 years old, and 4 spinsters in the 35–50 year-old age group; of the 13 in commerce there were 6 spinsters, 4 married women and 3 widows; of the 28 in manual occupations 26 were spinsters and only 2 were married women; whilst in the professional/clerical group there were

*Table 32: Occupations of Jewesses in Plymouth according to the Decennial Census, 1851–1881*

| | *1851* | *1861* | *1871* | *1881* |
|---|---|---|---|---|
| *Commerce* | | | | |
| Assistant | 6 | 2 | 5 | 6 |
| Dealer, | | | | |
| general | – | 2 | – | – |
| in shells | 2 | – | – | – |
| in toys | 1 | 1 | – | – |
| in curiosities | – | 1 | – | – |
| Draper | – | – | 1 | – |
| (?hawker) | 1 | – | – | – |
| Jeweller | 1 | 1 | – | 1 |

*Table 32 (continued)*

| | *1851* | *1861* | *1871* | *1881* |
|---|---|---|---|---|
| Newsagent | – | – | – | 1 |
| Pawnbroker | 1 | – | 2 | – |
| Stationery traveller | 1 | – | – | – |
| Tobacconist | – | – | 1 | – |
| | 13 | 7 | 9 | 8 |
| *Manual* | | | | |
| Bonnet maker | 2 | 3 | – | – |
| Clothes cleaner | 1 | – | – | – |
| Domestic assistant | 1 | – | – | 3 (at home) |
| Dressmaker | 6 | 5 | 2 | 1 |
| Feather dresser | – | 1 | – | – |
| Housekeeper | 1 | 3 | 2 | 2 |
| Lace, | | | | |
| Honiton manufacturer | 3 | – | – | – |
| manufacturer | 1 | – | – | – |
| transfer of | 1 | – | – | – |
| worker | 1 | – | – | – |
| Leather dresser | – | – | – | 1 |
| Mantlemaker | 1 | – | – | – |
| Milliner | 4 | 1 | 2 | – |
| Milliner's apprentice | 1 | – | – | – |
| Sempstress | 2 | 3 | – | 1 |
| Shoebinder | 3 | 2 | – | 2 |
| Tailoress | – | 3 | – | – |
| | 28 | 21 | 6 | 10 |
| *Professional/Clerical* | | | | |
| Daily Governess | 2 | 3 | – | – |
| Harpist | 1 | – | – | – |
| Optician's assistant | 1 | – | – | – |
| Photographic artist | – | 1 | – | – |
| Scholar | – | – | 3 | 3 |
| Schoolmistress | 1 | – | 1 | – |
| Teacher of Singing | – | 1 | 1 | – |
| | 5 | 5 | 5 | 3 |
| *Annuitant* | | | | |
| Pensioner | – | 4 | 2 | 1 |
| Lady | – | – | – | 3 (visitors) |
| Relief | – | – | 2 | – |

*Table 33: The Distribution of Jewesses according to Ocupational Class in the South West of England, 1851, Expressed as a Percentage of those Gainfully Occupied*

| | *Jewesses in the South West in 1851* % | *Male Jews in Plymouth in 1851* % |
|---|---|---|
| Annuitant | 20.7 | 6.8 |
| Commerce | 22.4 | 54.2 |
| Manual | 48.3 | 25.4 |
| Professional/Clerical | 8.6 | 13.6 |

(Source: 1851 census.)

2 married women and 3 spinsters. Most married Jewesses did not work for a living as it has already been noted that there were apparently 42 of them in Plymouth at the time of the census in 1851,[190] only 8 of whom were gainfully occupied.

To sum up, Jews in the South West in the eighteenth and nineteenth centuries followed quite a wide range of occupations, primarily in commerce, but also manual and to a limited extent, professional or clerical. They hardly engaged at all in the staple trades and industries of the area, such as farming, textiles and mining.[191] The proportion of annuitant men and women and the large number of men in commercial activities in the nineteenth century seem to indicate some degree of economic success.

In the early twentieth century,

> contemporary observers with no particular axe to grind were in no doubt at all that the immigrants were, when compared with their English neighbours, very unusual people indeed. . . . Whereas the English workman was content for the most part to remain an employee, the over-riding desire of the majority of the immigrants was to control their own destiny by becoming their own masters.[192]

Furthermore, the Jew who wished to remain part of Jewish society and attached to the synagogue needed more money than his non-Jewish counterpart. He needed to pay for his seat in the synagogue, not merely to have a seat in the synagogue but also to preserve his rights of burial; he needed extra money not only for his offerings in the synagogue which were socially, if not constitutionally required but also for the communal calls on his charitable purse; his *kosher* meat and food cost him more; he had to take a week off work when he was a mourner; he needed some days off work for at least the High Holydays, even if he did not strictly observe the Sabbath and other Festivals; if he kept the Sabbath, he had to pay a *shabbos goy* to bank up his fire in the winter; and if he wished to preserve any social self-respect in the Jewish community his wife had to have at least a charwoman to come in and scrub the floors and, perhaps more important, the front door step, every week. If he was self-employed he might, just might, be able to accomplish all this for himself. Perhaps he might be able to pay for a good education for his children and who knows, they could then aspire to white collar work or even a profession. The Jew's lifestyle, his whole *weltanschauung*, was geared to this aim. As a poor Jewish immigrant of no particular consequence told Her Majesty's Royal Commissioners in 1902, 'I neither smoke nor drink and believe in everything that will make me better off'.[193]

It was this desire for financial self-improvement which drove the immigrant to scour the streets, knocking on doors in all weathers, looking for customers who would buy his sponges, or who would pay him to mend their broken window panes. He might have earned more, and more steadily, working in the dockyard, but he and his children would never rise in the world that way. As late as 1955, the loan department of the London Jewish Welfare Board reported that

> the high wages [being generally offered] do not seem to lessen the number of prospective borrowers who prefer to be their own master, rather than take jobs which might make them more money.[194]

The traditional route in the nineteenth and twentieth centuries for the poor Jewish immigrant who came with a trade in his hand, usually as a tailor, cabinet maker or shoemaker, was to go to work for another Jew for a year or two, often in sweatshop conditions. Then, when he had found his legs and learned to speak English he would leave his employer and start up on his own. If he wanted to avoid the competition of the big city, then he would go to the provinces. If he had a wife, then he would soon open a shop, in a poor part of the town, with an upper part for accommodation for himself and his family, in a trade allied perhaps to the one he had learned in Eastern Europe, generally second-hand clothing or furniture. He would leave his wife in charge whilst he travelled around the area, buying and selling, making and repairing, until he became established. He would then go up-market, to the main shopping area. If he stayed in retailing he could earn a good living and live a comfortable lower-middle-class life. If he was more ambitious he would soon learn that it was more profitable to buy and sell houses than bedroom suites. By frugal living, he would save enough to put down a deposit on a house.

As he made his rounds, he would see an opportunity, buy a house, let it, and use the rent to pay the mortgage. After a year or two, he could repeat the process until he had several houses. In the meantime, if he had bought shrewdly he would sell a house here and there and begin to amass capital. After two or three decades, he, and his sons or sons-in-law if they chose to join him, would begin to develop property, buy land and build on it shops or offices or factories. Gradually, if they were wise, they would move away from the domestic market, for the landlord of a domestic tenant, as even town and borough councils have discovered, is not a popular figure, either to the tenant or to the public. The immigrant's success enabled him to provide his children with an education, for in those days there was no free education after the age of fourteen, except for the very few able to win scholarships. So the second generation might aspire to a profession, in law or medicine, or as ancillary to the business, in accountancy or estate management.

One can trace this progress trail from rags to riches of Jewish immigrants in the late nineteenth century and their descendants in the twentieth century by observing a number of Jewish families in Plymouth.

According to the 1871 census, the first Fredman to arrive in England appears to have been Jacob David. He and his wife Rachel were born in Russia. Their eldest child, Leah, was born in Birmingham in 1862. By the following year they were in Plymouth when their next child Phoebe was born. By 1871, he had set up shop as a clothes dealer in Queen Street, the shopping centre of Devonport. By now, he had brought over his brothers. One of them, Levy Fredman, was beadle of the synagogue. Another, (Samuel) Wolf, was away on the night of the 1871 census but his wife 'Freebey' Freedman,[195] is described as a 'traveller's wife'. He was probably travelling in the same line as the fourth brother, Lavine Fredman, later called Levin Fredman, who is described in the 1871 census as a hawker of sponges and leather. Levin, his wife and two children shared a house, 98 Pembroke Street, Devonport, with Jacob Roseman, also a hawker of sponges and leather, his wife and two children. Wolf, his wife and four children shared a house, 3 Canterbury Street, Devonport, with his married daughter Rachel and son-in-law Israel Roseman.[196] One could almost have guessed that Israel was also a traveller in sponges and leather. An Abraham Roseman, a young bachelor of 20, was a lodger, along with two other Jews, at 53 Mount Street, Devonport, and all three were travellers in sponges and leather. As might be expected, at this stage, none of them had a living-in servant.

According to the 1881 census, Jacob Fredman has moved to 34 Queen Street, Devonport, and is now a general dealer, and if nothing else he has had three more children since 1871. In 1882, his daughter, Leah, marries Jacob Israel Pollack of London, a corn merchant, whilst Jacob

Fredman is described as a shopkeeper.[197] In 1886 and 1888, still at the same address when his daughters Rebecca and Amelia marry, he is an outfitter.

In 1881, Levy Fredman was no longer a beadle but a general dealer at 11 Hoe Street, Plymouth. His eldest son, Nathan, aged 22, has married and set up home in a house with two other Gentile families at 66 King Gardens. He, too, is a general dealer. Financially and possibly socially, Levy seems to have improved himself. He has not got much time to further improve himself, for he will die in January 1886, Nathan having predeceased him in 1884.[198]

Woolf Fredman has moved to 12 Queen Street, Devonport, and he is still described as a traveller. His eldest daughter, Anne, aged 22, is not at home; perhaps she is married. In 1882, Sarah, then aged 20 and described as a hawker and the daughter of Samuel Woolf Freedman, also a hawker, married Barnett Cohen, a hawker.[199] As yet, Woolf does not seem to have made much progress. But better days are ahead. In 1886, he has moved to 38 Queen street, Devonport, and calls himself a traveller. His youngest daughter, Golda, marries Morris Sulski, a furniture dealer, the son of Jacob Sulski, a baker.[200]

Levin seems to have done well for himself in the decade that has passed since the 1871 census. He has moved to 4 James Street, Devonport, and he is no longer a hawker but is described as an outfitter. His eldest daughter, Amelia, 14, is a general assistant in his shop, whilst his wife, Hetty, is busy looking after their children, Myer, Aaron, Israel and baby Fanny, all born in Devonport. Most important, though, is that they have a living-in domestic servant. True, she is only 15 years old, but she marks the transition from migrant labourer and the working classes to middle-class respectability. When his daughter Amelia marries Myer Isaac Roseman, a furniture dealer, in 1890, he describes himself as a property dealer.[201] He has moved to 24 Catherine Street, Devonport, and he is still there trading as a house furnisher in 1894,[202] and as a furniture broker in 1896.[203]

Levin's son, Myer, continued the family business and occupied himself in Local Government, becoming Mayor of Devonport. When he died in 1927 he left a very considerable fortune. Levin's daughter, Fanny, married Reuben Lincoln, at one time minister of the Jewish Congregation in Bradford, and later minister of a New Jersey, USA, Congregation and eventually becoming a business man. Their children included Fredman Ashe Lincoln QC who served in the Royal Navy during World War II, rising to the rank of Captain; his son is a rabbi. One of Samuel Wolf's great-grandchildren became a conductor. Another was Arthur Goldberg, solicitor and Lord Mayor of Plymouth, whose son, David Goldberg QC, is a prominent barrister. The descendants of the Fredman brothers are today scattered across the world and to tell their story fully would require a book of its own.

Towards the end of the century a commercial directory notes in Plymouth and Devonport 19 Jews who were general dealers, house furnishers and furniture manufacturers, clothiers and outfitters, who were probably all much of a muchness, though some were no doubt more prosperous than others; 12 Jewish watchmakers, jewellers, and pawnbrokers; 2 Jews who were picture frame-makers; a Jewish financier, a diamond merchant, a hotel keeper, a tobacconist, a dairyman, a fish merchant, a bicycle dealer, and a musician.[204] These were all self-employed and had their own business establishments.

The entrepreneurs of the early part of the twentieth century followed a similar pattern. Thus, Ephraim (Frank) and Eva Holcenburg and their two daughters, Peggy and Gussie, arrived in England about 1909. They stayed in London for two years or so, moved to Exeter, and then on to Plymouth. He was a furrier, and using Plymouth as his base, he travelled the markets of Devon and Cornwall, and as far afield as Wales. Whilst he was travelling, selling and repairing furs, his wife and daughters opened the Imperial Fur Company in Frankfurt Street, Plymouth in 1913, moving to better premises in nearby George Street in 1925. Frank died in 1934. Business, however, prospered and the widow and her daughters opened a fashion shop, Woodhills, in 1936. The daughters married and the family, having made a tentative start in property, took part in the redevelopment of blitzed Plymouth after the war.[205]

Between the two World Wars, some of the more assimilated families began to take up a wider range of business activity. The Brock family, for example, was old-established in Plymouth. Eleazar (George) Brook (as the name first appeared) was a tailor in Plymouth when his son, George, a hawker, married Sarah, daughter of Lyon Levy, in 1838.[206] His son, Lewis Brock, was a hairdresser when he married Henrietta, daughter of Aaron Nathan the Plymouth constable, in 1860.[207] In the 1871 census he describes himself as a musician. Indeed, later, together with his five sons, Henry, Charles, Alfred, Jacob Nathan (John) and Ernest, they were known locally as Brock's Band. Charles became a bookmaker, with a share in a Plymouth nightclub and a clothing factory. By his will in 1947, he left some £25,000. Alfred, we have already seen as the first electro-plater in the West Country, had a jewellery shop, and left £1,400. Ernest became wealthy. He was a partner with his brother and another in a clothing factory which employed several hundred people until it was destroyed in the Blitz; he was a bookmaker, and he went into property. He was elected to the town council. When he died in 1950 he left £125,000. His widow, Lillian Ada, known as Cissie, a convert to Judaism and very proud of her new faith, also a

Town Councillor and prominent in local affairs, carried on his property business.[208]

A branch of a London Jewish family might settle for a while in Plymouth to look after expanding business interests. Members of a Smith family had been resident in Plymouth between the First and Second World Wars. A Caroline Smith died in 1933 and was buried in the Gifford Place cemetery.[209] In 1940, Mary Smith and her daughter Esther were the first civilians in Plymouth to be killed by enemy action. Mary's son, Nat, married a non-Jewish Plymouth girl. He was in the entertainment business, he had a small cinema and was a bookmaker; their son became a professor in America, patented a chemical process and became a multi-millionaire. A second son, Morry, had a barber's shop *cum* fruit shop *cum* bric-a-brac shop in Union Street. Another son, Hymie, settled in Exeter, where he was a bookmaker. Yet another son, Joe, was in bookmaking, but in a big way. He had business connections with the Stein family of London.[210] A daughter, Hetty, married Solly Silver who became a well-known bookmaker in Plymouth. Solly Silver introduced a Plymouth scrap-metal company, Davies and Cann, to a branch of the Smith family trading in non-ferrous scrap in London. In 1952, the directors of Davies and Cann sold their interest to the Smith's. So Jack Smith and his wife Evelyn and four sons, Malcolm, Harry, Derek and Colin, together with their six-weeks-old daughter Mary, came to Plymouth to manage the business. The business was expanded to deal with ferrous metals as well, and the sight of obsolete submarines being dismantled at Smith's Wharf, Pomphlett, became a familiar one to all who crossed Sutton Bridge. Jack and his family, strongly observant and closely identified with the Congregation, emigrated to Israel in the 1960s.[211]

Not all the Plymouth Jews were self-employed, some were employees, particularly after the Second World War when discriminatory practices were dropped. Mr Aloof, the Plymouth *shammas*, had three sons who all grew up and remained in Plymouth, Sidney, Lionel and Percy. Sidney was apprenticed to a Jewish watchmaker, King Field, and went into business on his own account, his son Brian remained in Plymouth, also in business on his own account. Lionel became manager of a jewellery shop belonging to a member of an old-established Plymouth Jewish family, the Nelson's; his childen moved to London where they married extremely observant Jews, his son, Martin, became a *hasid*, and is employed as a *shomer* (ensuring that food in shops licensed by the *Beth Din* is *kosher*). Percy, later to become president of the Congregation, when he came out of the army was employed by the Post Office in its then Telephone Service; his son, Marcus, qualified as a chef and is employed by a leading London hotel. Not all employees remained so attached to the Jewish community. One, employed in the dockyard, drifted away. So did another, employed as a conductor on the Corporation buses. These two returned in their old age. Others have disappeared without trace.

From the above account it is clear that in the twentieth century, as in former times, a few Jews prospered and left considerable fortunes, many lived their lives comfortably, earning a little more than they spent. When they died some surviving relative took out Letters of Administration to wind up their small estate of two or three hundred pounds.[212] Most, however, just kept their heads above water, lived in rented accommodation, and left only a few personal effects which needed no formal legal procedure to liquidate, and few records or traces of themselves.

## CHAPTER FIVE

# Cemeteries, Synagogues and Other Buildings used by the Jewish Communities in the South West of England after 1740

Jewish communities in England have always emerged after a process of growth, one or two Jews settling in a town, to be followed by a few more, so that over a period of years there would be enough to form a community.[1] Accordingly, services were first held in the house of an early established Jew, who had a room large enough to accommodate a dozen or so worshippers. Then a larger room was rented, and after that as the numbers grew too large to be conveniently seated in a hired room or small hall, land was acquired and a synagogue built. But long before a permanent house of prayer was built somebody would have died and a duly sanctified burial ground was required. Indeed, the acquisition of a cemetery is often the first communal act of a newly formed community.[2]

The Jewish communities of the South West developed on just these lines. It has already been shown that a small number of Jewish families settled in Plymouth in the early 1740s. About this time, apparently, a local Jew died and it was not convenient to take him (or her) to London or to the nearest established community, Portsmouth. Almost certainly this Jew was buried in a garden on the Hoe which, in 1744, was either already in the possession of Sarah Sherrenbeck or was bought for this purpose, because this land ultimately formed part of the first Plymouth Jewish Cemetery (see Map 5). In 1752, this land was held by Joseph Jacob Sherrenbeck in trust for his wife Sarah.[4] In the next five or six years the

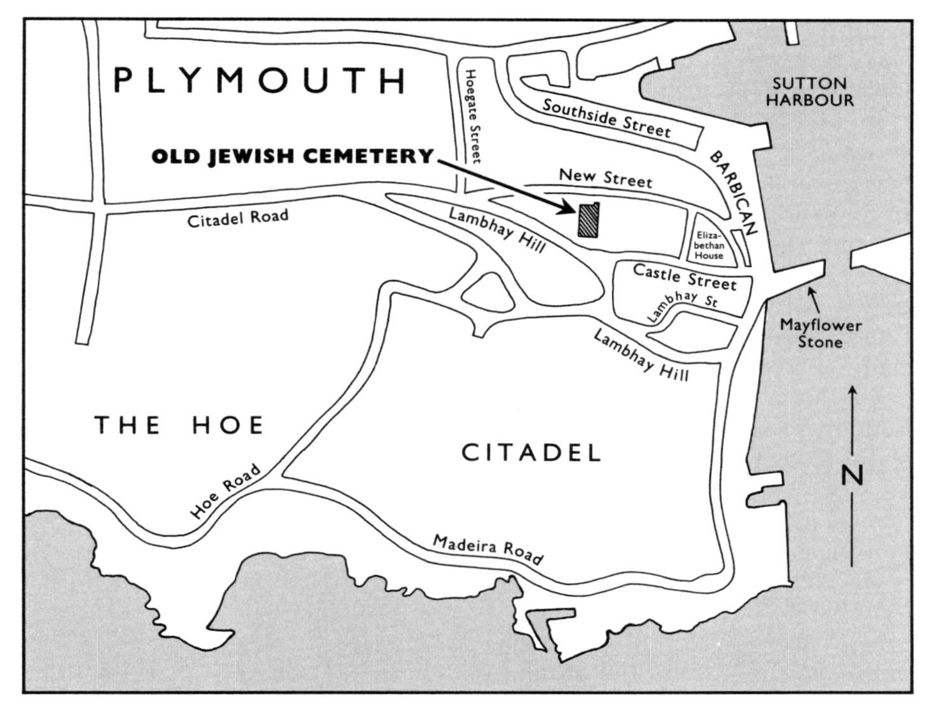


*Map 5*

The site of the old Jewish Cemetery on Plymouth Hoe.

community apparently expanded and required a proper burial ground with a small chapel for laying out the dead, and Sarah Sherrenbeck's original garden which had served its purpose as a kind of private burial plot was no longer large enough. In June 1758, therefore, another quarter acre of ground and a summer house, with its garden near to, and probably adjoining the garden owned by Sarah Sherrenbeck, was bought by three prominent London Jewish merchants, on behalf of the Congregation.[5] The reason for the involvement of the London merchants was probably that the community was still not sufficiently well established—it did not yet possess its own synagogue building—for its members to be certain that the Congregation was a permanency. They therefore arranged for London merchants, who probably advanced much of the necessary cash as well, to be nominal owners, to ensure that whatever happened the cemetery would be looked after and the graves cared for. In 1811, when the Congregation was flourishing and a further piece of land was acquired, nobody outside Plymouth was involved.[6] This time the land was conveyed to three Plymouth Jews and a non-Jew,[7] the latter in case Jews were not legally entitled to hold land.[8]

It is typical of the early development of Anglo-Jewry that there were a series of *de facto* situations which time legalized. In the same way as there was never any formal declaration by Cromwell permitting Jews to resettle in England (and hence no legislation about them for Charles II to repeal)[9] so, too, there was no formal application for permission to establish a Jews' cemetery in Plymouth. Private ground already in Sarah Sherrenbeck's possession was used for a burial ground, and only fourteen or fifteen years later in 1758 was the purpose for buying more ground 'officially' disclosed in the lease.

The lease and release of 1758 relates that in consideration of £40 the London Jewish merchants were to have a garden and summer house about 55 feet in length and 35 feet in breadth

> to permit and suffer all and every such persons as profess the Jewish ceremonies and religion and who now reside in or near the borough of Plymouth or at any time thereafter . . . to be used as a place of burial . . .[10]

Similarly, the lease of the ground bought in 1811 declares that it is to be used as a 'burying place for the Society of Jews in Plymouth'.[11] There is no similar statement in the lease of Sarah Sherrenbeck's ground, but the mere fact of its inclusion with the other burial ground leases on the Hoe as well as the contiguity of the plots makes it almost certain that it was indeed the first Jewish burial ground in Plymouth and the South West. Continued expansion of the Plymouth Congregation necessitated a further purchase in 1868 of land in Compton Gifford adjoining the Old Plymouth Cemetery near Central Park.[12]

Turning to Exeter, the first physical evidence of a Jewish cemetery there was a decision by the Exeter City Council to grant a lease on 28 March 1757 to,

> Abraham Ezekiel for a term of 99 years determinable on three lives in late Tanners plot in the parish of the Holy Trinity at the yearly rent of 10*s*. 6*d*. for a burial place for the Jews . . .[13]

The plot was '22 feet towards Maudlin Street and in depth backwards 80 feet'.[14] The lease was made from 18 May 1757 for a consideration of five shillings on the lives of Abraham Ezekiel himself, then aged 31, Rose his daughter, aged 2, and Israel Henry, the son of Israel Henry, also aged 2.[15] The lease on this plot of ground was renewed on 7 January 1803 for a consideration of five shillings and double the rent. This time the lessee was Moses Mordecai and the three lives were Solomon Ezekiel, then aged 17, and Simon Levy and Jonas Jonas who were then both 12 years old.[16] It seems as though the Exeter Congregation was 'outgrowing' its cemetery, because only four years later on 23 June 1807 Moses Mordecai took out a new lease for the original ground and also for a plot adjoining it.[17]

In 1845, the Exeter Congregation reported to the Chief Rabbi that,

> two out of the three lives have dropped and the third is very aged, which leaves us in a very precarious state. The trustees have offered to receive a piece of ground in lieu of it and to convey our present ground to us as freehold which will cost us £300. We have applied for subsriptions to the parties who have relatives lying there but have not succeeded.[18]

The freehold was eventually purchased, and the burial ground is still in use.[19]

It has not been possible to trace any lease of the Falmouth Jewish cemetery. It is situated on the Penryn Road midway between Falmouth and Penryn, and was thought to have been presented to the Jewish community by Lord de Dunstanville towards the end of the eighteenth century.[20] In 1913, however, it was up for sale as the lease had expired. It was bought by a local Jew, A.A. de Pass, and it is now cared for by the Board of Deputies.[21] It was already in use by 1791, as the oldest surviving tombstones date from that year,[22] though B.L. Joseph, about 1850, apparently read the date on one stone as 5534 (= 1774).[23] It is reasonable to suppose that the Falmouth cemetery was in use some decades before the date of the oldest surviving tombstones because the Plymouth Jewish cemetery was in use by 1750, and the oldest tombstone there is dated 1776,[24] whilst the Exeter Jewish cemetery was opened in 1757 and its oldest readable tombstone is dated 1807.

When the Penzance Congregation built its first synagogue in 1807, it already had a cemetery at the back of Leskinnick Terrace.[25] There, too, the oldest stone is dated 1791, and it may well be that the cemetery was in use some time before that date.

The Exeter Jewish cemetery has a small chapel still in use where the dead are washed in the last purificatory rites; at Plymouth's old Jewish cemetery and at Falmouth, only a fireplace against the wall indicates where the chapel stood. The Plymouth Congregation's new cemetery at Gifford also has a chapel which was rebuilt in 1958, as well as a caretaker's house.

After the purchase or leasing of a cemetery the next concern of a growing Jewish community was to engage itself in building a synagogue.

The synagogue buildings of the four South-West Congregations are all extant.[26] Those of Plymouth and Exeter are still in use as synagogues, Falmouth's was sold about 1900 and is currently used as a furniture warehouse, and Penzance's was sold in 1906 and is at present used for other purposes.

The most detailed records have survived for Plymouth's synagogue. On 27 April 1762 the mayor and commonalty granted a lease to one Samuel Champion[27] for 99 years determinable on the deaths of George and John Marshall and Joseph Jacob Sherrenbeck, at a yearly rental of £2 12*s*. 0*d*. of 'a garden in St Katherin's Lane . . . to erect or build any houses or edifices thereon'.[28] It is noteworthy that there is no mention of the purpose for which the 'edifice' was to be built. Just one month later Samuel Champion signed a deed in which he declared that he held the lease in trust for Joseph Jacob Sherrenbeck and Gumpert Michael Emden of Stoke Damerel, shopkeepers, elders of the Synagogue.[29] The lease was renewed in 1786 and again in 1797.[30] In 1834, the mayor and commonalty transferred the freehold of the synagogue's land to seven members of the Plymouth Congregation acting as trustees, for a consideration of £100.[31]

Besides the main synagogue in Plymouth, there were at least two buildings in Plymouth Dock (now called Devonport) used as a 'branch' synagogue, or 'Minyan Room'. The first was in operation before 1810, but heads of agreement were signed by a committee of Plymouth and Dock residents on Tuesday, 10 July 1815.[32] It was probably dissolved by 1844, when Mrs B. Moss presented to the Plymouth synagogue a spice box used in the *havdalah* ceremony, which she had previously given 'to the Plymouth Dock Minyan Room'.[33] A second Minyan Room was opened about 1907, according to the late Mr H. Greenburgh of Plymouth, although it is first recorded in *The Jewish Year Book* of 1914. It was bombed in the blitz on Plymouth in the Second World War, its scrolls of the Torah being rescued from the rubble by F. Ashe Lincoln QC.

The lease for the ground of the Exeter synagogue in St Mary Arches was granted on 5 November 1763 to William Luke,[34] Abraham Ezekiel and Kitty Jacobs, at an annual rental of two guineas.[35] An inscription in a Breeches Geneva Bible gives the exact date and order of service of the consecration and first use of the synagogue ten months later:

> August the 10th, 1764; This day the Jews Consecrated their new Synagogue in the City of Exeter. The service began with prayers for the King and Royal Family. As soon as those prayers were ended, the musick (which consisted of two violins, a heautboy and bassoon) played God save the King. They then carried the Law of Moses seven times round their reading desk and between the times of carrying it the following psalms were sung in the following order, 1st—5, 91; 2nd—30; 3rd—24; 4th—84; 5th—122; 6th—132; 7th—100.

> After these psalms and some prayers they stopped about half an hour and then said the first service of their Sabbath, the whole lasted above three hours.[36]

The Falmouth Congregation first erected a synagogue in 1766 in Hamblyn's Court, later known as Dunton's Court, and then moved in 1806 near to Smithick Hill.[37]

The Penzance synagogue was built after a lease granted to Hart Woolf and others on 11 December 1807.[38]

In common with most eighteenth-century synagogues, and nonconformist meeting houses for that matter, all the South-West synagogues had plain exteriors to avoid unwelcome attention and envy. For these reasons the entrances of many eighteenth-century synagogues were tucked away from main thoroughfares.[39] Both the Plymouth and the Exeter synagogues have only one entrance and that fronts on to a narrow pavement which is used only as a pedestrian short-cut, whilst those at Falmouth and Penzance were hidden away in the back streets.[40] The Exeter Synagogue like that of Hull, has no windows and is lit through the roof, possibly to prevent noise annoying the church next door or to give an extra sense of security to congregants.[41] There may be an additional reason for not making the entrance on the main road. Ideally, the synagogue should be built with an east–west orientation, so that the worshipper on entering the synagogue would be facing the ark and at the same time facing Jerusalem. An inscription placed over the inner doors leading into the Exeter synagogue at its refurbishment in 1836 displays three biblical verses quoting Jerusalem, the chronogram for 1836 based on the Hebrew word for Jerusalem, and a specific injunction (in translation): Pray according to the law towards Jerusalem.[42]

The South-West synagogues were austere not only in their outward appearance, the interior was also plain with bare plaster walls coloured by water-paint, plain ceilings, roughly planed wooden floors, and cheaply-constructed wooden benches. According to a craftsman employed in the restoration of the Plymouth synagogue in 1965, the benches display joints typical of eighteenth-century naval craftsmen and were probably made by dockyard carpenters. Apart from the *bimah* (the central dais) and the ark on the east wall, the four synagogues in Devon and Cornwall could equally well have been nonconformist meeting houses. It is not considered likely that architects were employed for provincial synagogues, and usually their design was taken by the builder, himself usually a nonconformist,[43] from his local chapel.[44] Both the Exeter and Plymouth synagogues are listed as buildings of special architectural or historic interest under section 30 of the Town and Country Planning Act, 1947, and they are strikingly similar. The ark in the Exeter Synagogue is a simplified copy, reduced in scale, of the one in Plymouth.[45] In the Plymouth Synagogue the ark is the most ornate item of furniture. It might have been imported from Germany because there are indications that it has been cut down to size.[46] It figured in the schedule of fittings which were mortgaged with the synagogue in 1770:

> One Alter[47] and one Tabernacle.[48] Three large Brass Chandeliers,[49] Eight large Brass Candlesticks,[50] Five setts of the five books of Moses engrossed on parchment in the Hebrew language.[51] One clock, seats or chairs or other furniture.[52]

The architect responsible for the restoration of the Plymouth synagogue in 1965 has described the fittings in these terms:

> The *Bimah* must have been a very beautiful and dignified platform, with the eight great candles alight, and the richly polished woodwork which would reflect the light of the candles contained in the three great Dutch type brass candelabra which were no doubt suspended from the centre of the three circular ventilation rilles in the lofty ceiling. The great Ark which closely resembles the one in the old Synagogue in Venice, with its pediment cartoushe at the top, its beautiful cornice and carved decorations in the Roman Corinthian order, was completely covered with silver and gold leaf. The columns, capitals, cornices, flowers and mouldings were all in gold leaf, while the plain surfaces which are now painted dull red, were in silver leaf.[53]

It has been pointed out that,

> whereas carved detail in other types of buildings sometimes becomes coarser and less plentiful as the surfaces recede in perspective from eye level, richness was evenly spread over the Ark front and, if anything, tended to become more elaborate towards the top. A possible explanation lies in the traditional arrangement of the synagogue, the high balcony fronts used for the screening of women cut off from view the lower half of the synagogue and concentrated attention on the upper part of the Ark. Furthermore, as Georgian synagogues were generally small, the detail of the Ark had to satisfy closer scrutiny.[54]

The eight large brass candlesticks with their wide sconces and candle holders are replicas of those used at the Bevis Marks Synagogue, London, and symbolize (according to the beadle there in 1963) the three requirements of the *havdalah* service, light, wine (the candle holder has the shape of a traditional wine cup), spices (the long column of the candlestick has

ק״ק פלימוט

# PLYMOUTH'S HISTORIC SYNAGOGUE

## BUILT 1762

*Ancient Menorah, used in Plymouth Synagogue for some 200 years*

# THE PLYMOUTH SYNAGOGUE

*"Blessed be he that cometh in the Name of the Lord.,..".*

There appear to have been contacts between the Jews and Phoenicians on the one hand and Devon and Cornwall on the other more than 2,000 years ago. There was a well established Jewish community in Exeter from 1181. By the middle of the 18th century there was a Jewish community settled in Plymouth. They came mostly from Central Europe, one family called Emden, whose descendants still live and trade in Southside Street, emanating from the town of that name, came via Amsterdam and in 1761 had a "Goldsmith's shop known by the sign of the tea-Kettle and Lamp, near the Gate, Plymouth Dock." By *1745* they were holding regular services in their own houses and then in rented raoms. In *1759,* they planned this Synagogue, and on 27 April 1762 the mayor and commonalty granted a lease to one Samuel Champion (a non-Jew, because there were doubts then whether or not a Jew might own or lease land) of "a garden in St Katherin's Lane ... to erect or build any houses or edifices." In 1834, the Congregation bought the freehold. In 1864, Mr. Leon Solomon extended the Ladies' Gallery by adding the two wings along the length. This appears to be the only major alteration since the synagogue was built in 1762.

Let us now take a tour through our beautiful Synagogue, the oldest in continuous use for the Ashkenasi rite in the English speaking world.

Above the entrance doors is a foundation stone which reads in translation: Holy to the Lord. This holy and honourable house was founded and built in the year, 'Come let us worship, bow down and bless before the Lord' (Psalms 95,6 with slight changes, and the chronogram gives *5522* (= 1761/1762).

In the vestibule is fixed a small silver shield with the names of the Vestry members who in 1784, were apparently responsible for the completion of the Synagogue's interior furnishings and repayments of the mortgage. Here, also, is the prayer for the Royal Family painted on canvas. It was promised to the Congregation and Sabbath in 1759 at a cost of 8 guineas and bore the names of King William IV and Queen Adelaide. The monarchs' names were updated from time to time.

As you enter the Synagogue, you will see the original windows to left and right, though the warm coloured glass is comparatively recent. These stained glass windows portray aspects of the Sabbath and the Festivals - New Year, Day of Atonement and the Three Pilgrim Festivals. The 2 windows on the east wall were cut into the wall after 1874.

On the east wall is the Holy Ark. It may have been made locally, more likely it was imported from the Continent. Local craftsmen made a smaller, less ornate, replica of it for the Exeter Synagogue, built in 1764. Above the Ark are the 2 tablets with the first words of each of the Ten Commandments. In front of it is the *Ner Tamid,* the Everlasting lamp, reminding us that one lamp in the seven branched lamp of the Tabernacle and Temple was always alight

In the Ark, are the sacred Scrolls of the Torah, the Pentateuch, adorned on special occasions with ancient religious appurtenances.

In the centre is the ***BIMAH* - *a*** Greek world meaning "a speaker's tribune from which the Torah is read every week in an annual cycle (*cf.* Acts 15,21). Around it are the eight great brass candlesticks which featured in the original mortgage deed along with the building, the Ark, and the *BIMAH*.

The central chandelier, the *Ner Tamid,* and the silver plated eight branched candelabrum, the *Chanukiyah,* are all modern, but the little brass offertory box facing, is original and is still in use!

Come and sit on the old pine benches. Generations of worshippers have given them the patina of age. Here sat Solomon Hart, well known Victorian artist and Librarian to the Royal Academy; Abraham Joseph, friend of kings whose descendants played a notable part in the development of the Commonwealth; Phineas Levi, one of the first Jews to be elected to public office in England; Jacob Nathan, one of the largest benefactors of Plymouth's poor in the 19th century; the writers Grace Aguil and Amy Levy; Abraham Daniel, one of England's great miniaturists; and a long line of men active in local government, including Abraham Emdon and his son Eliezer Emdon, Myer Fredman who became Mayor of Devonport, Ernest and Cissie Brock, Isidore Joseph who became Mayor of Torquay, and Arthur Goldberg who became first Jewish Lord Mayor of Plymouth.

This gem of a Synagogue has seen many happy occasions, but has also passed through many troubled times. Jewish Plymothians served in the Volunteer Companies recruited in 1798 to repulse any Napoleonic invasion, and the Congregation gave its sons and daughters in two world wars.

The glorious traditions of the past are kept alive by the small community of the present. Long may it be so.

Thank you for your visit, may you derive knowledge and understanding from it and may the Blessings of the Almighty accompany you.

".... WE BLESS YOU OUT OF THE HOUSE OF THE LORD"

**Rabbi Dr. B. Susser**

**The oldest Ashkenazi Synagogue in the whole of the English speaking world.**

Printed by the Plymouth Hebrew Congregation. Charity No. 220010
Catherine St. PL1 2AD tel. 01752 301955
email: info@plymouthsynagogue.co.uk
WWW.plymouthsynagogue.co.uk

*Illustration 10*

The interior of the Plymouth Synagogue.

*Illustration 11*

Detail of the Plymouth Synagogue's Ark.

**Plymouth Hebrew Congregation.**

Honorary Officers:

Mr. T. BRAND, President. Mr. D. JORDAN, Treasurer.
Mr. M. FREDMAN, Burial Warden. Mr. H. ORGEL, Secretary.

**זמירות ושירות**
**לשיר במקהלות**

ביום חנוכת בית הכנסת בק"ק פלימותה יע"א

ביום א׳ ה"י אדר ראשון "אחרי הרחיבם את הבית" שנת סתרי לפ"ק

ORDER OF SERVICE

AT

**The Re-Consecration**

OF THE

SYNAGOGUE,

CATHERINE STREET, PLYMOUTH.

ON

SUNDAY, FEBRUARY 27th, 5670—1910.

The Service will be conducted by the
Revs. D. JACOBS and A. K. SLAVINSKY
and Choir.

E. W. RABBINOWICZ, 91, High Street, Whitechapel, E.

*Illustration 12*

Plymouth Synagogue 1910 Order of Service of Re-Consecration.

*Illustration 13*

Chief Rabbi Sir I. Brodie, followed by Rabbi B. Susser and Revd S. Ginsburg taking in the Torah Scrolls at the service to commemorate the bi-centenary of the Plymouth Synagogue, 1961.

the shape of a 'spice tower'). All the joinery work, such as seating, balustrading, etc., which is in pine, was stained and polished into a rich tone to contrast with the pale tints of the decorations.[55]

Various alterations and additions have been made to the Plymouth synagogue since it was first built. As female worshippers in the synagogue are generally restricted to a balcony there is seldom sufficient room for them. Originally there was a balcony only across the west wall,[56] and this was later extended across the north and south walls.[57] Iron pillars had to be installed to support the gallery in 1807.[58] Two of these were placed under the balcony at the entrance to the synagogue. Their place there might be structurally necessary, but equally well they may symbolize *Jachin* and *Boaz*, the columns associated with Solomon's Temple.[59] The synagogue itself had to be extensively repaired in 1795 when a carpenter was engaged to carry out repairs at a cost of £50,[60] and additional seats for children were added in 1811 on the west (rear) wall.[61] In 1863, Leon Solomon 'unsolicited enlarged the gallery, painted and redecorated this synagogue at his sole expense'.[62] There were further extensive refits of the synagogue in 1910.[63] and again in 1965.[64] The Exeter synagogue underwent a thorough repair and refurbishment in 1854,[65] in 1905 in memory of Revd S. Hoffnung and his eldest son,[66] and again in 1962.

Surprisingly, in view of the importance of the institution, there are few references in the surviving minutes of any of the South-West Congregations to a *mikveh*, either to the building of one or to its repair.[67] There was, however, a *mikveh* attached to each of the South-West synagogues.[68] The ruins of the *mikveh* at Falmouth are still to be seen, but there is no trace of the one at Penzance. The *mikveh* at Exeter was built in the synagogue, possibly at the time when the synagogue was built in 1764, at a cost of £84. The Congregation informed Chief Rabbi Adler that the *mikveh* had to be built on the second floor and the apparatus for heating the water was above that. The difficulty of obtaining a proper water supply and the injury to the premises from the steam and wet led the Congregation to abandon the use of the *mikveh* in 1844. The report continues:

> consequently the public baths are now resorted to where there is a bath constructed which on investigation is found to be within two inches of the prescribed rule for size as *kosher*. But we regret to add that on account of a trifling extra expense it is not generally used.[69]

The earliest reference to a Plymouth *mikveh* is in 1821, when there appears to have been a '*mikveh*house',[70] though the *mikveh* was probably in the Congregation's house adjoining the synagogue which was used for the beadle's residence and for a schoolroom. It is not known how long this *mikveh* laster, there are references to one in 1833,[71] and again in 1854.[72] New arrangements were made in 1910 for the Plymouth Corporation to provide a *mikveh* at the public baths. It cost £50 and the Congregation paid £25 per annum for the use of it.[73] This arrangement was subsequently abandoned and until 1974 the *mikveh* was situated in the Congregation's vestry house.[74]

The vestry house itself was built in 1808.[75] It was rebuilt in the latter part of the nineteenth century to provide classrooms for the *cheder*, a flat for the caretaker or beadle on the ground and upper floors, and a vestry room for Congregational meetings together with a classroom on the middle floor. It was again rebuilt in 1975, this time providing additionally a flat for the resident minister, or after 1981 a visiting minister.

The only other building known to have been used for Jewish communal purposes in the South West was the Jacob Nathan School, founded in 1869, which met at 69 Well Street, Plymouth, until after the First World War.[76]

## CHAPTER SIX

# The Communal Organization of the Jewish Communities of the South West, 1750–1900

## *Part 1*

### *Lay and Religious Leadership*

Until the Second World War the constitutions of the four South-West Congregations, like all the historic congregations of Anglo-Jewry,[1] were essentially oligarchic in character. In this regard they conformed to the general pattern of closed municipal corporations of eighteenth- and nineteenth-century England, with the status of privileged membership (*Hezkat HaKehillah*) corresponding to that of the freedom of the corporation which was available by purchase, inheritance, or apprenticeship. V.D.Lipman has pointed out even closer parallels between the London synagogal organization and the close vestries of the parishes where there was even identity of nomenclature.[2] All the Congregations made a clear distinction between full members—*Baalei Batim* enjoying *Hezkat HaKehillah* (congregational rights or vestry membership), and the renters of seats—*Toshavim* or seatholders. Outside these two classes, all others were regarded merely as *Orchim*, strangers or guests, even though they may have been resident in the town for many years.[3]

The original *Baalei Batim*, or vestry members as they were later called when the congregations translated their traditional terms,[4] were the wealthier members of the nascent community who were willing and able to shoulder the first expenses. Possibly they allowed the poor but exceptionally learned to join their ranks.[5] The group of *Baalei Batim* governed the affairs of the Congregation in the form of an executive committee known as *Kohol*.[6] Once the Plymouth Congregation's *Kohol* had been constituted, vestry membership was automatically granted to the sons and sons-in-law of *Baalei Batim* on payment of half a guinea,[7] whilst others if they obtained a majority of votes in their favour after being proposed and seconded in *Kohol* would be granted the same rights on payment of two guineas.[8] A similar arrangement was in operation in Exeter:

> The son of a vestry member having had a seat two years, paying the full amount due to the *Kehillah* by him to the day of his admission, and of a good moral character, may be admitted a member of *Kohol*, if married at twenty-one years of age and if not married at twenty-five years of age; on admission to pay half a guinea. If a person should marry the daughter of a vestry member he must be Twenty-one years of age, pay One Guinea admission and be subject to all the aforenamed conditions. A Seatholder having had a seat for three years may be proposed for a vestry member but must be twenty-one years of age if married, and if not married Twenty-five years of age, on admission to pay one Guinea, such person can only be elected at the Annual Meeting and must be proposed at a Quarterly Meeting previous.[9]

The same system operated in Penzance. In 1844, when other small provincial congregations were abandoning the oligarchic system and there were murmurings against it in London,[10] the tiny and declining Congregation there revised its regulations and still maintained the class distinctions. The Congregation forming the Penzance community was classed thus:

> *Baalei Batim*, members having *Hezkat HaKehillah*, that is to say, being possessed of all rights and privileges appertaining to them as established Members of the Congregation.
>
> *Toshavim*, persons having a seat in the synagogue for twelve months.
>
> All other descriptions of persons are called *Orchim* or strangers.[11]

The *Baalei Batim* formed about one-third to a half of the total Jewish community, as Table 34 illustrates.

Among what may be termed the religious privileges of the *Baalei Batim* were those of officiating as *Hatan Torah* and *Bereishit*;[12] to officiate as *Segan* on the Sabbath preceding the wedding of one of his children, or on the Sabbath when one of his sons was *Bar Mitzvah*, or if his wife attended the synagogue for the first time after child birth, or if his son was circumcised that day; of being given an *Aliyah* on Festivals, on the Sabbath before and after the marriage of a child, the circumcision of a child or *Yahrzeit*, on the day of his wife's first appearance in synagogue after her confinement, the

*Table 34: Baalei Batim, Toshavim, and Orchim in the South-West Congregations in 1845*

| | *Baalei Batim* | *Toshavim* | *Orchim* |
|---|---|---|---|
| Exeter | 14 | 8 | 20 (estimate) |
| Falmouth | 9 | 3 | 8 (estimate) |
| Penzance | 11 | 0 | 9 |
| Plymouth | 19 | 33 | 30 (estimate) |

(Source: MSS 104, Chief Rabbinate Archives, pp. 37, 47, 135, 139.)

day of his son's *Bar Mitzvah*, or when he was obliged to *Bentsch Gomel*; to have the attendance of the cantor and beadle on the occasions of a circumcision or mourning; to lead congregational prayers except on Sabbath and Festivals. The privileges of the *Baalei Batim* extended to the very portals of the next world. They were also entitled to

> burial on the high ground belonging to the Congregation, free of expense for himself and wife, and right of ground free of expense for his children, parents, brothers and sisters.[13]

The *Baalei Batim* strenuously defended their privileges and preserved their status even when they left town, paying half a guinea to retain their rights.[14]

The *Baalei Batim* had not only religious privileges, they also had control of the governing body of the congregation, *Kohol*, as well as the power to assess contributions and control financial outlay. It is fair to add that they also carried the lion's share of the financial outlay. In 1815, for example, the average annual per capita income from the Plymouth Congregation's *Baalei Batim* was £10 12*s*. compared to £4 5*s*. from the seatholder.[15] The four Congregations in the South West had a very similar hierarchy. A *Parnas*—the president or warden; a *Gabbai Zedakah* (literally, charity collector)—the treasurer; a *Gabbai Beth Hayyim*—an overseer of the burial ground; a committee of the Five Men or the Three Men; and *Kohol*, the vestry.[16] All of these were recruited only from the ranks of the *Baalei Batim*, who formed a social elite.[17]

It is convenient to describe the rights and duties of the members of this hierarchy as they were found in the Plymouth Congregation, as this will serve as a model for the other Congregations as well. The most powerful individual in the Congregation was the *Parnas*. He had 'the general superintendence of all the affairs of the Congregation, whether relative to the state of the community in general or to the Synagogue in particular.[18] He acted as *Segan* unless that office was otherwise disposed of, presented all *mitzvot*, sending out notices each Thursday telling which seat-holders had to attend on Sabbath for an *aliyah*. Marriages could not be celebrated without his permission nor burials, nor could announcements be made in the synagogue or notices displayed there without his special licence. Moreover, the *Parnas* could empower the *Gabbai* to give not more than one guinea to any one necessitous person.[19]

The *Gabbai* had the management of all receipts and expenditures of the Congregation, and was responsible for distributing casual relief to poor applicants, provided he did not give any individual more than five shillings in any one month, and also *matzot* to the poor. The seating arrangements in the synagogue were under his care and the letting of seats under his hand. It was also his province to order and superintend repairs to the synagogue or its ancillary buildings and to purchase whatever was needed, but he was not to spend more than two pounds without prior consent of the vestry (*Etrogim*[20] and candles excepted). It was naturally the duty of the Treasurer to keep the books, inspect the accounts of the collector, and to render an account to the vestry in the month of *Heshvan* (= October, i.e. after the High Festivals).[21]

There was, and is, a special enclosed seat, 'the box', in front of the *Bimah* in which the *Parnas* and *Gabbai* had to sit,[22] and Honorary Officers are still colloquially known as 'the box'.

The third member of the executive was the overseer of the burial ground. His primary duty was to make all the necessary arrangements for funerals and to see that the burial ground was kept in proper order. To this latter end in his own discretion he could expend not more than half a guinea. Monies received from funerals or laid out for cemetery expenses were kept in a separate account called the *Tikkun Beth Hayyim*, and he had to give an account of this fund to the vestry at the appropriate time (*Heshbon Zedek*). Like his two colleagues he was also concerned with welfare work. He had to arrange for donations collected at funerals[23] to be distributed immediately to the poor. The relief of poverty due to sickness was his special province, and he could give up to ten shillings and sixpence to any one case. Moreover, he was obliged to visit the sick and arrange a rota of every member of the Congregation, whatever his status, to attend the sick and render such help as was necessary.[24]

But however influential the Honorary Officers were, it was the vestry, *Kohol*, which was the ultimate source of power, both making laws and enforcing them. The vestry was composed of the *Baalei Batim* who had to attend all meetings, unless unwell or out of town,[25] and who were obliged to vote yea or nay on each proposition.[26]

The Committees of Three[27] or Five[28] men had very little executive power, their function being largely advisory.[29]

It is rather strange that they did not have a committee of seven, styled in Hebrew, the *shiva tuvei ha'ir* (the 'seven good men' of the Talmud and the *Responsa* literature), but possibly such a large subcommittee would have been unwieldy in relation to the size of the *Kohol*.

Another office of some importance, as it could only be served by *Baalei Batim* of three years standing, was that of Treasurer of the Perpetual Lamp. Only Plymouth of the South-Western Congregations appears to have maintained this office, a nonexecutive one, and even there only in the eighteenth century.[30] Election to executive office was dependent on a form of apprenticeship. The Plymouth Congregation's rules stipulated that no person could be 'elected to the office of *Parnas* unless he has first served or paid fine for the office of *Gabbai* nor can he be elected to the office of *Gabbai* unless he has served or been fined for the office of overseer of the cemetery fund'.[31] Even this last and lowly office could only be served by a *Baal Habayit* of three years standing.[32] In the eighteenth century, bachelors were not admitted to executive office though they could exercise the other rights and privileges of *Baalei Batim*.[33]

The oligarchic type of constitution remained in force in the Falmouth, Penzance and Exeter Congregations until their disbandment in the late nineteenth century.[34] In Plymouth, too, it remained rigidly in force. Until the Second World War, the vestry members wore silk top hats at Sabbath and Festival services,[35] but ex-Servicemen returning after the war insisted on one type of membership with equal rights for all members.[36] This democratization did not, however, extend to women, who, even if members in their own right (such as widows or spinsters), never had a vote nor could they be elected to any executive office. From time to time since 1960, there have been attempts by some of the members of the Plymouth Congregation to secure voting rights for women, but these were vetoed by the religious leadership. Eventually, a resolution giving women the vote and the right to serve on the General Purposes Committee but not be chairman or treasurer, was passed in 1975.

Side by side with lay leadership, the South-West Congregations had at various times differing degrees of religious leadership. Traditionally, a well-organized Jewish community needs a rabbi, a *shochet*, a *mohel* and a teacher; it is also desirable to have a cantor and a beadle.

The functions of a rabbi are esentially judicial, with an independent jurisdiction, a function in Anglo-Jewry which is nowadays exercised by a *dayan* of a *Bet Din*. He gives rulings on the requirements of Jewish law both for the community as well as for individual Jews. As Judaism is an all-embracing way of life, the rabbi's authority extends to the very warp and woof of a Jew's life, both religious and secular. Not only religious services, observance of the Sabbath and dietary laws and other religious precepts, but also business dealings such as contracts, loans and their repayment, credit purchase and interest agreements, as well as matters affecting personal status such as marriage and divorce, contraception and abortion, organ transplants and the like, are the legitimate province of the rabbi, who advises how these matters may be carried out consonantly with Jewish law and who settles any disputes which may arise. Theoretically, and largely in practice, in Jewish communities throughout the world at least until the seventeenth century, the rabbi was the head and leader of the Congregation or town which appointed him.

In the Anglo-Jewish community from the re-settlement in the mid-seventeenth century until towards the end of the nineteenth century few Congregations elected a rabbi of the type just described.[37] For the most part, both the London and provincial Congregations were content to utilise the rabbi of the Great Synagogue, London, who was often styled in the eighteenth century 'the High Priest', and afterwards 'the Chief Rabbi'. Several factors account for this centralization of rabbinical functions. In the first place, newly emerging congregations were rarely sufficiently well financed to be able to afford the 'luxury' of a rabbi. Particularly, as between them the congregants had a fair knowledge of the requirements of Jewish law in most day to day situations. Furthermore, it is probable that Jewish immigrants to England in the eighteenth and nineteenth centuries, as well as the twentieth, were not averse to relaxing strict rabbinical supervision of their lives. Indeed, it may well be that the absence of such supervision and the general lack of social pressure to conform with Jewish religious requirements prompted some immigrants to leave their strictly ordered lives in their native town and settle in the more liberalized atmosphere of England. As one Toynbee Hall, London, resident put it, 'It is a common saying amongst the foreign Jews that England is a *freie Medinah*—a country where the restrictions of orthodoxy cease to apply . . .'.[38]

Then again, financial control and hence ultimate power was vested in the lay leadership which was loath to share its authority or possibly lose it altogether. For all these reasons, it was far more convenient for a provincial Congregation to recognize the Chief Rabbi[39] in London as its spiritual head and to submit to him peripheral points which affected in the main only the externals of Judaism, whilst retaining local autonomy to deal with many matters as it thought fit, rather than to appoint a local rabbi.[40] It has also been argued that after 1870 the Chief Rabbi's authority was strengthened by Anglo-Jewish communities which wanted an equivalent of the Archbishop of Canterbury.[41]

In recognizing the Chief Rabbi in London as their rabbinical authority and eschewing a local rabbi, the Congregations of the South West conformed to the pattern followed by most other provincial Anglo-Jewish Congregations in the eighteenth and nineteenth centuries. The Plymouth Congregation entered into its Book of Records written in 1807, a special page entitled (in translation):

> And these are the names of the Gaonim who had Rabbincal Authority here in the holy congregation of Plymouth, may our city be speedily rebuilt, amen, [and there follow the names] the late Rabbi Zvi Hirsch; the late Rabbi David Tevele Schiff, righteous priest; Rabbi Solomon, may his light shine, the son of the aforesaid Gaon Zvi Hirsch.[42]

The superscription is so couched that at first reading it might well be thought, and indeed it once was, that Rabbi Tevele Schiff and the others had actually been the local rabbis in Plymouth.[43] This is, however, impossible, as the career of the last named is too well known to admit of the possibility that he had ever been a rabbi in Plymouth. The authority of the London Chief Rabbi was amply acknowledged by the Plymouth Congregation. Its regulations of 1779 compel every person called to the Torah to mention the name of 'the Gaon, the Head of the *Beth Din* of the Great Synagogue in the holy congregation of London'.[44] The same rule was incorporated in the Penzance Congregation's regulations in 1844, by which time, however, it was omitted in Plymouth.[45] A special commemorative prayer was made in Plymouth on behalf of the departed Chief Rabbis of England.[46] A manuscript prayer book written specially for the Plymouth Congregation in 1805 commemorates the following:

> Shraga ben Naftali [Aaron Hart, in office 1709–56]
> Rabbi Jonathan ben Nathan
>
> Rabbi David ben Solomon the Priest [David Tevele Schiff, 1756–64]
>
> Rabbi Moses ben Meir [Moses Myers]

and, added in a later hand,

> Rabbi Solomon ben Zvi [Solomon Hirschell, 1802–42][47]

In the 1779 regulations it was also enacted that in the event of a dispute between members of the Congregation,

> they shall not go to the Gentile courts but it shall be dealt with by our vestry here. If the matter is difficult then they should bring it to the Priest, the Head of the *Beth Din*, the Gaon of the Great Synagogue.[48]

ספר

מזכיר נשמות

שהלכו לעולמם

אשר הם קרובים יושבים

ד׳ד׳ עקסיטאר

והם נדרם לצדקה

נכתב על ידי ה״ק אלכסנדר בר אליעזר

בשנת

תהיינה נפשותם צרורות בצרור החיים

*Illustration 14*
Title page of the Exeter Hebrew Congregation's necrology, 1858.

Similarly an authorization from the Chief Rabbi was required before any marriage was solemnized.[49] A minute of 1802 shows just how seriously the Congregation took its responsibility in this matter:

> Behold, there is a certain man here, and his name is *Elimelech ben Rabbi Moses YZV*, and it is in his mind to arrange a wedding for his wife's daughter, the maiden *Pessela bat Nathaniel*, may his lamp shine, together with the bridegroom *Zelig ben Asher*, and he has asked us to send on his behalf for authorization of the wedding ceremony. However, these men have no portion or inheritance in our Congregation YIA, neither do we know anything at all about the bride or groom as to who they are. Therefore we have withheld our hand from writing in this matter except by consent of our Teacher, may his lamp shine.[50]

Moreover, from an early date the Congregation would not appoint a *shochet* unless he was duly certified as competent by the Chief Rabbi, even if he had been so certified by well-known rabbis.[51] The control was very tight. When it was discovered in 1764, during the interregnum that followed Hart Lyon's[52] retirement, that *Moses ben Uri Hamburger*, one of the London *shochetim*, had given a licence to a certain Hirsch Mannheim of Plymouth, the officers of the Great Synagogue, London, insisted that his position should be regularized, until a new Chief Rabbi was elected, by the three official London *shochetim*.[53] It may also be observed that when important agreements between various groups within the Congregation were made, they were entered into under the auspices of the Chief Rabbi who was the natural arbitrator when any dispute about the interpretation of the agreement arose.[54]

In the early part of the nineteenth century the Plymouth Congregation paid what amounted to an annual retainer to Solomon Hirschell, the Chief Rabbi in London. In 1808, for example, the vestry voted ten pounds to be sent to him as 'head of the *Beth Din* of London and State'.[55] Similarly, in 1811 it was decided 'to send a present to the Rav, the Gaon, our master and teacher, the sum of £15 for the past three years'.[56] Apparently the rate was fixed at about five pounds per annum. The relationship between the Congregation and the Chief Rabbinate was formally recognized at the election of Chief Rabbi Nathan Adler in 1844, when Plymouth had two votes by virtue of making an annual contribution of ten guineas.[57]

The relationship between the Exeter Congregation and the Chief Rabbis of London was never as close as that of the Plymouth Congregation. There are no indications in the surviving accounts of the Exeter Congregation of any 'presents' to him, there was no obligation to mention the name of the Chief Rabbi when called to the Torah, nor are the names of the Chief Rabbis commemorated in its necrology.[58] In 1844, the Exeter Congregation declined to contribute to the support of the Chief Rabbi and consequently had no delegate or vote at the election of Nathan Adler in 1844.[59] Its attitude on this occasion seems to have stemmed not so much from the fact that it was a small and declining community—so were the Falmouth and Penzance Congregations which each sent a delegate—but rather due to a traditional disinterest in the London Chief Rabbi. The Exeter Congregation and its members did, however, call on the services of various Chief Rabbis particularly in matters concerning personal status such as conversions and marriages but also, especially in the 1850s and 1860s, to settle disputes amongst themselves. There is no clear cut reason why Exeter, apparently alone of the South-West Congregations, had this distinctly cool attitude to the Chief Rabbinate. Possibly, they were fortunate in having cantor/*shochetim* of high calibre who were able to assert themselves as *de facto* rabbis. Indeed it does appear as though the Exeter Congregation referred to its cantor in the first half of the nineteenth century as the rabbi,[60] and the right to *pasken* (give decisions based on Jewish religious law) in the synagogue, a purely rabbinical function, was exclusively reserved to him.[61]

The Plymouth Congregation, on the other hand, appears to have had only one rabbi, *qua* rabbi, throughout its history.[62] He was Rabbi *Phineas ben Samuel* who was appointed for one year on 22 June 1800 as 'Preacher and to give ritual decisions at £45 per annum plus lamp for lighting and coal for fire'.[63] There must have been some conflict of authority, perhaps between him and Chief Rabbi Hirschell, because in 1801 his unfettered right to celebrate weddings was restricted so that he could only perform them at the discretion of *Kohol*.[64] *Phineas'* appointment was renewed for a further year on 14 June 1801 but his name then drops from the Congregational records in Plymouth.[65] Apart from *Phineas ben Samuel* there was no other appointment of a rabbi as such in Plymouth,[66] nor was the engagement of one contemplated, there being no rules at any time governing the appointment or conduct of a rabbi. Nor was there at any time in any of the South-West Congregations a seat for the rabbi, another indication that such an appointment was never considered on a permanent basis.

Although all male Jews aged thirteen years and over may lead congregational prayers, in practice most Congregations engage a professional singer, called a cantor, who conducts the main services, often with a (male) choir.[67] A top flight cantor, internationally known in Jewish communities, has always been able to command a high salary and is called upon to conduct services on Sabbaths, Festivals, and special occasions, and he would have few if any other duties. But most Congregations generally engaged men with a pleasant voice who were able to perform other functions as well, particularly that of *shochet*, or beadle. Indeed, in the Plymouth Congregation, the duties of cantor and beadle were regarded as reciprocal, if either was absent the other had to perform his duties.[68]

About 1823, when the Plymouth Congregation was in a poor state, both financially as well as numerically, officials were no longer appointed primarily on their cantorial prowess, but rather as *shochetim*[69] in the first instance, who could also act as a *Baal Tefillah*, i.e. one with a pleasant voice who was able to lead the prayers and read the scriptures, but not a professional singer.[70]

At most periods in the South West, each Congregation had only one cantor. Plymouth, however, from 1796 until 1816 had a second cantor known as the *chazan sheni*. The second cantor's duties were, in the main, the same as the first's.[71] The appointment of a *chazan sheni* was probably due to the advancing age of the first cantor, *Jacob Judah ben Benjamin*, who was appointed cantor about 1770 and remained in office until his death as a near centenarian in 1829.[72]

The cantor, unlike a rabbi in the performance of his rabbincial duties, was subject to the control of the *Parnas* or *Gabbai*. This was spelt out in the Exeter Congregation's regulations of 1833:

> he shall at all times and all places in his official duties be under the direction of the *Gabbai* or person acting as such, excepting at funerals when he must conform to the orders of the *Gabbai* of the Cemetery Fund.[73]

The cantor had to be present at all synagogal services whether he led them or not, and dressed in his proper attire.[74] This attire was specified in 1794 in Plymouth as consisting of '*kregil* and mantle'.[75] The Exeter Congregation in 1823 similarly insisted on a proper uniform: 'the cantor must not be in synagogue in time of service without his Mantell, Biff and Hat, subject to a fine of 2/6*d* for each offence'.[76]

This preoccupation with a uniform, which was later to be called 'canonicals', the very term redolent of the church, has remained a feature of Anglo-Jewish synagogal officials until the present day. When Mr H. Aloof answered an advertisement in the *Jewish Chronicle* in 1924 inviting applications for 'a *shammas*, Reader and Collector' in Plymouth the one requirement beyond carrying out his duties was that he had to wear his 'uniform'—silk top hat and clerical gown—at all Sabbath and Festival services.[77] Until World War II, almost every single Jewish minister in Britain wore the Christian clergyman's badge of office, the 'dog-collar'. This went out of fashion after the war, though the author was asked whether or not he would wear one in Plymouth when he applied for the position as late as 1961. Lay insistence that rabbis and ministers should wear 'canonicals' has become a point of issue in many British synagogues.[78]

The cantor was expected to conduct the entire service, though on Sabbaths and Festivals, when the service generally lasts two and a half or three hours, the beadle helped out by saying the first part of the prayers, until *shochen ad*.[79] The three South-West Congregations whose rules have survived[80] all insisted that it was the cantor's 'positive duty to attend in the synagogue on the day prior to every Sabbath and Festival for the purpose of rehearsing the portion alloted for the occasion, and to be careful in noticing and correcting any error that may have occurred in the Manuscript of the Scripture, which might altogether desecrate the Scroll of the Law or require correction'.[81]

The cantor in the South-West Congregations was also expected to act as a book-keeper and collector of monies,[82] as well as a secretary to keep minutes and the registers of births, marriages and deaths.[83]

It appears that of the four South-West Congregations only Plymouth was ever able even to contemplate engaging a top flight, internationally famous cantor. In 1815, *Yedidiah Naftali Hirtz ben Moses* of (?)Lichtentam came to Plymouth for an audition over Passover and for five Sabbaths, at a period when the Congregation was at its most prosperous. His wages were to be £100 per annum, but evidently this was not sufficient for him, or perhaps the number of worshippers was not great enough to satisfy his artistic ego, because he declined the appointment, the Congregation observing 'and the worshippers had much pleasure from his singing, therefore the cantor and choristers shall have £10 from the charity box'.[84] After this failure to secure a top flight cantor the Congregation went back to its pedestrian cantors who could and did combine their artistic talents with the more prosaic ones of ensuring a supply of *kosher* meat or teaching children.

Another synagogal office was that of the beadle, called in Hebrew the *shammash*, though more generally pronounced *shammas*, the jack of all trades and maid of all work. Whereas the rabbi and, to a lesser extent, the cantor occupy an office of leadership which tends to distance them from the ordinary congregant, the beadle is close to the ordinary worshipper both physically as well as metaphorically. Besides doubling up for the cantor he had to open and close the synagogue and superintend its cleaning.[85] Nowadays, the *shammas* sets out the prayer books and *chumashim* (copies of the Pentateuch) for each service. He shows the untutored the place, and helps mourners who have not previously attended weekday services to put on the phylacteries and to recite the *kaddish* at its appropriate times. He ensures that the Torah scrolls are rolled to the right place in readiness for the next service, and that the right boards are slotted into the wall display notice-board announcing the portion of the week, or special liturgical insertions. He prepares a white curtain in front of the Ark and white mantles to cover the Scrolls in readiness for the High Holydays and, in Plymouth, for all the days on which *Yizkor* is recited. To the *shammas* was entrusted the task of maintaining decorum during the services. This was the acid test of his authority. Often, he would cultivate a steely glance which was sufficient to quell a city magnate in his seat. In Plymouth he had to make announcements in the synagogue,[86] and in every Congregation he was the general factotum of the community, carrying messages, taking round the *lulav* and *etrog* to those unable to come to the synagogue so that they could say the blessing over them at the festival of Tabernacles, announcing births, inviting people to religious festivities, and generally making himself, if he was at all an able person, an almost indispensable part of the community. When he was efficient services ran smoothly, he would inform the wardens which worshippers were due to be called to the Torah and made sure that they were called to a passage consonant with their dignity, potential disputes were settled by a tactful word before they flared up, and the community flourished. The beadle has always been the lynchpin of a congregation.[87]

The beadle, like the cantor, was expected to have other skills. He was invariably the second *shochet*, a *mohel*, and/or a teacher. In 1802, in Plymouth, for example, the beadle *Joseph ben Judah* was also a *shochet* and was recognized as sufficiently learned to conduct marriage services; his wages were £40 per annum.[88] In 1816, Hayyim Issacher's basic wage as the beadle to the Plymouth Congregation was £50 per annum.

There was one other official essential for the convenience of the community—the *shochet*. Jewish requires that food animals for consumption be killed by a trained man using a traditional technique known as *shechitah*, which avoids unnecessary pain to the animal. This man, whose training usually takes two or three years,[89] is called a *shochet*. The carcase then has to be examined for signs of disease, the man doing this job being called a *bodek*. The two jobs are invariably combined, and the man is then called a *shochet ubodek*.[90] If a community had no *shochet*, *kosher* meat had to be brought in from a neighbouring community, an expensive and highly inconvenient expedient at any time.

It has already been mentioned[91] that the *shochetim* in England at least from the early nineteenth century were licensed, after an examination, by the Chief Rabbi in London. Each *shochet* receiving a licence from Chief Rabbi Hirschell was obliged to give an undertaking, besides the usual obligations of the Jewish religion, not to shave with a razor and not to drink the wine of Gentiles.[92]

To ensure that the *shochet* had not forgotten the laws relating to the slaughter of animals and the examination of the carcase, and also that he had not lost his skill in setting the knife to exquisite sharpness, he had to report to the Chief Rabbi from time to time for re-examination. For this purpose the Plymouth Congregation sent Hayyim Issacher to London on Sunday, 27 February 1814, to be examined by Rabbi Solomon Hirschell. The Congregation allowed Hayyim eight pounds for the expense, which was perhaps calculated on a generous scale.[93] In 1837, B.A. Simmons went to London from Penzance to be examined at the request of Rabbi Hirschell 'and is to be allowed six pounds for expenses going up there and coming home'.[94]

The *shochet* in Penzance had to kill twice a week in winter and thrice a week in summer.[95] In Plymouth, with its larger community, he was busier, and besides killing for the butchers he had to attend the homes of those who required poultry to be slaughtered.[96] Besides the inspection of the carcase, the *shochet* had also to 'porge' the meat,[97] i.e. to remove certain veins, fat and sinews forbidden by Jewish law,[98] and to ensure that it was watered within three days of slaughter.[99]

Rabbi Hirschell licensed not only professional *shochetim*, but also private individuals who were going to remote parts where there was no supply of *kosher* meat,[100] and possibly also individuals who preferred to slaughter for themselves to abide by various stringencies which the regular *shochet* did not observe. In the case of individuals the authorization only extended to the slaughter of poultry. This would explain an otherwise strange reference in the Plymouth Congregation's rules of 1835 to 'persons duly authorized by the Chief Rabbi to kill poultry for themselves only' who were permitted to slaughter even without the express permission of the officers of the Congregation, and even when the Congregation had its own *shochet*.[101]

A man's income is an important factor in his life and his family's happiness, often affecting his attitude to his work and the length of his service. How well off were the Jewish officials in the four Congregations in the South West? Until about 1830, wages were not ungenerous, and taking into account perquisites and income from secondary activities, officials seem to have lived comfortably, and if length of service is any guide, to have enjoyed their work. From about 1840 until the end of the century their basic wages remained much the same, with even a tendency to some decrease.

The best paid official was the cantor. In the Plymouth Congregation, *Jacob Judah ben Benjamin*, for example, was receiving fifty guineas per annum in 1807[102] which was increased to six guineas per Jewish month in 1814.[103] The second cantor, *Lima ben Ze'ev*, started at £25 per annum in 1796, which was increased to about £40 in 1800, and £50 in 1816.[104] In

1815, a *Nahman ben Isaac* was appointed as first cantor, though without the title, as *Jacob Judah ben Benjamin* was still alive, at £60 per annum.[105] He was able to augment his salary with a further £20 per annum by acting as *shochet* for the Dock Congregation.[106]

The basic wage of a *shochet* approximated to that of the cantor. In 1805, the *shochet*/beadle received £36 per annum for his services in Plymouth and a further £10 for his work at Dock, whilst a year later he got a composite salary of £50 per annum.[107] At this time his colleague the cantor was getting £52 10*s*. per annum. But the *shochet* was probably financially better off as he also received a fee from the housewife each time he slaughtered a bird. In 1822, for example, these payments were 4*d*. for a goose or turkey, and 3*d*. for a pair of fowls, ducks, or pigeons.[108] Unfortunately, these fees being a private arrangement between householder and *shochet*, there is no record of just how much they amounted to. It may be surmised that they approximated to a further £25 per annum. The *shochetim* traditionally also received certain parts of the animal for their own use,[109] or had an allowance of free meat in lieu thereof.[110]

As the century wore on the Congregations of the South West expected the cantor to double up as the *shochet*, but the wages, if anything, declined. In 1844, the cantor/*shochet* in Exeter got 19*s*. 6*d*. per week,[111] whilst his counterpart in Penzance received only 14*s*. 6*d*. per week.[112] There are no figures available for the wages of the Plymouth Congregation's officials in the latter part of the nineteenth century, until 1865 when it advertised the vacant post of *shochet* at 23*s*. a week.[113] In 1884, the Revd A. Spier, general factotum to the Jewish community in Plymouth, was paid 40*s*. a week.[114] It was little enough, but at the same period, the Penzance cantor/*shochet* had to manage on just half that amount.[115]

Apart from the basic wage, the communal officials in the South-West Congregations, as elsewhere, had various opportunities to augment their income in one way or another. In the eighteenth and early nineteenth centuries, for example, legacies were sometimes left to the officials, usually to recite memorial prayers where the legator did not have male issue to say them. Examples of such bequests include one by E.A. Ezekiel who left three guineas in 1806 to Moses Levy 'teacher of the Synagogue, Exeter, for which he shall read the usual Lectures to my memory and to say *Kaddish* for me during eleven months';[116] and another by Jacob Jacob who left

> £5 to Mr Ephrim to say a certain portion of the Holy Scriptures, *Torah*, as a prayer for me on every Saturday, also £5 to my brother-in-law Rabbi Simon for saying a prayer called *Kaddish* for me in the Synagogue every day.[117]

Salaried officials could also expect to earn a few extra shillings each year by entering births, marriages and deaths in a register. Hayyim Issacher's contract of reappointment in 1822 specified:

> The beadle shall keep a register book of all the children both male and female . . . also a memorial of all weddings, may they be for good luck, and also a memorial of may we live and not die . . . The beadle to have one shilling for each . . . name inscribed.[118]

Until the early nineteenth century, officials in the South-West synagogues frequently engaged in trade to a lesser or greater extent, and there does not seem to have been any opposition on the part of the South-West Congregations to this.[119] In the Plymouth Congregation, Levi Benjamin, cantor for more than 60 years until his death in 1829, was an umbrella maker,[120] Hayyim Issacher, beadle, was a slop-dealer,[121] Falk Valentine, *shochet*, acted as an agent for a London money changer, travelling through Devon and Cornwall buying up gold for paper money with fatal consequences to himself.[122] In the Penzance Congregation, the general factotum B.A. Simmons, had a lucrative sideline selling crockery as well as bones, the later presumably in connection with his *shechitah* activities.[123]

In the latter part of the nineteenth century when the members of the Congregation in the South West began to demand that their officials should behave in a 'professional' way,[124] trading by officials was frowned upon if not forbidden. In 1840, the Exeter Congregation noted that

> Mr Green having opened a shop in opposition to the wishes of the members—unless he give it up in three months, that he have notice to quit the situation . . . Mr Green's answer is: I have no intention of giving up my shop.[125]

None of the South-West Congregations, however, seems to have objected to its officials teaching on a freelance basis, though the Exeter Congregation in 1851 did attempt to regulate the fees paid by parents.[126] There has long been a Jewish tradition for parents who could afford it, and many who could not, to provide 'private' lessons for their children. The Plymouth Congregation itself in 1812 arranged with *Simeon ben Nathan*, its teacher, who had been appointed in 1806 at £42 per annum,[127] that he would have £15 per annum for teaching poor children plus a further two guineas per annum for each additional child. At the same time he must have received at least two and probably three guineas per annum for each child of parents who could afford to pay.[128] Even the *shochet* when he was able to do so, augmented his income by giving lessons, in one instance to Christian clergymen. Such a one was Michael Solomon Alexander who gave private lessons in Hebrew and German in 1824 to the Revd B. Golding of Stonehouse,[129] with far reaching consequences which will be discussed in Chapter 10. At a later period in Exeter, Revd M. Mendelssohn, 'teacher of Hebrew and German', had three pupils boarding with him and advertised in 1863 that he had room for another nine.[130]

Undoubtedly, the officials of the South-West Congregations received gifts in cash or kind, from time to time, as, for example, when they officiated at a wedding, circumcision,[131] funeral, tombstone consecration, or sold *hametz* at Passover.[132] But no evidence has survived to show the extent of such gifts. Until the twentieth century it was customary in many communities for those called to the Reading of the Law to announce a donation to the cantor and beadle, and to the rabbi where there was one, in addition to donations made to the synagogal or other communal charitable funds. Surviving records do not suggest whether or not this system was ever in operation in the South-West Congregations.

Accommodation was provided by the Plymouth Congregation for its beadle in the house owned by it and which was adjacent to the synagogue.[133] Other officials in the South West almost certainly had to provide their own accommodation.

It is not possible to assess the importance of these secondary sources of income in relation to the basic salary. Twentieth-century practices suggest an order of 10–15 per cent. On the other hand, there is some evidence that in the mid-nineteenth century some of the well established officials in Plymouth and Exeter lived on a rather better scale than their basic salaries would imply. Myer Stadthagen with four children at home in 1841 kept two living-in servants,[134] perhaps indicating an income of some £500 per annum on Marion Lochhead's scale referred to before. Revd Hoffnung and Revd Mendelssohn in Exeter each had one servant in 1851 and 1861 respectively, as did B.A. Simmons in Penzance in 1851, possibly suggesting an income of £150–£300 per annum.[135] Stadthagen left a not inconsiderable sum after his death in 1862. His estate was valued at under £1,500, and was comprised of a house and about £1,000 in cash and shares.[136] It must be observed that it was unlikely that Stadthagen saved all this out of his salary. He had married Arabella, daughter of Judith Moses and Isaac Joseph. Isaac was a comfortable merchant, and Judith was the granddaughter of Moses Jacob who was well settled in Redruth by 1767 and great-granddaughter of the founder of the Falmouth Congregation around 1750,[137] and she might well have had some inheritance of her own.

The basic incomes of the Jewish officials in the South West, however, compare poorly to those of their Christian counterparts in both the Established as well as the Nonconformist Church. According to Trollope, Mr Quiverful and his brood of fourteen were poverty stricken on £300 per annum, and at that salary Mr Arabin would not contemplate the responsibilities of a family. A curate's average wage in 1837 was £81, and in 1897 £145, whilst in the early days of their existence, when pluralism was being restrained, the Ecclesiastical Commissioners made up livings with 500–1,000 inhabitants to £100. In 1893, when pluralism was rare, the average annual wage for incumbents was £246.[138]

Amongst the Nonconformists, the Presbyterians, the richest of the dissenters in Exeter, in 1818, paid their minister £200 per annum and in 1847, £343.[139] The Baptists had two churches in Exeter after 1817. In the South Street church, where membership fluctuated from about 60 to 200, the minister's salary rose from £80 in 1834, to £120 in 1846, and £200 in 1865. At the other Baptist church in Bartholomew Street, where membership fluctuated between 100 and 250, the salary gradually fell from £142 in 1830, to £120 in the period 1840–60, to £100 in 1864 and down to £80 in 1870. In Topsham, the Congregationalist minister, though he never had more than 50 members, received £70 in 1872, £100 in 1892 and £120 in 1911.

In the light of the salaries paid to Christian clergymen, those paid to the Jewish religious officials in the South West look decidedly anaemic. Indeed, the Jewish clergy in the four Congregations in Devon and Cornwall would hardly have been much worse off financially had they been agricultural labourers, who, for most of the century after 1796, received between 10*s*. 6*d*. and 18*s*. per week.[140] From Table 35 it will be seen that for the most part, from 1796 until 1885, the cantor/*shochet* in the Congregations of the South West received just a few shillings a week more than an agricultural labourer and about half or two-thirds of a skilled London artisan's weekly wage.

It should be borne in mind, however, that the figures shown in Table 35 do not include other forms of monetary (and non-monetary) payment received by the Jewish officials, as also by agricultural labourers, for example. This does mean that such comparisons as these need to be considered with some caution.

In the nineteenth century, Jewish religious functionaries could expect higher wages in larger Congregations elsewhere, both in England and overseas.[141] The desire on the part of some of the Jewish clergy to better themselves probably accounts in part for their comparatively short tenures of office, whilst the other attractions of larger Congregations no doubt also played a part. It is perhaps not surprising that after B.A. Simmons retired in 1854 after 43 years of service[142] there was a succession of 'ministers' in the Penzance Congregation with its low emoluments, and also in the Exeter Congregation after Moses Horwitz Levy died in office

*Table 35: Comparison of Weekly Wages of Religious Officials in the Jewish Congregations in the South West of England with the Average Wages of an Agricultural Laborer and a Skilled London Artisan*

| *Year* | *Congregation* | *Cantor/ Shochet* | | *Agricultural Laborer* | | *London Compositor* | |
|---|---|---|---|---|---|---|---|
| | | *s.* | *d.* | *s.* | *d.* | *s.* | *d.* |
| 1796 | Plymouth | 10 | 0 | 10 | 6 | 30 | 0 |
| 1800 | Plymouth | 14 | 6 | 12 | 6 | 30 | 0 |
| 1802 | Plymouth | 16 | 2 | — | | — | |
| 1805 | Plymouth | 17 | 8 | 15 | 6 | 33 | 0 |
| 1807 | Plymouth | 20 | 2 | 16 | 0 | 34 | 0 |
| 1813 | Plymouth | 21 | 11 | — | | 36 | 0 |
| 1814 | Plymouth | 29 | 1 | 15 | 6 | 36 | 0 |
| 1823 | Plymouth | 15 | 6 | 11 | 6 | 33 | 0 |
| 1844 | Penzance | 19 | 6 | 11 | 6 | 33 | 0 |
| 1847 | Exeter | 14 | 6 | 11 | 3 | 33 | 0 |
| 1853 | Exeter | 14 | 0 | 13 | 6 | 33 | 0 |
| 1863 | Exeter | 20 | 0 | 13 | 9 | 33 | 0 |
| 1864 | Exeter | 16 | 0 | 14 | 0 | 33 | 0 |
| 1865 | Plymouth | 23 | 0 | — | | 33 | 0 |
| 1867 | Exeter | 20 | 0 | 14 | 0 | — | |
| 1875 | Penzance | 20 | 0 | 18 | 0 | 36 | 0 |
| 1884 | Plymouth | 40 | 0 | — | | — | |
| 1885 | Penzance | 20 | 0 | 14 | 0 | 38 | 0 |

(Sources: *Encyclopedia Brittanica* (1951), 23, 270; PHC Min. Bk II, *passim*, PHC A/c. Balance Sheet, 1884; EHC Min. Bks 1838–45, 1848–54, 1860–97; PenHC Min. Bks 1843–63, 1829–92.)

in 1837 after 44 years of service.[143] Tables 36, 37 and 38 show the length of service of the known religious officials in the Plymouth, Exeter and Penzance Congregations in the period under study as a whole.

From Tables 36–8 it is clear that after the early prosperous years of the Exeter and Penzance Congregations[144] when the incumbents served in office for 44 and 43 years respectively, most of their successors remained for comparatively short periods.

Apart from the small sizes of the Jewish communities in the South West, the main reason for the comparatively low rates of pay for the Jewish officials was probably that the supply of officials nearly always outstripped the demand. Throughout the nineteenth century until the end of the First World War there was an apparently inexhaustible supply

*Table 36: Period of Service of Religious Officials in the Plymouth Hebrew Congregation, 1764–1981*

| *Name of incumbent* | *Main duty* | *Year of appointment* | *Year of leaving* | *Number of years* |
|---|---|---|---|---|
| Hirsch Mannheim | Shochet | 1764 | Not known | — |
| Joseph ben Joseph Meir | Cantor | Before 1779 | 1784 (died) | at least 6 |
| Mordecai | Shochet | Before 1779 | Not known | — |
| Moses Isaac | Beadle/teacher | Before 1778 | 1803 (died) | at least 25 |
| Levi Benjamin | Cantor | Before 1774 | 1829 (died) | at least 55 |
| Joseph Levy | Shochet | 1795 | *c.* 1821 | *c.* 26 |
| Phineas ben Samuel | Rabbi | 1800 | 1803 | 3 |
| Hayyim Issachar | Beadle | 1813 | 1830 | 17 |
| Michael Solomon Alexander | Shochet | 1823 | 1824 | 1 |
| H. Harris | Shochet | 1829 | 1831 | 2 |
| Myer Stadthagen | Cantor | 1829 | 1862 (died) | 32 |
| Leopold Stern | Shochet | 1864 | 1865 | 1 |
| Joshua Levy | Shochet | 1865 | 1867 | 2 |
| L. Rosenbaum | Minister | 1863 | 1893 | 30 |
| Moses Lewis | Shochet | 1884 | 1885 | 1 |
| A.N. Spier | Shochet | 1886 | 1893 | 7 |
| J. Posner | Shochet | 1896 | *c.* 1903 | *c.* 7 |
| M. Berlin | Minister | 1896 | 1906 | 10 |
| E. Jaffe | Shochet | 1896 | Not known | — |
| D. Jacob | Minister | 1903 | 1914 | 11 |
| A.K. Slavinsky | Cantor | 1909 | 1918 | 9 |
| M. Zeffert B.A. | Minister | 1918 | 1928 | 10 |
| S. Wykansky | Cantor | 1920 | 1932 | 12 |
| H. Aloof | Beadle | 1924 | 1959 (died) | 35 |
| W. Woolfson | Minister | 1928 | 1944 | 16 |
| M. Goodman | Cantor | 1933 | 1959 | 26 |
| S. Susman | Minister | 1944 | 1952 | 8 |
| D. Josovic | Minister | 1954 | 1956 | 2 |
| I. Broder | Minister | 1959 | 1960 | 1 |
| B. Susser B.A. | Minister | 1961 | 1965 | 4 |
| A. Ginsburg | Minister | 1965 | 1975 | 10 |
| B. Susser Ph.D. | Rabbi | 1977 | 1981 | 4 |

(Sources: Minute Books and Marriage Registers of the Plymouth Congregation; *Jewish Year Book*.)

*Table 37: Period of Service of Religious Officials in the Exeter Hebrew Congregation, 1792–1917*

| *Name of official* | *Year of appointment* | *Year of leaving* | *Number of years* |
|---|---|---|---|
| Moses Horwitz | 1792 | 1837 | 44 |
| M.L. Green | 1839 | 1841 | 2 |
| S. Hoffnung | 1841 | 1853 | 12 |
| B. Albu | 1853 | 1854 | 1 |
| Myer Mendelssohn | 1854 | 1867 | 13 |
| S. Alexander | 1867 | 1869 | 2 |
| Joseph Lewis | 1869 | 1870 | 1 |
| David Shapiro | 1870 | 1871 | 1 |
| S. Bach | 1871 | 1874 | 3 |
| Harris | 1874 | 1874 | 3 months |
| Manovitz | 1875 | 1876 | 1 |
| Lazarus | 1876 | 1878 | 2 |
| I. Litovitch | 1895 | 1897 | 2 |
| Rittenberg | 1898 | 1898 | 6 months |
| H. Bregman | 1898 | 1904 | 6 |
| Rosenberg | 1908 | 1913 | 5 |
| Shinerock | 1916 | 1917 | 1 |

(Sources: Minute Books of the Exeter Congregation.)

*Table 38: Period of Service of Religious Officials in the Penzance Hebrew Congregation, 1811–1887*

| *Name of official* | *Year of appointment* | *Year of leaving* | *Number of years* |
|---|---|---|---|
| B.A. Simmons | 1811 | 1854 | 43 |
| Solomon Cohen | 1854 | 1857 | 3 |
| B.A. Simmons | 1857 | 1859 | 2 |
| Hyman Greenburgh | 1859 | 1863 | 4[a] |
| Spiro | 1863 | 1863 | 1 |
| Bischofswerder | 1874 | 1886 | 12 |
| I. Rubinstein | 1886 | 1887 | 1 |
| Michael Lankion | 1887 | 1887 | 1 |

(a) In his article on Penzance Jewry, Roth refers to a factotum called A. Lupshutz who acted for a year in 1861 (C. Roth, 'Penzance', *JC Supplement*, June 1933, p. ii).

(Sources: Minute Books of the Penzance Congregation.)

of recent immigrants from Eastern Europe, who were willing to take any form of employment provided it offered a wage, however meagre,[145] to use it as a stepping-stone to obtain a better job elsewhere. Thus, the temporary resurrection of the Exeter Congregation from 1895 until 1917 may be attributed not only to the presence in Exeter of East European Jews in somewhat larger numbers but also to the willingness of young Jewish men, particularly bachelors, from Poland and Russia, to become *shochetim*, cantors, and teachers, at bare subsistence wages.

Owing to the relatively poor emoluments and short service of most of the nineteenth-century officials in the South-West Congregations, it may well be that there was little true religious leadership. Leadership of a high order could perhaps hardly be expected from a cantor who was merely expected to lead prayers, nor from a *shochet* who was expected to kill efficiently sufficient animals for the Jewish community's needs, and it has already been shown that few rabbis or ministers were appointed in the Congregations of the South West. Nonetheless, some sort of spiritual lead was expected from these officials, however ill-equipped they were to give it. Myer Stadthagen in Plymouth, for example, who was originally appointed as *shochet* and porger in 1828, eventually became the cantor and then quasi-minister. In the latter capacity, he was responsible for sick and prison visitation, as well as preaching on special occasions. English was not his mother tongue and preaching must have been difficult for him. There is no indication, indeed, that he preached at all until he had been with the Congregation for many years. So much so that sermons were often delivered, when they were given, by members of the Congregation. In 1842, for instance, a Sabbath discourse delivered by Mr Levy, Junior, was sent to the *Voice of Jacob* which regretted 'that a young man of such good parts should not possess the opportunity of qualifying himself for the important office to which he aspires'.[146] In a situation where congregants were often as learned, and sometimes more so, than the officials, and, moreover, where the officials were lowly paid, it is not surprising to find that some officials were sometimes treated as mere paid employees, with all that that implied in the nineteenth century.

The difficulties were understood by Lemon Hart, former warden of Penzance, when on behalf of the Penzance Congregation he engaged B.A. Simmons in London as a *shochet* for Penzance in 1811. Hart sent two letters with the youthful aspirant to office: one, a recommendation from 'The High Priest Solomon Hirshell', and the other was a hope that the community 'would behave to him properly, for you may rest assured those articles are very scarce in this Market'.[147] Hart's pious hopes were not to be fulfilled. The earliest extant minutes of the Penzance Congregation tell of charges against Simmons and counter-charges by him, followed by threats of dismissal, fines, and reports to the Chief Rabbi in London.[148]

There is evidence to indicate that a similar state of affairs prevailed from time to time in Exeter and Plymouth in the nineteenth century.[149] Frequently the pages of the Exeter Congregation's surviving minutes deal with the 'misdemeanors' of its officials, and the Chief Rabbi subsequently called in both by the Congregation and the official. One example of the more serious disputes may be given as an illustration. In 1851, the Chief Rabbi was writing to Revd Hoffnung saying that he would support Hoffnung's candidature elsewhere, and whilst still in Exeter he should try to act as moderately as possible.[150] A day or so later Chief Rabbi Adler wrote to him,

> Alexander (the Parnas) has written to me that you broke into a terrible passion in Synagogue and said to Marks, "May you not reach the end of your journey". Is this true?[151]

There is no record of Hoffnung's reply, but Adler was soon writing to Alexander that he deprecated the Exeter Congregation's treatment of Hoffnung: 'I cannot suppress stating that no respectable Reader will go now to Exeter to be exposed to such insults'.[152] At the same time he wrote to Hoffnung: '. . . stay away next Sabbath pleading indisposition so that there should be no scenes in the Synagogue'.[153] One would have thought that the Congregation would be glad to see such an incumbent leave. The Exeter Congregation, however, delayed paying Hoffnung's salary so that he could not travel elsewhere to apply for another position.[154]

The Falmouth Congregation in 1860 also delayed payment of the *shochet*'s salary. He invoked Dr Adler's aid and on 18 April 1860 the Chief Rabbi wrote to the President of the Falmouth Congregation:

> Mr. Herrman is leaving for Sheffield and complains you object to him leaving. I think you have no right to form an impediment to the man's promotion in life.[155]

Revd Stadthagen in Plymouth also had his share of troubles. In spite of having given the Plymouth Congregation a quarter of a century's service, various members made his life so miserable that he tendered his resignation in 1856.[156] Once more Dr Adler championed the cause of his subordinate. He wrote to the wardens: 'In former times your Congregation was one of the most united and disciplined, now it has become a byword throughout the Empire'.[157] The trouble seems to have flared up in January in 1855 when Stadthagen sued Aaron Levy for libel, though the case was dropped on Adler's intervention.[158]

Unfortunately, similar disputes between Congregations and their officials were by no means infrequent in the Anglo-Jewish community as a whole, as a perusal of the *Jewish Chronicle* or published records of other provincial Congregations indicates.[159]

On the other hand, Congregations did show appreciation for long service by paying pensions[160] and erecting tombstones for former officials.[161] Moreover, long silences in the minutes probably indicate that officals were satisfactorily serving their Congregations, and undoubtedly there were many officials in the South-West Congregations who gained the love and respect of most, if not all, of their members.[162]

Theoretically, a Congregation could dismiss any official, as the terms of the usual contract, at least in the eighteenth century, enabled the official to leave or the Congregation to dismiss him provided in either case three months notice was given.[163] In practice, few Congregations ever dismissed an official, unless he was guilty of gross dereliction of duty. The factors which fettered the legal right of the Congregation to dismiss an official were the influence of the Chief Rabbi who was usually able to make peace; the difficulties and expense of obtaining a better replacement; and, deep down, a respect for the office of the official, influenced for the good, perhaps, by the contemporary attitudes of his flock to the Christian clergyman.

## *Part 2*

## *Synagogal Finances*

The Anglo-Jewish community has never had a central fund from which subsidies could be given to local Congregations to enable synagogues to be built or to help pay officials' wages,[164] so that provincial Congregations, at least, which rent or build houses of worship and engage officials must devise some machinery to raise the money to cover expenses. In this chapter an attempt will be made to describe the methods employed to finance both capital and recurrent expenditure.

At first, the typical eighteenth-century Anglo-Jewish Congregations' financial needs were small. The few worshippers met in one of their homes. Each Jew provided his own prayer book and ritual regalia. The most expensive item at the nascent stage was the Scroll of the Torah. This was usually provided by a wealthy individual,[165] or possibly loaned by a neighbouring Congregation.

Usually, a Congregation first became involved in recurrent expenditure of any size when it rented a room or hall for its services. Now, regular weekly expenditure had to be matched by regular weekly income. This was achieved in two ways: seat rentals which were the mainstay, and contributions which those called to the reading of the Torah were expected to give. In most Anglo-Jewish Congregations there was also a supplementary income from sale of synagogal honours, fines from those refusing honours, entrance fees from those wishing to become vestry members, and bequests. Nonrecurrent expenditure, such as the purchase of land, buildings, or Scrolls of the Torah, major repairs, arranging the defence of Jews whose conviction would bring the Congregation into disrepute, or contributions to Government War Appeals, was covered by appeals to the members of the local and other Congregations or by special levy.[166]

Turning first to seat rentals, in Plymouth these were fixed in 1760 at two guineas per annum for a vestry member and one guinea or half a guinea for seatholders.[167] More than a century later, the seat rentals had increased by one hundred and fifty per cent, the rentals in 1883 being two shillings weekly for the vestry members, and a shilling for the seatholders, or sixpence for those in straitened circumstances.[168]

The other major source of income was that from offerings, either those made to secure the privilege of performing certain rites in the synagogal service, or donations, generally subject to a minimum amount, made after receiving such an honour.

Thus, the Plymouth Congregation's regulations of 1779 specify that one called to the Torah on Sabbath or Festival must offer three donations to the synagogue, charity, and cemetery funds, the minimum to each being three pence.[169] The Bridegrooms of the Law had each to offer 7*s.* 6*d.*[170] One could buy the honour of standing *Segan,* i.e. to decide who should be called up, for sixpence.[171] The other South-West Congregations had similar rules.[172] These offerings formed a considerable proportion of a Congregation's total income. In 1760, for example, before the Plymouth Synagogue was built, the offerings amounted to £66 6*s.* 11*d.* from the 25 resident donors, who subsequently formed the nucleus of the Congregation.[173] In 1765, that is three years after the Plymouth Synagogue was built, the total income from fourteen contributors (the accounts may be deficient) was £81 11*s.* 3*d.* of which only £22 11*s.* 0*d.* came from seat rentals whilst £59 0*s.* 3*d.* came from donations.[174] This was possibly an unusually high proportion as just over £31 came from two donors.[175] Unfortunately, the next sufficiently detailed accounts of the Plymouth Congregation which have survived, so that a comparison of seat rentals with donations may be made, are those of 1883. In that year seat rentals brought in to the Plymouth Congregation £155, and donations from members were £54 15*s.* 0*d.*[176] The balance sheet for the year ended 3 September 1885,[177] conveys a similar picture:

| | £ | s. | d. |
|---|---|---|---|
| Members' seat money | 162 | 13 | 0 |
| Members' offerings | 58 | 16 | 0 |
| Strangers' offerings | 8 | 9 | 6 |
| Collections (on Purim, at a *Brit Milah*, etc.) | 3 | 4 | 0 |

At the outbreak of the First World War, seat rentals doubled but the offerings remained approximately the same; if anything they dropped, certainly the proportion they formed of the total income dropped:[178]

| | £ | s. | d. |
|---|---|---|---|
| Members' seat money | 326 | 14 | 0 |
| Members' offerings | 49 | 0 | 6 |
| Strangers' offerings | 13 | 6 | 0 |

The tendency for voluntary donations to form a decreasing proportion of the Plymouth Congregation's income may be explained in the following way. There was an ever decreasing number of regular worshippers, so that the same peope were called to the Torah more and more often. These regular worshippers decreased their donation on each occasion so that their total for the year remained constant. These voluntary offerings fluctuated throughout the year and therefore did not form a reliable method of financing the Congregation's expenditure, so it was exceedingly difficult to budget ahead. The drawbacks of relying on voluntary donations were already manifest by the early nineteenth century when Treasurers frequently had to lay monies out of their own pockets until their accounts were finally paid.[179] To ensure a ready supply of money to pay bills on hand, the South-West Congregations resorted to a tax on meat (which the observant Jew could only buy under Congregational auspices), either a farthing or halfpenny on each pound weight. This tax was collected by the butcher, remitted to the Treasurer and then credited to the member's account to *contra* his seat money or offerings.[180]

One form of supplementary income which substantially helped the Plymouth Congregation, and to a lesser extent the other three Congregations in Devon and Cornwall, was that which came from legacies. These were often spent as they came in, except in Plymouth where legacies from the mid-nineteenth century were invested so that only the interest on the investments was used. According to the necrology of the Plymouth Congregation, some £2,500 was left to the Plymouth Synagogue between 1780 and 1900, and this by no means represents all legacies, as the munificent legacies of Jacob Nathan in 1868, which include £1,000 to the

Synagogue, were for some unknown reason, not incorporated in the necrology. Many of the legacies were left on the express understanding that a *Yizkor* (Commemorative prayer) should be recited at the appropriate time on Festivals.[181]

At least one legator was afraid that his wishes would be neglected. He was Jacob Jacobs, Plymouth silversmith, goldsmith and Navy Agent, and the relevant portion of his will reads as follows:

> I bequeath to the Jewish Synagogue of Plymouth of which I am by the blessing of God a member £100 to be invested in five per cent annuities the interest to be given to the poor Jews of Plymouth on every year in the month of *Ellul*. On condition that at the normal times of the commemoration of souls my late father Mordecai Jacobs and myself be commemorated for ever. In default of which my heirs are to sue the Congregation in payment of £100. £10 towards 10 poor people for making competent meeting for prayers every Saturday and £5 for Mr. Ephrim [sic] to say a certain portion of the Holy Scriptures, *Torah*, as a prayer for me on every Saturday, also £5 to my brother-in-law Rabbi Simon for saying a prayer called *Kaddish* for me in the Synagogue every day.[182]

The Plymouth Congregation's far sighted decision to invest legacies and use only the interest has provided it with a substantial 'unearned' income. In 1884, for example, the interest from four legacies amounted to over £60 out of an income of £320, nearly one-fifth. In 1913, seven legacies provided £62 out of a total income of £538, and in 1963 the interest from 43 legacies amounted to £424 out of £3,341, or about 12 per cent.[183] If the Congregation continues to decline numerically this investment income will play an increasingly important part in its finances.

The regulations of the South-West Congregations will be discussed in detail towards the end of this chapter. At this stage it is apposite to mention one aspect of them, and that is that they, in common with other Anglo-Jewish Congregations, freely prescribed fines. To give just a few examples: those who did not accept office,[184] or absented themselves on a Sabbath or Holyday after receiving notice that they would be called to the Torah,[185] or who made a disturbance during meetings or prayers—or allowed their children to do so,[186] or who were contumacious or rebellious,[187] and paid officials who neglected their duties,[188] were all liable to pay fines ranging from as low as a few coppers to sums as grand as five guineas.

It is clear that the fines were intended to be penal rather than fiscal, that is to say, they were designed to encourage members and officials to do their duty rather than to increase income. In Exeter, for example, fines of £12 3*s*. were imposed over a six year period from 1827 to 1833, but were rarely paid.[189] Fines often proved to be a source of discord, those who felt aggrieved paying reluctantly or not at all.[190]

Entrance fees to the Congregations of the South West were payable only by aspiring *Baalei Batim*, the vestry members with superior rights and privileges. The married son or son-in-law of a vestry member could, on election, become a vestry member on payment of a half guinea, others had to pay one and a half guineas.[191] In Penzance the same privileges were acquired by an entrance fee of two guineas,[192] and in Exeter for one guinea.[193] Again, these entrance fees can hardly be regarded as an income-producing institution as there were rarely more than one or two new vestry members admitted in any one year.

Reference has already been made to buildings and land which were owned by the Congregations of the South West. At this stage it is appropriate to discuss the ways by which these Congregations funded capital projects. Invariably, the first financial step taken by any Anglo-Jewish Congregation in the eighteenth and nineteenth centuries towards paying for large capital expenditure was to launch an appeal to local and distant Jews.[194] If the appeal did not bring in sufficient funds then a mortgage was raised, which was later generally repaid out of current income.

No records of any appeals by the South-West Congregations have survived which relate to the purchase of lands or buildings in the eighteenth century. Indeed, apart from the one instance of the Plymouth Synagogue, no financial details relating to eighteenth-century capital expenditure by the South-West Congregations seem to have survived.

Examining the records which relate to the building of the Plymouth Synagogue in 1762, it is apparent that the Congregation did not have sufficient funds in hand to complete the building. Accordingly, on 3 September 1762 the elders of the Congregation through a Gentile, Samuel Champion, to obviate possible legal complications, had to borrow £300 from a Mrs Elizabeth Aven of Plympton 'to complete the buildings edifices and erections now building and erecting thereon and which is designed for a Jewish Synagogue or place of worship for persons professing Judaism'.[195] The Congregation was unable to pay even the interest on this sum, as in 1770 there was £336 owing.[196] Mrs Aven having died, her executor called in the mortgage but not only was there insufficient to pay that, but the Treasurer was short of £34 to pay Raymundo Putt and Edward Stephens, who were probably the builders.[197] The Congregation's difficulties were temporarily solved by a loan of £370 from one Christopher Harris who advanced £220 on the security of the Synagogue and £150 on a joint bond of the elders. The debt was steadily paid off; the interest was not unreasonable being five per cent of the outstanding balance, which on 19 April 1780 was £100.[198] The mortgage was finally paid off on 6 November 1783, twenty-one years after the synagogue was built.[199]

The Plymouth Synagogue's opening brought a number of large gifts. Joseph Sherrenbeck gave £52 and then £30 on Passover and Tabernacles in 1760 in anticipation of the opening,[200] as well as £10 10*s*. 3*d*. for the *Almemor* (platform in front of the ark) in 1762;[201] *Mordecai ben Yehiel* gave £33 on Pentecost 1762,[202] the cost of Ark; and at the same time the Prayer for the Royal Family, sign-written on canvas, at a cost of eight guineas.[203]

A major repair programme was carried out to the Plymouth Synagogue in 1805 for which an appeal was made on the last day of Passover and which brought in promises of £141.[204] There were 87 subscribers to the appeal, two of them—Joseph Joseph and Abraham Emanuel of Plymouth Dock—gave nearly half, £40 6*s*. and £20 4*s*. respectively. Friends, relatives or former members who resided elsewhere were among the contributors: one each from Exeter, Basingstoke and Bristol, two from Portsmouth and five from London. Seven widows and a spinster were amongst the contributors.

Yet another inscription, this time dated 1813, records the names of 26 donors who gave £160 1*s*. for major repairs to the Plymouth Synagogue. The one shilling was given by the beadle, Hayyim Issachar.[205] A large appeal in 1858 brought in £1,035 for the new Burial Ground at Compton Gifford.[206] There were 83 donors to this appeal, all local; more than half was given by Jacob Nathan, the Plymouth Congregation's Mycenas, who gave £555. There is nothing to suggest in any of these appeals that the members were levied according to their comparative wealth or by any arbitrary assessment. Each donor gave according to the generosity of his or her heart. Undoubtedly, similar appeals were made for similar objects in the other South-West Congregations but no record of them appears to have survived.

Where the good name of the Anglo-Jewish community was endangered the Plymouth Congregation did not hesitate to pay for legal representation, and if necessary, to apply to the rest of Anglo-Jewry for help. Some trouble seems to have befallen one *Solomon ben Hayyim* in 1804 when the Congregation granted his wife fourteen shillings a week, and the same amount to himself. Within a few months it was decided 'that the vestry should hire a counsellor and lawyer' for him out of Congregational funds.[207] Another occasion arose in 1817. The dry entries of debits and credits in account book are suddenly illuminated by the following (translated) account:

> Tuesday, 18 Shevat, pericope Yitro, 5577
> [4 February 1817]
> For our many sins Judah ben Hayyim Mannheim was forced to go in chains of iron on his feet and hands to stand his trial at March

> Assizes, may they come to us for good, for there is a State Prosecution against him to condemn him because they found that he forged on the Greenwich Navy Agent.
>
> 17 and 18 March 1817
>
> The aforementioned was tried and acquitted and came out free. His attorney was Alley Isaac of London.
>
> | | £ |
> |---|---|
> | For his expenses with the counsellor . . . | 25 |
> | and with other expenses here at Exeter . . . | 6 |
> | and the attorney here when he was arrested . . . | 3 |
>
> Through the applications to the Rav[208] [the Chief Rabbi] from London and other communities £46 was made. So the surplus of money has been given to the parents of the aforesaid.[209]

The murder of Isaac Falk Valentine in Fowey in 1811 involved the Plymouth Congregation in heavy expense which was met by a levy on the local Jews and an appeal to the Jewish community throughout England. Valentine, appointed *shochet* to the Plymouth Congregation in August 1811, acted as an agent for a London money changer, buying up gold for paper money.[210] One Wyatt lured Valentine to Fowey and drowned him. Subsequently, a roll of sea-soaked notes amounting to £260 was found in a pile of dung in Wyatt's stables. Wyatt was hanged for this crime.[211] For some reason that is not now clear the Plymouth Congregation was saddled with an obligation to pay legal expenses on account of Valentine. A committee was set up on 2 February 1812, before Wyatt's trial, 'to remedy the state of affairs'.[212] It reported on 7 June 1812 that the 'legal expenses' were £275.[213] The committee recommended that each individual member be levied a year's seat money, payable in three instalments.[214] The levy would only bring in £150,

> which leaves a deficiency of £125 which is too onerous for our Congregation to raise. It is therefore agreed that letters should be sent to the Congregations of England, and one also be sent to the Rav, the Gaon, the head of the Beth Din of the holy Congregation of London, and one also to request from the Rav, may his lamp shine, a testimonial which we might send to the Congregations that they should help us.[215]

One more example of raising funds by levy, this time for a national emergency, may be given. The relevant minute vividly gives the circumstances:

> This day, Sunday 11 *Ellul* 5539 [= September 1779].
>
> All the vestry was gathered and it was agreed . . . that the vestry should give a donation of fifty pounds to the Mayor at the Guildhall for the war. And everyone shall immediately pay what he has been debited. It is confirmed that everyone who separates himself from the community and does not pay according to the list shall be put in prison. The President and Three Men must help him[216] for the sake of peace in the vestry.[217]

To what extent were the various methods of financing the South-West Congregations effective? From one point of view the Congregations were always financially solvent. Income nearly always exceeded expenditure, and if for one or two years it did not, then expenditure was reduced.

This may be seen from the early minute books of the Plymouth Congregation. From 1788 until 1813, only the accumulated balance each year is recorded. From 1814 until 1834, however, the actual income and expenditure each year is minuted, and the relevant details are given in Table 39.

From Table 39 it appears that there was a drastic fall in the income of the Plymouth Congregation after the end of the Napoleonic Wars. There are indications in the Congregation's minute book that already in 1814 'it found that the Congregation is in a very poor state'.[218] By 1816, the fall in income made it necessary to budget very carefully:

> from today [3 November 1816] the expenses of the community are to be:

| | £ |
|---|---|
| Wages of Cantor Limma | 50 per annum |
| Wages of the beadle Reb Hayyim | 50 per annum |
| To Eliezer ben Joseph | 15 per annum |
| Mistress Davis | 10 per annum |
| Yehiel ben Naftali | 10 per annum |
| Zelvelcher [= Samuel Simon] | 10 per annum |
| Miss Benjamin | 6 per half year and she must continue to make wax candles as previously. |

Unfortunately, neither the minute books nor the surviving account books indicate how the Congregation managed to reduce expenditure from £747 14*s*. 8*d*. in 1815 to rather less than half that amount in 1816.

One balance sheet of the Plymouth Congregation of the early nineteenth century has fortuitously survived, that of 1824. It gives a picture of the type of expenditure incurred by a Congregation at that period, and, as there is nothing in the minutes of that year to indicate the contrary, was probably typical. In the original spelling and style it runs:

Anno 5584 [= 1824]

Exps

| | £ | s | d |
|---|---|---|---|
| Mr Alexander[a] | | | |
| 55 weeks @ 21/– | 57 | 15 | 0 |
| do. for stationary | 1 | 5 | 0 |
| Cantor 13 months | | | |
| @ 66/8d | 43 | 6 | 8 |
| Judah Moses[c] | | | |
| 55 weeks @ 5/– | 13 | 15 | 0 |
| Mrs Abrahams | | | |
| 11 months @ 21/8 | 11 | 18 | 4 |
| Mrs Davis | | | |
| 13 months @16/8 | 10 | 16 | 8 |
| Do. Ralph | | | |
| 17 weeks @ 5/– | 4 | 5 | 0 |
| 3 years Ground Rent | 7 | 16 | 0 |
| Do. Insurance | 4 | 0 | 0 |
| Wax candles 6. 17. 7 | | | |
| Tallow Do. 5. 9. 6 | 12 | 7 | 1 |
| Old Debt | | 14 | 9 |
| Strange and | | | |
| resident poor | 8 | 0 | 6 |
| do. to Moses for eating | 1 | 11 | 0[d] |
| Moatzers & | | | |
| gave the Poor | 9 | 11 | 6[e] |
| Printing | 3 | 6 | 6 |
| Hambley | 1 | 16 | 6 |
| Incidental Expenses | 7 | 16 | 6 |
| Error in Casting | 1 | 16 | 2 |
| Balance | | | |
| last year's acct | 25 | 1 | 10 |
| do. do. paint bill | 3 | 3 | 10 |
| | 232 | 14 | 5 [sic] |

Receipts etc

| | £ | s | d |
|---|---|---|---|
| By Mr Alexander | | | |
| from Emden[b] | 26 | 1 | 10 |
| do. from Heard | 36 | 2 | 7½ |
| do. from Butlands | 62 | 14 | 6 |
| do. from Smith | | | |
| & Toll | | 8 | 9½ |
| By offerings and | | | |
| seats | 97 | 13 | 7 |
| By error in casting | 1 | 16 | 2 |
| By Balance due | 7 | 17 | 5 |
| | 232 | 14 | 11 |

(a) This was Michael Solomon Alexander who later became Bishop of Jerusalem, see below, p. 241ff.

(b) This and the following three items probably represent monies collected by Gentile butchers on behalf of the Congregation.

(c) This and the next three entries refer to pensioners of the Congregation.

(d) The poor received cash donations, and were also allowed to eat at Judah Moses, who was paid for his expense.

(e) Matzot and special help to the poor at Passover.

*Table 39: Annual Income and Expenditure of the Plymouth Hebrew Congregation, 1814–1834*

| *Year* | *Income* | | | *Expenditure* | | | *Balance* | | |
|---|---|---|---|---|---|---|---|---|---|
| | £ | *s.* | *d.* | £ | *s.* | *d.* | £ | *s.* | *d.* |
| 1814 | 654 | 17 | 0 | 596 | 1 | 10 | +88 | 15 | 2 |
| 1815 | 774 | 3 | 9 | 747 | 14 | 8 | +26 | 9 | 0 |
| 1816 | 287 | 13 | 0 | 304 | 14 | 10 | −17 | 1 | 10 |
| 1817 | 312 | 0 | 7 | 330 | 14 | 5 | −18 | 13 | 10 |
| 1818 | 168 | 8 | 5 | 204 | 2 | 2 | −35 | 13 | 9 |
| 1819 | 187 | 18 | 0 | 158 | 6 | 0 | +29 | 12 | 6 |
| 1820 | 230 | 10 | 8 | 213 | 16 | 10 | +16 | 13 | 10 |
| 1821 | 172 | 12 | 8 | 163 | 7 | 5 | + 9 | 5 | 2 |
| 1822 | 188 | 14 | 6 | 184 | 5 | 5 | + 4 | 9 | 1 |
| 1823 | 157 | 18 | 6 | 183 | 0 | 5 | −25 | 1 | 11 |
| 1824 | 224 | 17 | 0 | 232 | 14 | 5 | − 7 | 17 | 5 |
| 1825 | 246 | 17 | 10 | 234 | 16 | 10 | +12 | 1 | 0 |
| 1826 | 217 | 8 | 5 | 202 | 15 | 10 | +14 | 12 | 7 |
| 1827 | 212 | 9 | 6 | 202 | 16 | 5 | + 9 | 13 | 1 |
| 1828 | 235 | 14 | 6 | 193 | 18 | 9 | +41 | 5 | 7 [sic] |
| 1829 | 317 | 11 | 3 | 290 | 9 | 9 | +29 | 1 | 6 |
| 1831 | 300 | 18 | 4 | 256 | 1 | 4 | +44 | 17 | 0 |
| 1832 | 178 | 12 | 10 | 162 | 15 | 6 | +15 | 17 | 4 |
| 1834 | 286 | 13 | 6 | 239 | 9 | 2 | +47 | 4 | 4 |

(Source: PHC Min. Bk I, II, *passim*.)

When there were too few members to shoulder the responsibility of paying salaries, rates, repairs and other items of expenditure necessarily incurred in maintaining a Congregation, then it had to close down. This was the fate of the Falmouth and Penzance Congregations.

## *Part 3*

## *Communal Discipline within the Congregation*

Societies, complex or simple, require a constitution, written or oral, by which they regulate their conduct. Congregations and other Jewish societies are no exception.

Various sets of regulations have survived from the Plymouth, Exeter, and Penzance Congregations, though those of Falmouth have disappeared. The earliest rules of any Congregation in the South West are those of Plymouth and date from 1779. They are handwritten in Yiddish at the front of a book entitled (in translation):

> *Pinkes* of the regulations of our community, the holy Congregation of Plymouth. In the year "The work of righteousness is peace".[219]

There follow 64 rules covering the gamut of congregational life, from birth to death, dealing with the synagogue and the Jew's relations with his fellow Jews and with non-Jews. Various additions were made from time to time at vestry meetings, and these rules served until 1796 when it was minuted, still in Yiddish:

> Since in the regulations of former days in our community there are many matters which do not suffice for this time and also they are not kept as they ought to be, the undersigned . . . are to review the regulations . . . so that "peace be within thy rampart, prosperity within thy palaces".[220]

The Plymouth Congregation next considered its regulations on 3 November 1834, when it was 'resolved that a Select Committee . . . be appointed to alter, amend and add such rules as seem proper, for the better government of the Congregation . . .'. The result of the Committee's deliberations was a booklet of 26 pages printed in English in 1835 by John Wertheimer of London.[221] There were probably revisions of this rule book in the latter part of the nineteenth century, but the next one to be published was in 1949.

The Exeter Congregation's original regulations were in Yiddish but these have disappeared. They are referred to in a manuscript volume which starts:

> Exeter, Sept 20, 1823.
>
> The *baalei batim* of the Kehillah of Exeter, taking into consideration the state of their *Takkanot* [= regulations] are unanimously of Opinion, they require translating into English, revising and adding many new *Takkanot* . . .[222]

The Congregation was able to manage on 48 rules until these were revised in 1833.[223]

A similar situation prevailed in Penzance where revised regulations were made in 1884,[224] no less than 95 laws being necessary to maintain harmony amongst its eleven seatholders![225]

The regulations of the Plymouth Congregation in 1835 are typical of those of the other Congregations of the South West, as well as broadly representative of earlier as well as later sets of rules in force in Congregations in England in the nineteenth century, and can now be considered in detail.

They open with a reminder that 'it is the duty of every Jew regularly to attend Divine Service . . .'. They then specify the various Honorary Officers of the Congregation together with their mode of election, privileges and responsibilities. Similarly, the rules specify the different types of membership, the mode of becoming a member, together with the rights, duties and privileges of the various kinds of member. The rules govern the calling and conduct of meetings, as well as the ways and means of financing the Congregation's budget. There is a special section governing the conduct of Divine Service, as well as sections which specify the day-to-day responsibilities of the paid officials. A variety of other matters and situations which had probably been the subject of disputes in the past are also dealt with. These include the recitation of *Kaddish*, the making of *Mi Sheberach* prayers, the election of *Hatan Bereishit* and *Hatan Torah*,[226] the legal preliminaries to the celebration of marriages, and burials.

It is easy enough to make rules and regulations, it is another matter to enforce them. The usual method was to fine an offender, and fines ranging from coppers to two guineas were imposed.

The following representative fines are taken from the Plymouth Congregation's Regulations, 1779:

| OFFENCE | FINE |
|---|---|
| Cantor omits Chief Rabbi's name in Mi Sheberach | 6*d*.[227] |
| Refusing office as President or Treasurer | 2 guineas[228] |
| Taking out writ before Mayor without vestry's consent | 39*d*.[229] |
| Refusing office of Hatan Torah or Hatan Bereishit | 10/6*d*.[230] |
| Not helping to bake Unleavened Bread | 10/–*d*.[231] |
| Leaving synagogue to avoid being called up | 39*d*.[232] |
| Singing too loud, thereby confusing the cantor | 2/6*d*.[233] |
| Coming to synagogue in long boots | 1/6*d*.[234] |
| Chewing tobacco in synagogue | 2/6*d*.[235] |
| Arranging minyan at home without prior permission | 2/6*d*.[236] |
| Men entering Ladies Gallery on Rejoicing of the Law | 10/6*d*.[237] |

(Source: PHC Min. Bk I, Regulations 3, 4, 8, 12, 23, 30, 46. For the Hebrew terms see Glossary).

It was realized, however, that these fines would not prevent a troublesome member from making a nuisance of himself, hence a rider to the rules that the Executive had the power to increase or diminish any fine according to circumstances.[238] Furthermore, it was envisaged that contumacious members would not pay their fines or determine to have the satisfaction of flouting the authority of the Honorary Officers even if payment of a fine was entailed. In such cases the Honorary Officers could forbid a man to be called to the Torah and ban his children from *cheder*.[239]

The traditional abhorrence of the Jew in exile (and indeed any member of a persecuted minority) of the informer is expressed forcibly in the following terms:

> If any Jew . . . gives information to the authorities about another Jew and causes him harm either financial or bodily, then the informer shall forever be barred from our Congregation and for all his days shall never be called to the Torah.[240]

Outrageous behaviour called down the ultimate sanction of excommunication, the dreaded *Herem*.[241] Only one instance of this has been noted in the South West, when it was put into operation against a certain Leib Shtievel of Plymouth who shaved in the nine days of mourning[242] in 1776 and was fined,

> nine and thirty three-penny bits, that is, as many as the maximum number of flagellations permitted by the Rabbis from Holy Writ. But he regarded not the vestry . . . and for spite shaved during the intermediate days of Tabernacles, and he boasted that he had done so deliberately. Therefore he is to be excommunicated all the days of his life, and he is not to come to any holy matter, small or great, nor is any member of the Congregation to assist him whether in joys or troubles, in health or sickness, and when he dies then none of our members shall perform the last rites.[243]

Such stringent measures were effective as long as the local Jewish community was tightknit and all members wanted to use facilities which only the Congregation could provide. Once a situation prevailed, as it did in the late nineteenth century, where there was a substantial number of Jews in the Plymouth Congregation who did not care whether or not they attended synagogue, whether they ate *kosher* or *treifah* meat, whether they were buried or cremated, then the sanctions lost their force and were gradually dropped. By the time the Russo-Polish Jewish emigration arrived in the South West in the latter part of the nineteenth century, such sanctions were no longer fashionable and they never reappeared.

## CHAPTER SEVEN

# Jewish Religious Life in Devon and Cornwall

The religious life of most Jews in the period under study, at least until the First World War, was led in both synagogue and home. More can usually be said about the synagogal life of the Jew than his home life, for the former was more often recorded. There is, however, by no means an entire silence concerning Jewish life outside the four walls of the synagogue, and both aspects as reflected in the lives of the Jews living in Devon and Cornwall after 1750 will now receive attention.

Consideration will be given first to the most obvious aspect of synagogue life—Divine Service. The liturgy of the South-West Congregations when they were first founded must have reflected that in operation in Germany, from whence the original members emigrated. It soon settled down to *Minhag Polin*,[1] that is to say, the form adopted by Polish Jews of the non-*Hasidic* variety, with one or two additions peculiarly English, notably the hymns recited at the end of the Sabbath services, *Yigdal* and *Adon Olam*.[2] There does not appear to have been any direct *Hasidic* influence in the communal life of South-West Jewry.[3] Services in Plymouth are still read from an *amud*, a prayer stand at the front of the synagogue. The inauguration of the Sabbath service[4] is read from the *bimah* in the centre of the synagogue.[5] Other vestiges of the original eighteenth-century German liturgical rites have survived in Plymouth. One is the recitation of the Hymn of Unity[6] on Sabbaths. The other is possibly unique in all Anglo-Jewry. Mourners, or those keeping a *yahrzeit*, come to the front of the synagogue, to the very steps leading up to the Ark, wearing a *tallit* (prayer shawl), even at night services when it is not customary to do so, in order to say the *Kaddish*.[7]

Public services in synagogues are generally held daily, in the morning and the evening. There are many incidental references to Sabbath and

Festival services in all the synagogues of the South West, but nothing is recorded which suggests that these were markedly different to those which take place nowadays in any traditionally inclined Anglo-Jewish synagogue.

Curiously, there are very few references to weekday services. A 1779 regulation which gives relief from a fine for the worshipper who arrives at the synagogue in long boots when he has just come from a journey,[8] seems to imply a reference to weekday services. But this is not conclusive as it is possible that the regulation refers to services which were held on special weekdays, such as Mondays and Thursdays,[9] on days when a member had *yahrzeit*,[10] or New Moons.[11]

There is one direct reference to daily services in the will of Jacob Jacobs who died in 1811. He left £5 to his brother-in-law Rabbi Simon, 'for saying a prayer called *Kaddish* for me in the Synagogue *every day*'.[12]

It is indeed possible that by 1835 daily services were no longer being held in Plymouth because in that year the regulations specify that the synagogue was to be open on the following weekdays:[13] Purim, Hanukah, Fasts, Intermediate days of Festival, Penitential days (from before New Year to Day of Atonement), New Moon and its eve,[14] *yahrzeit*,[15] when the Torah is read, and *Omer*.[16] As the synagogue had to be open only on certain weekdays, it follows that on all other weekdays it did not have to be open. This seems to be a very strange conclusion, yet it is partially confirmed by the comparatively small number of worshippers in the Plymouth synagogue on a typical Sabbath in 1851, only 45 according to the religious census of that year.[17] At a Monday morning service in 1856 in the Plymouth synagogue there were only twelve men present.[18]

Apart from the normal weekday, Sabbath and Festival services, special services were also held from time to time in the synagogues of the South West to mark national events, synagogal anniversaries and occasions of thanksgiving, as well as special charity performances.

At the coronation of George IV, for example, there was a special service in the Exeter synagogue at which 'an excellent lecture was delivered by Rabbi Levy'. The coronation service at the Exeter cathedral was followed by a dinner for the poor at Heavitree. In the synagogue, however, it was 'the elders and chief part of the Congregation who sat down to a handsome dinner'.[19] The death of Prince Albert was on 23 December 1861 duly marked in the Exeter synagogue which was 'draped in black and had a most mournful appearance' by a 'most impressive discourse from Rev. M. Mendelssohn'.[20] Queen Victoria's Jubilee was celebrated in the Plymouth synagogue in 1886, the Order of Service leaflets costing £1 11*s*. 3*d*., the telegram of congratulation 2*s*. 11*d*., and the tea that followed cost the Congregation one pound.[21]

There were other services of intercession at times of need and thanksgiving in the nineteenth and twentieth centuries. Sometimes no local record of them has come to light. Such an instance must have occurred when Chief Rabbi Adler wrote to the Plymouth Congregation asking them to hold a service in connection with the Indian Mutiny.[22] In many synagogues worshippers have a box under or by their seat in which to keep their prayer books. After a special service some worshippers place the Order of Service for the occasion in their box, a fruitful source for later historical research. Although worshippers in the Plymouth synagogue, other than the wardens, do not have such boxes, the synagogue itself has capacious boxes beneath the seats on the *bimah* and at the back of the synagogue. A search revealed Orders of Service for the Coronation day of King George V and Queen Mary on Thursday, 22 June 1911; for the 'Restoration of Peace after the Great War' in 1918;[23] various Armistice day services; the Silver Jubilee of King George V's accession to the throne, on Sabbath, 11 May 1935; a memorial service for him on the day of his funeral, Tuesday, 28 January 1936; the Coronation of King George VI and Queen Elizabeth in 1937; 'a service of Prayer, Intercession and Thanksgiving in connection with Britain's Fight for Freedom', Sunday, 23 March 1941; services to mark the second, third and fifth anniversaries of the 'Outbreak of Hostilities', 1941–1944; 'Praise and Thanksgiving for the Victories of the Allied Nations in the World War', in 1945; a 'National Day of Prayer by command of King George VI', on Sunday, 6 July 1947; a Memorial Service for King George VI, in February 1952; the Coronation of Queen Elizabeth II.

The vicissitudes of the Jewish people in the twentieth century were also noted in the Plymouth synagogue. Amongst the Orders of Service found in the synagogue was a 'Memorial Prayer for the Victims of the Massacres in the Holy Land', when Jews were slaughtered by Arabs, particularly in Hebron in 1929. Even more ominously was the 'Service of Prayer and Intercession for the Jews in Germany', Sunday 10 November 1938, after the rape of German Jewry in the Crystal Night pogrom; and an 'Order of Service on the day of Fasting, Mourning and Prayer for the Victims of Mass Massacres of Jews in Nazi Lands', Sunday, 13 December 1942. All these Orders of Service have survived by chance. Undoubtedly there were many services of which there is now no trace.

Services were held to celebrate landmarks in the lives of the South-West Congregations. The opening of the Penzance synagogue was marked by a special service on Friday, 12 Shevat 5563 (= 4 February 1804).[24] So was the re-opening of the Exeter Synagogue in 1854. Of the latter, a detailed account is given in the *Jewish Chronicle*.[25] Three boys carrying wax candles led a *chuppah* (wedding canopy) followed by a Scroll of the Torah into the flower-bedecked synagogue. Seven circuits of the synagogue were made, and a prayer, specially composed by the Chief Rabbi, was recited. The bicentenary of the Exeter synagogue was marked by a special service conducted by the Revd B. Susser of Plymouth on Wednesday, 2 September 1964.

Ceremonies were held in the Plymouth synagogue after extensive repairs and renovations. An inscription hanging in the synagogue records its renovation in 1864. The title-page of the 'Order of Service at the Re-Consecration of the Synagogue' on Sunday, 27 February 1910 informs us that the Honorary Officers were: T. Brand, President; D. Jordan, Treasurer; M. Fredman, Burial Warden; and H. Orgel, Secretary. The service was conducted by 'the Revs. D. Jacobs and A.K. Slavinsky and Choir'. A service was held in the Plymouth synagogue to mark the bicentenary of the Congregation at a 'Re-Consecration of the Synagogue on Sunday, 25 May 1952'. It seems that some Honorary Officers and ministers are not loath to have special services, for the Plymouth Congregation celebrated another bicentenary just nine years later. This time it was the two hundredth anniversary of the building of the synagogue. The cover of the Order of Service informs us that the service was 'held on Wednesday, 14 June 1961 and was attended by the Chief Rabbi, The Very Rev. Dr Israel Brodie and The Lord Mayor of Plymouth, Alderman Arthur Goldberg. Officiating Clergy: The Chief Rabbi, and the Reverend B. Susser. Israel B. Black, President; Maurice Overs, Treasurer; Brian I. Pearl, Secretary'. A fortnight or so earlier, there was 'a Civic Service of Prayer and Thanksgiving attended by the Lord Mayor and Lady Mayoress, Alderman and Mrs Arthur Goldberg, with the Aldermen, Councillors and Corporate Officers of the City Council on Saturday, 3 June 1961 conducted by the Lord Mayor's Chaplain, The Rev. B. Susser, B.A.'. This Civic Service followed the precedent set when 'His Worship the Mayor of Devonport, Alderman Myer Fredman J.P. accompanied by the Magistrates and Members of the Corporation' visited the synagogue on Sunday, 19 November 1911.

A rather extraordinary series of services took place in the Exeter synagogue in 1815. Messrs Lightindale, Solomon, and Braham officiated at the Sabbath services towards the end of May, probably travelling around the country as a trio. Gentiles, particularly the musical public, were invited to be present. *Trewman's Flying Post*[26] went into enthusiastic raptures. So great was the impression made by 'the sweet singers in Israel' that they were prevailed upon to give six further services, on Friday and Saturday, 9 and 10 June, and on the following Tuesday, Wednesday and Thursday at 9 a.m. and 6.45 p.m. each day.[27] Still the 'musical amateurs' of Exeter were not sated, and tickets were sold for the trio's final Sabbath service, the proceeds in aid of the Devon and Exeter

Hospital.[28] The services were a financial success, Mr Ezekiel sending £30 to the hospital.[29]

Services in most English synagogues have always tended to be somewhat informal. Even when the Plymouth congregation was on its very best behaviour, at the Civic Service in 1961, the Plymouth Coroner, walking afterwards in procession to a reception in the Plymouth Guildhall, remarked to the minister, 'Do you know what impressed me about the service?' 'No,' replied the minister, expecting a compliment on the decorum. 'Well, it was the delightful informality of it all!' There are certain days, such as Purim and the Rejoicing of the Law,[30] when there is a licensed or tolerated revelry with banging, clapping, dancing and drinking. Even ordinary Sabbath services tend to have a relaxed atmosphere.

From the early nineteenth century onwards there was a growing movement in Anglo-Jewry to introduce greater decorum into synagogue services.[31] This culminated in a brochure issued by Chief Rabbi Adler in 1847: *Laws and Regulations for all the Synagogues in the British Empire*,[32] which made elaborate arrangements for improving decorum, notably by cutting down 'the prolonged *mi-sheberach*'. The *Mi Sheberach* was a blessing recited by the cantor after each man was called to the Reading of the Law. In each *Mi Sheberach* it was customary for the man called to the Torah to offer a sum of money to the synagogue or its ancillary charitable societies and at the same time invoke Divine blessing on his family, the dignitaries of the synagogue, and his friends and neighbours. As it takes about a minute to recite the *Mi Sheberach*, it follows that if ten men (which is not an unduly large number even nowadays) were called to the Reading of the Law and for each man seven *Mi Sheberach*'s were recited (and this number, as will shortly be shown, was apparently not unusual), then the service was prolonged for well over an hour. The general trend in Anglo-Jewry to improve decorum was closely paralleled in the Devon and Cornish Jewish communities. Early attempts to maintain decorum are apparent in the Plymouth Congregation's regulations of 1779.[33] There was an unsuccessful attempt to keep the *Mi Sheberach* problem under control, limiting the number to six for each person. But though the title of the rule is in the index, there is a blank left in the text itself.[34] Riotous behaviour on the Rejoicing of the Law was to some extent tackled: 'both in the evening and in the day, none of the young boys—and how much more so none of the men—shall enter the women's section'.[35]

In Exeter the attempt to improve decorum was spelt out in greater detail. The regulations of 1823 empower the Honorary Officers to give a spot fine to anyone creating a disturbance and, if necessary, to 'call a Constable and treat him or them as the Act directs, being protected by the Bishop's license'.[36]

At the High Festivals of New Year and the Day of Atonement, when the synagogue was particularly full, no boys under six years of age were allowed in, nor girls under the age of 12. In both cases parents and friends had to keep children in order under pain of fine.[37] Whereas the Plymouth Congregation was interested in keeping the men out of the women's gallery, in Exeter the rule was 'no females at all to be suffered in the men's synagogue'.[38] Once again the *Mardi Gras* days of the Rejoicing of the Law and Purim were regulated. On the former day, children under *bar mitzvah,* that is below the age of 13, were allowed to have flags and banners but no lights fixed to them:[39] nothing was to be thrown down from the ladies' gallery: no liquor, wine, or cake was to be given out in the synagogue: children under *bar mitzvah* could dance around but adults had to keep to their seats.[40] Although, children were permitted to make a noise during the reading of the scroll of Esther;[41] but

> no person above *bar mitzvah* is to make any noise in the synagogue neither with hammers nor in any other way whatever under fine of Five Shillings.[42]

The 'any other way whatever' may refer to fireworks. Mayhew interviewed a Gentile firework seller who told him that 'Poram fair, sir, is a sort of feast among the Jews, always three weeks [actually four], I've heard, afore their Passover; and then I work Whitechapel and that way'. This particular seller cleared 13*s.* 6*d.* 'last Poram fair'.[43]

Even before Chief Rabbi Adler's attempt in 1845 to streamline services, there had been previous attempts to shorten the inordinately long services on the Day of Atonement. These were particularly prolonged by the Penitential prayers (*Selihot*) recited on that day. As early as 1830, printed slips were handed to worshippers indicating the particular prayers to be recited. These slips presumably emanated from Chief Rabbi Hirschell and were probably used in Plymouth, as a copy was found in a festival prayer book used there.[44]

In an effort to improve decorum the South-West Congregations attempted to regulate the type of attire worn to services. Plymouth Jews in 1779 were fined for coming to services in long boots and a minute of the Exeter Congregation stipulated that no-one was to come to services in a 'Jim Crow or dustman's hat'.[45] Many of these rules were adopted from one Congregation to another in the eighteenth and nineteenth centuries. In Liverpool (1799), for example, 'none called to the Torah dare wear Jack boots outside his trousers, nor a coloured handkerchief around his neck, nor may he chew tobacco'.[46]

It is customary nowadays in most Anglo-Jewish Congregations for all worshippers saying *Kaddish* to do so simultaneously, either in their seats,

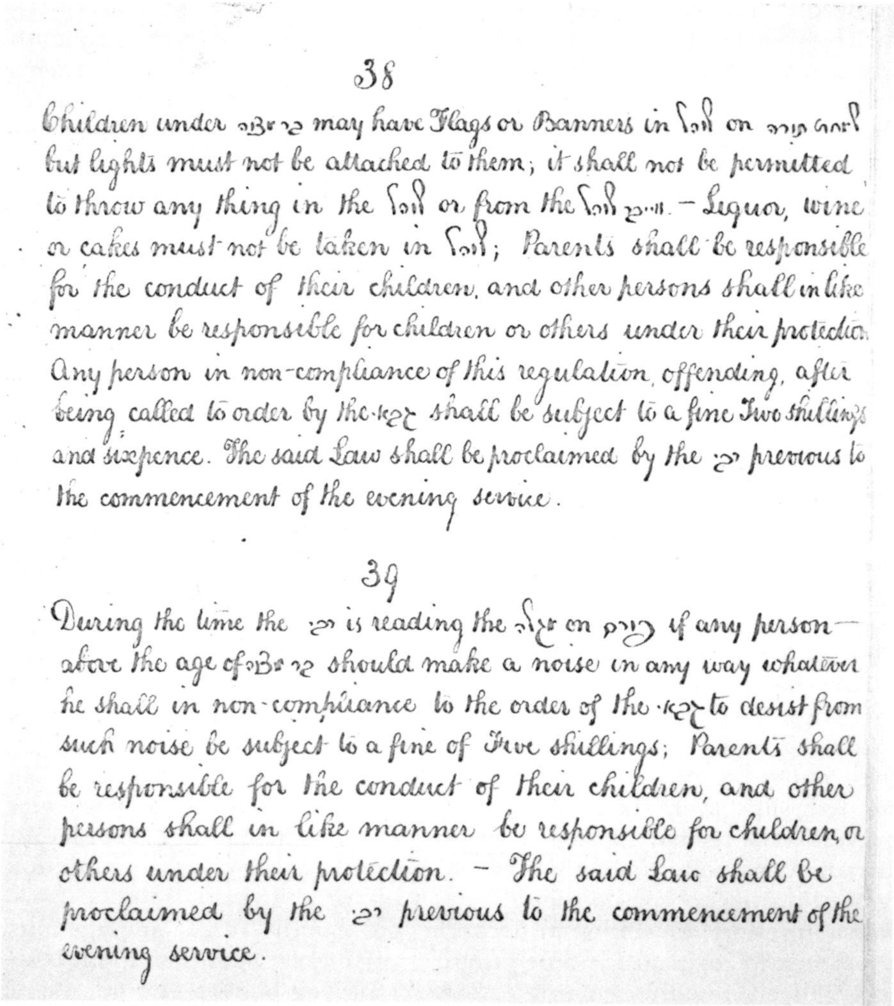

38

Children under בר מצוה may have Flags or Banners in שול on שמחת תורה but lights must not be attached to them; it shall not be permitted to throw any thing in the שול or from the שול בה״כ. — Liquor, wine or cakes must not be taken in שול; Parents shall be responsible for the conduct of their children, and other persons shall in like manner be responsible for children or others under their protection. Any person in non-compliance of this regulation, offending, after being called to order by the גבאי shall be subject to a fine Two shillings and sixpence. The said Law shall be proclaimed by the חזן previous to the commencement of the evening service.

39

During the time the חזן is reading the מגלה on פורים if any person above the age of בר מצוה should make a noise in any way whatever he shall in non-compliance to the order of the גבאי to desist from such noise be subject to a fine of Five shillings; Parents shall be responsible for the conduct of their children, and other persons shall in like manner be responsible for children, or others under their protection. — The said Law shall be proclaimed by the חזן previous to the commencement of the evening service.

*Illustration 15*

The Regulations of the Exeter Hebrew Congregation referring to decorum, 1833.

at the *bimah*, or before the Ark. In the Plymouth synagogue, at least until 1835 and probably until the East European influx, each *Kaddish* was recited by only one person, and he was chosen according to a very strict order of precedence.[47] The rules are set out in great detail in an illuminated manuscript prayer book written in 1805 and kept up-to-date until the mid-twentieth century.[48] Just how important the order of precedence in saying *Kaddish* was to the eighteenth-century Jew is demonstrated by a letter sent by Chief Rabbi Hirschell in 1832 informing the Plymouth Congregation that he and his *Beth Din* had converted the children of Abraham Franco. He reminded the Congregation that the girls could not marry *cohenim* (priests), a very usual reminder which is given to converts to this day, but he also stipulated that the boys had no precedence whatsoever in the saying of *kaddish*,[49] a stipulation that no English *Beth Din* would make nowadays.

It is not unlikely that the rules of precedence for saying *Kaddish* were dropped because too many arguments arose from their application. Much the same applies to the system of obligatory *aliyot*. There were various occasions when a vestry member in the South West could demand to be called up to the Torah, whilst others were called at the Honorary Officer's discretion.[50] Principally these were: the Sabbath before and after the marriage of a child; a circumcision, or *yahrzeit*; the day of his wife's appearance in the synagogue after her confinement; the day of his son's *bar mitzvah*;[51] when he had to *bentsch gomel*.[52] Although not mentioned in the rules of any South-West Congregation it was also customary to call a mourner after the week of mourning, as will be seen shortly.

Difficulties arose when the *Segan* (the presiding warden responsible for allocating honours during the service) favoured his own relations and friends on the occasion of some rejoicing in his family and overlooked the rights of others. The Exeter Congregation, no doubt in response to a situation which had once arisen, made a rule that if the *Segan* neglected to call someone entitled to an *aliyah*, then the *Gabbai* had to order the cantor to call up the man for the seventh portion.[53] In most Congregations to the present day there are worshippers who set great store by the position of their *aliyah*. In their view, the third[54] is the best, followed by the sixth.[55] To offer the penultimate *aliyah* was often regarded as an insult.[56] In 1855, one member felt impelled to write to the *Jewish Chronicle* that the Plymouth Honorary Officers only called up their friends '. . . causing great noises, when the police are called in, and summonses taken out to appear before the magistrates'.[57] Matters reached a head a year later when Josiah Solomon summoned Solomon Solomon, Joseph Joseph, Hyman Hyman, Woolf Emden, and Abraham Ralph. The dispute arose because when Josiah was *Segan* he did not call the son of a deceased Mrs Joseph on the Sabbath after the week of mourning when the synagogue was well attended, even though the mourner had been called to the Torah the previous Monday, but then there were only twelve men at the service.[58] Hyman Hyman, seeing that Josiah called a Joseph Roborichkey as the last *aliyah* so that there was no chance for the mourner to be called, shouted out: 'Shame, iron hearted fellow, tyrant'.[59] Abraham Joseph came down to Plymouth from London to try to patch things up, unsuccessfully. When the case came to court the plaintiff addressed the magistrates: 'If I cannot have justice as an Israelite let me have it as a Christian'!![60] Eventually the solicitors and the magistrates' clerk effected a settlement by which the plaintiff paid all costs and the defendants apologized.[61] For months the community was riven on account of the case.[62]

On one occasion, a brawl in the Plymouth Synagogue, in 1858, attracted the attention of the local press.[63] Judah Solomon Lyon claimed that he was flung from his seat in the synagogue on the first day of Tabernacles (= 23 September 1858) by William Woolf: 'I went into the vestry with my sister and he squared his fists and said, "Let me get at him". He seized a candlestick and attempted to strike me'. The *Plymouth Herald*'s comment was: 'The Jews are a small body in Plymouth and should be able to live in harmony as they have hitherto done with their Gentile neighbours'.

The four Jewish communities in the South West undertook a number of responsibilities in order to make the celebration of Festivals easier for their members. Passover, which occurs in the spring, comes first and with it the need for *matzot*, the unleavened bread. It is exceedingly difficult to prepare *matzot*, as the dough must be baked within 18 minutes of the flour being wetted, and all the crumbs from one batch of dough disposed of lest they leaven the next. For the Jews of Devon and Cornwall, obtaining *matzot*, the staple diet on Passover, was a perennial source of anxiety.

In 1779, the Honorary Officers of the Plymouth Congregation were obliged to ensure a proper supply of flour and appropriate utensils so that the *matzot* could be baked locally.[64] A special oven was kept for the purpose and at one time the oven used by the Falmouth Congregation was on display at the Falmouth Museum.[65] In 1801, pursuant to the twenty-two years' old regulation the Plymouth Congregation decided to bake *matzot* at 7½*d*. per pound. The baking was done on two days and every seatholder with a seat on the right hand side of the synagogue had to help on the first day and those who sat opposite them had to help the next day. Anyone desiring *matzot* but who did not help, had to pay a penal price of 2*s*. 6*d*. per pound.[66]

Apparently the communal baking was not too successful, as in 1803 it was decided that the *matzot* should be ordered from London.[67] One can imagine the difficulties of transporting some half a ton of brittle *matzot* over nearly 200 miles of bone-rattling roads, unless they came by ship, which would have posed its own problems, and the state in which the *matzot* arrived. It is not surprising that the community decided to bake the *matzot* locally again. This time, in 1805, the baking was put out to tender, and a consortium of three members of the Congregation—the brothers Eleazer and Phineas Emden and one *Abraham ben Isaac*—agreed to bake the *matzot* at their own expense and provide them to local Jews at 8*d*. per pound.[68] This last arrangement seems to have been more satisfactory as it lasted for 30 years or more. The contract was renewed in 1814 when the vestry ordered 1,300 pounds from Eleazer Emden at 9*d*. per pound,[69] and in 1833 when it ordered 220 pounds of *matzot* and 20 pounds of flour at 5*d*. per pound.[70] The last reference to the baking of *matzot* in the Plymouth Congregation's Minute Books was in 1834 when there were three tenders for the baking, at 4¾*d*., 5*d*., and 5½*d*. per pound.[71] In 1884, the Plymouth Congregation purchased £8 19*s*. worth of *matzot*, these were presumably for the poor.[72]

An unusual Passover celebration occurred in 1962 when the Lord Mayor and Lady Mayoress of Plymouth (Alderman and Mrs Arthur Goldberg) held a civic *seder* for more than a hundred Jewish personnel of the American Task Force 'Bravo', which was in Plymouth on Nato exercises. The *seder* was attended by the Honorary Officers of the Congregation and chaplains of the American fleet. It was conducted by the Lord Mayor's chaplain, Rabbi B. Susser.

At Pentecost, which occurs in May or June, it is traditional to decorate synagogues with flowers and shrubs. This custom was presumably maintained in the four synagogues of the South West throughout their existence. There are direct references to it in Exeter in 1833,[73] and in Plymouth where 11*s*. 9*d*. was spent on flowers in May 1884.[74] Another custom associated with Pentecost is for some men to remain awake all night spending their time in study. This tradition was kept in Exeter, and there is a reference to it in the Congregation's minutes of 1833.[75] Almost certainly it was also kept in other years, as well as in the other Congregations of the South West, although no reference to the custom has been noted in their surviving records.

Again, with regard to the Festival of the Day of Atonement, which occurs in October, there is a custom that each household lights a large candle which burns for 25 hours. This candle was often taken to the synagogue. Records of purchases of wax for the Day of Atonement occur in every year for which detailed accounts survive.[76] The beadle stayed up in the Plymouth Synagogue on the night of the Day of Atonement,[77] and it is likely that he did so in order to watch the candles in case of fire.

There are two rites associated with the Festival of Tabernacles, which is celebrated each year in October. One is to wave the Four Species, i.e.

the palm, bound with myrtle and willow twigs, and the citron, during the recitation of Psalms, 113–118.[78] Each person ought to buy his own set, but where this is not possible it was customary for the members of a Congregation to pay into a special fund in order to purchase a set. There is evidence that this custom was maintained in Plymouth and Exeter in the nineteenth century. In Exeter, members paid 1*s.* each every year to a communal fund for the *'etrog'*, i.e. Four Species.[79] In 1884, the Plymouth Congregation bought four citrons and three palms paying £1 8*s.* for them, as well as 2*s.* 6*d.* for carriage, 5*s.* for dressing them (i.e. binding the myrtles and willows to the palm), and 5*s* to one Frank 'for taking Louliff[80] round' to people in their homes.[81]

The other ceremonial observed at the Festival of Tabernacles is the building of a booth, known as a *succah*, covered with a leafy roof in which the Jew eats and, subject to limitations of climate, lives. Apart from exceptional circumstances each Jewish family should build their own booth. Nowadays, when many Jews neglect their obligation to build a *succah*, it is customary for Congregations to erect one. This was not always the case. The *Voice of Jacob* in an editorial in 1845[82] praised the Bevis Marks Synagogue, London, for having built a *succah*, lamenting that for many years it had been the only Congregation to do so, and expressing its pleasure that Crosby Square Synagogue had followed suit. The editorial went on to exhort all synagogues to build a *succah*. It is therefore most significant to find that the Exeter Congregation at least as early as 1823 required its Honorary Officers to 'provide and fit up a decent *succah* in the synagogue yard at the the Congregation's expense'.[83] The implication would seem to be that Exeter Jews were not punctilious about making a *succah* in their own homes. This impression is somewhat confirmed in Plymouth where the members of the *Meshivat Nefesh Society* arranged a function at which coffee was served at the London Inn on an intermediate day of Tabernacles in 1799.[84] They could not have been over-particular about the precept of *succah*. A congregational *succah* is first mentioned in Plymouth in 1884,[85] but there was possibly one there very much earlier.

It may be observed that there is some evidence that the custom of learning throughout the night of *Hoshannah Rabbah* (the last intermediate day of the Festival of Tabernacles) was also observed in both Plymouth and Exeter from time to time during the nineteenth century.[86]

Since 1842, when the West London Synagogue was established, there has been an active Reform movement in the Anglo-Jewish community. The Jewish Reform movement has modified or abolished many traditional practices and ceremonial observances of Judaism, particularly the dietary and Sabbath laws. In its synagogues, men and women sit together, an organ is played during services, and the use of the *tallit* (prayer shawl) and head coverings for men are optional. None of the

*Illustration 16*

Rabbi B. Susser with Mr and Mrs M. Overs in the *Succah* in the Plymouth Synagogue.

changes introduced by the Reform movement were ever put into effect by any of the Congregations in the South West.

It is clear from surviving records that the major religious ceremonies and practices of Jewish life attendant on birth, religious majority, marriage and death were all carried out in traditional style in the Devon and Cornish Congregations throughout the period of their existence.

The rite of circumcision at eight days old or as soon thereafter as is medically practicable, for example, an invariable step for a boy of the Jewish faith, was carried out in the Jewish communities of the South West. There is, however, no evidence to suggest whether the circumcisions took place at home or in the synagogue. When the infant is brought in for the operation, he is first placed on a seat known as 'Elijah's Chair'. There is no record of such a chair in any of the Congregations of the South West,[87] though there is a record of a *brit milah* (circumcision) cushion in the possession of the Plymouth Congregation.[88] Each *mohel* (circumciser) had his own instruments. Women are ineligible to perform the rite, and as he had only daughters Revd Stadthagen bequeathed his instruments to any grandson who would become a *mohel*.[89] The two silver goblets used by Joseph Joseph during the circumcision ceremony may be seen at the Jewish Museum, London. One was used to hold the wine for the blessing, the other held whisky or brandy to wash out the mouth to disinfect it before performing the action of *metzitzah*,[90] (sucking the wound to stem the flow of blood).

There used to be a beautiful custom associated with the birth of a boy which has fallen into almost complete disuse in Anglo-Jewry.[91] The mother, perhaps whilst resting and recuperating, would lovingly embroider a binder for the Scroll of the Torah. One example of this has survived from the Plymouth Congregation. It was embroidered by Mrs Brimay Joseph,[92] on the birth of her firstborn. Her embroidery reads, in translation:[93]

> *Abraham ben Nathan Nata KZ* born, for good luck, on Thursday 28 *Kislev* 5562 [= 3 December 1801]. May the Lord rear him to the Torah, to the *Chuppah*, and to good deeds. Amen. Selah.

After the covenant of circumcision, which takes place eight days after his birth or as soon after that time as is practicable, the next important milestone in every Jewish boy's life is his *bar mitzvah*.[94] Technically, 'bar mitzvah' means that henceforth the boy is obliged to carry out the religious duties of an adult male Jew, and this takes effect when the boy is thirteen years and one day old. Nowadays, this 'coming of age' is widely celebrated in most Jewish communities by calling the boy to the Torah, when he generally reads the *Haftarah* (portion from the Prophets) on the first Sabbath after his thirteenth birthday, and celebrating with a party or even formal banquet, very much as if he were a bridegroom on his wedding day. The records of the Congregations of the South West show that a boy's *bar mitzvah* was an important day in his life. Henceforth he was eligible, at least in the eighteenth century, to become a *baal habayit* (vestry member)[95] with the privileges of that status.[96] It is not now possible to tell whether boys read the whole pericope on their *bar mitzvah*, or just the *Haftarah*. In 1821, Zvi son of Samuel Hyman of Plymouth read the whole of his *bar mitzvah* pericope the year after his *bar mitzvah*, so it was probably the custom for able boys, at least, to read the whole pericope.[97] On one occasion in Plymouth, in 1821, a boy under *bar mitzvah* read the *Haftarah*[98] but the records are too sparse to indicate whether this was usual or unusual.[99] The occasion of a *bar mitzvah* would often bring Gentile friends to the synagogue. On an unusual occasion where there were two *bar mitzvah* boys on the same Sabbath, in 1890, a correspondent wrote, 'the [Plymouth] synagogue was crowded with Christians'.[100]

A girl becomes obliged to carry out her duties when she is twelve years and one day old, i.e. when she is *bat mitzvah*. Traditionally, there were no celebrations corresponding to those of the *bar mitzvah* boy. Reform Congregations marked the occasion for girls with a 'Confirmation' service at or about her sixteenth birthday. After the Second World War, Orthodox Congregations began to hold special *bat mitzvah* services. These were designed for young teenage girls, often for several at a time, attending a course of instruction emphasizing the duties of a Jewish woman. The first such ceremony in Plymouth took place on Sunday, 21 June 1964 when Caroline Peck and Andrea Lewis celebrated their *bat mitzvah*. This was followed by a similar celebration by Sharon Aloof and Monique Hirshman on 3 December 1978.

Another major religious ceremony, affecting Jewish men and women, is that of the wedding. There is no reason to believe that wedding ceremonies performed under the aegis of the Congregations of the South West were at any time different to those performed by similar Congregations elsewhere. It is interesting to note, however, that there does seem to have been a change of venue for the actual ceremony. Until the nineteenth century, Ashkenasi weddings were never solemnized in the synagogue itself.[101] This procedure was followed in Plymouth until 1874. Prior to 1874, all 68 weddings registered in the marriage register of the Plymouth Congregation took place at halls or private homes, though whether indoors or outdoors is not stated.[102] In 1874, a wedding was celebrated for the first time in the synagogue itself. Thereafter until the First World War, 35 weddings were celebrated in the synagogue and 51 elsewhere.[103] The first wedding celebrated in the Penzance synagogue took place in 1839, much earlier than in Plymouth. From 1841 to 1892 there were a further 14 weddings, of these half took place in the synagogue and half elsewhere.[104] In Exeter, weddings were celebrated in the synagogue from an early date, 1838, yet of the 32 weddings celebrated from 1838 until 1907 only 11 were in the synagogue, the rest being celebrated elsewhere.[105]

The other matter related to marriages where there appears to have been a change in the practice in the latter part of the nineteenth century concerns levirate marriage. According to Jewish law when a married man dies childless, his widow is regarded as married to his brother, unless the brother frees himself of his obligations to her and cuts the bond binding them together by performing a ceremony known as *halitzah*,[106] and thus enabling her to remarry outside the family. There are unscrupulous men to the present time who refuse point blank to perform the ceremony out of spite, or will do so only after receiving comparatively large sums of money. In order to prevent brothers blighting their widowed sister-in-law's remarriage prospects it was customary, until the twentieth century, for European Jews to give the bride a document called a *shetar halitzah*, some time prior to the ceremony. In this document, brothers undertook, in the event of the bridegroom dying childless, to perform the ceremony of *halitzah* with the widow. The undertaking was made in the form of a most solemn oath with provisions for the payment of punitive damages in case of non-fulfilment. Only one example of a *shetar halitzah* in the South West, dated 1840, has come to light, but undoubtedly it was the custom to give one until it fell out of general use in Anglo-Jewry towards the end of the nineteenth century. The relevant document begins (in translation):[107]

> For a memorial of what was before us on Wednesday, 23 *Sivan* 5600 [= Wednesday, 24 June 1840] according to the counting we count here in the City of Exeter, how the bachelor brother, *Isaac ben Moses*[108] . . . promised *Ella bat Zvi*[109] known as Hirsch who is to be the wife of *Abraham ben Moses*[110] . . . that if, Heaven forfend, my brother shall die without leaving holy seed . . . then I will give proper *halitzah* without charge immediately after three months . . . And if I do not do so within six months of being asked to do so then I shall pay her damages of fifty pounds sterling . . .
>
> Winesses: *Benjamin ben Samuel SGL*[111]
> John Levi[112]

This agreement was made, possibly at some ceremony of *tannaim* (engagement) three years before the actual marriage ceremony, which took place on 21 June 1843.[113] Only one other reference to the writing of a *shetar halitzah*, also in Exeter and about February 1867, has been noted. Chief Rabbi Adler wrote to Solomon Elsner of Exeter at that time asking him 'have you a brother or brothers who will attend wedding to write *shetar halitzah*?'.[114]

Some light is thrown on the social aspect of a wedding in upper-class Jewish society[115] in Exeter in the early part of the nineteenth century, by the invitation to the wedding of Ellen Levy's parents in 1810. The forms of etiquette were observed, the wedding reception took the form of a tea; and *mirabilé dictu*, no presents were accepted.

At the wedding of Mr S. Elsner and Miss Silverstone in the Exeter synagogue in 1853, there was 'a numerous attendance (including non-Jews) most of whom sat down to a rich dinner'.[116] A year later when Chief

Rabbi Adler happened to be in Torquay, probably on holiday, he came over to Exeter to marry Revd B. Albu to another Miss Silverstone. After the service they returned 'to an inner chamber of the synagogue . . . where dejeuner . . . for about 70 persons was served'.[117]

Traditionally, Jewish couples remain in town for the week following their wedding, and a special grace is recited in the presence of the bridal couple, known as *sheva berachot* (= seven blessings). No evidence has come to light to show whether the Jews of the South West maintained this beautiful tradition.

The rites and ceremonial associated with death play a large part in the life of the Jew. It would appear that throughout the eighteenth and nineteenth centuries Jewish laws and customs in respect of the last rites in sickness, funerals and burials were maintained by the Congregations of the South West.

Poor Jews in the South West who became seriously ill could expect to be looked after and to be kept company. The Exeter Congregation's Regulations of 1823, for example, stipulated that the Charity Warden should discharge all expenses occasioned by the sickness of members.[118] Wealthy individuals must have paid their own way, but the poor were provided with a nurse and even, during the cholera epidemic of 1832, brandy.[119]

In Plymouth, as early as 1779, when any Jew attached to the Congregation fell dangerously ill, then the Cemetery Fund Treasurer had to place the names of all members into a ballot box and draw out two of them. These two had to keep vigil and one had to move into the invalid's bedroom 'provided he is a suitable person to be with an invalid'.[120]

The present practice is to use only linen shrouds. Rachael Benjamin who made her will the day before her death on 11 March 1817 left instructions 'my shroud to be made of wollen if it can be got, if not of linen', whilst Eleazar Emdon, a humble and religious man, specified that his shrouds should be of linen.[121] Friends, relatives, or fellow members of Friendly Societies such as the *Meshivat Nefesh* or *Bikkur Cholim* sat up at night watching the corpse at home[122] in those cases where the burial was delayed to the next day.

Jewish law requires that a corpse should be buried as soon as possible, preferably on the day of death.[123] As far as can now be ascertained about one corpse in five in the old Jewish cemetery on Plymouth Hoe was buried on the day of death.[124] There are 37 tombstones now extant in that cemetery on which the dates both of death and burial are inscribed. Nine people died on the Sabbath, on which day burials may not take place. Sixteen people were buried on the day following death, in five cases there was a somewhat longer delay, particularly where death took place on a Friday.[125] In only seven cases was the person buried on the day of death, and one of these was a cholera victim.[126] It seems that as the nineteenth century wore on there was a tendency to perform funerals on the day after death. Indeed, in the last decade of the century, Chief Rabbi Adler wrote in peremptory terms to Revd Spier at the Plymouth Congregation because he buried a child soon after death: 'It is highly necessary to prevent a repetition of a proceeding which may lead to grave scandal or danger'.[127]

Even more necessary than watching the corpse before burial for kabbalist or other reasons, was the watch maintained after burial to prevent body-snatching. Precautions were taken by both the Plymouth and Exeter Congregations for the post-funeral protection of corpses. The Plymouth Congregation's Regulations in 1835 specify:

> To guard against the robbery of graves, watch shall be kept on the ground for three nights after burial, by three Jews[128] each night, to be decided by a ballot of every person in the congregation: in case of refusal to be fined Five Shillings.[129]

The Exeter Congregation with a small cemetery, required only two men to stand guard and these were presumably hired professional guards as each member was charged one shilling to pay the expense.[130]

In view of these precautions against body-snatching there is a certain grim irony that in early nineteenth-century Devon,

> one of the largest dealers [in bodies] was Israel Cohen, commonly called Izzy, a Jew, well known to surgeons and sextons. By the surgeons he was patronized, of the sextons he was patron: and so complete the understanding that the interest of all three was advanced by coalition. He was a square built, resolute ruffian, with features indicative of his Hebrew origins, black whiskers and a squint.[131]

It may be noted that the name of Israel Cohen does not occur in the records of the South-West Congregations.

Standing guard at the Jewish cemeteries involved not only physical discomfort but also some anxiety occasioned by superstitious beliefs. Shemoel Hirsch, then a pedlar down on his luck, made his way from the South West to Ipswich about 1823:

> The Rabbi asked me to watch at a grave of a dead Jewess who had asked for two people to watch the first four nights. I said I had no objection if he would lend me a prayer book to use during the night. (Note: The Jews consider the burial ground a most solemn and awful place, and would never pass one at 12 o'clock at night . . . as they believe the dead awake at that time and go to prayers in the synagogue. Their fear is diminished if they have *tsitsith* and proper prayers.) Good *tsitsith* I had, but I wanted a prayer-book.[133]

In most Jewish communities there is a society, generally known as the *Hebrah Kaddishah*, whose members, male and female, attend to the needs of the corpse, doing in a voluntary capacity the work which is generally done by the undertaker in Gentile society. There is no doubt whatsoever that such a society existed in each South-West Congregation, and the overseer or warden of the Cemetery Fund, whose duties have been discussed above, was almost certainly its chairman. Nonetheless, there appear to be no records of any *Hebrah Kaddishah* until the *Bikkur Cholim* Society of Plymouth founded in 1901. The primary objects of this society were:

(a) Personally to visit the sick;
(b) To secure watchers (of the sick) when required;
(c) To provide watchers after death (before burial);
(d) To perform the last rites, viz. *Tahara* (washing, purifying and dressing body) and *Levoyah* (funeral), and to visit the mourners.[134]

The warden of the Cemetery Fund was also probably the chairman of the *Hebrah Kaddishah*.

There is no fixed custom nowadays in Anglo-Jewry as to where the last rites of purification and clothing the body in shrouds take place. In some communities these rites are performed at the place of death, be it at home or in a hospital at the mortuary, in others in the chapel (*ohel*) at the cemetery. It would appear that in the Exeter Congregation, at least in the mid-nineteenth century, these rites took place in the deceased's home. The relevant Exeter rule reads (in a mixture of English and Yiddish):

> Agreeable to the custom of London by order of the Chief Rabbi, *Alle koved was onbelongt zu der meis* [all honours due to the dead] must take place in the house where the *meis* [corpse] is *mitaher* [washed and purified] and not at the cemetery.[135]

This rule implies that not only the last rites of purification took place in the house but also funeral orations were given and certain prayers were recited.[136]

From the proximity of certain tombstones in the old Jewish cemetery on Plymouth Hoe, it seems that, sometimes at least, coffins were buried on top of one another.[137] This is apparently the purport of a letter from the Chief Rabbi in London to the Congregation referred to in a minute of 1825: 'it was agreed to take up the ground in the new cemetery and to cover in the old cemetery some graves which have been reserved . . .[138]

In the Jewish cemeteries at Exeter and Plymouth places were reserved 'beyond the boards' for Gentiles married to Jews, as well as for Jews married to Gentiles, and for suicides.[139] In 1838, the Exeter Congregation wrote to Dr Adler asking whether a memorial prayer could be said in the synagogue for the late Mrs Jonas 'in as much as it was said that she was not made a convert by sanction of the Chief Rabbi and Beth Din of London'. The letter went on to say that 'the said Mr and Mrs Jonas were interred in our ground in the *regular row* . . .'.[140] It is clear from this last remark that those who were certainly not Jewish were buried outside the normal plot in the Exeter cemetery. Similarly, Abraham Joseph wrote to Dr Adler in 1867 asking him to ensure that if a Samuel Ralph, married once or twice in church, *was* buried in the Jewish cemetery in Plymouth, it should not be in the main part.[141]

From the early part of the eighteenth century and possibly earlier, the corpse whether in its permanent coffin or a temporary one, or on a bier covered with a shroud, was conveyed to the cemetery in a hearse. This was certainly the case in the South-West Congregations and in London. So important was this 'kindness to the dead' that in Exeter in 1823 the Cemetery Fund Warden 'where parties defunct are *known* to be poor' had to provide a hearse at the Congregation's expense, as well as a mourning coach.[142] In Plymouth, the Congregation apparently had its own hearse about 1835. There are references to it at a time when its upkeep was becoming expensive and every seatholder was expected to pay 2*s.* 6*d.* per annum towards its upkeep.[143]

B.H. Ascher in his *Book of Life*[144] intimates that in Germany Jews customarily set a tombstone a year after death at the *yahrzeit*, whereas 'in England, more especially in London, the tomb is set after the thirty days of mourning' or even sooner. At some time, possibly after the influx of East European Jews after 1880, the prevalent custom throughout England was to set the stone about a year after the burial. There is no evidence to indicate which custom was followed in the South-West Congregations, all that can be said is that tombstones were usually erected by the family or friends of the deceased. If there were no family or friends, and a dead man left sufficient money to pay for it, the Plymouth Congregation arranged for a stone to be set. A loose piece of paper now bound into the Plymouth Congregation's second minute book notes the expenses incurred by the Congregation on behalf of one Jacob who died in February 1803:[145]

| | £ | *s.* | *d.* |
|---|---|---|---|
| All the expenses during the whole time he lay ill until after his burial | 4 | 16 | 11 |
| Paid for his debt to Joseph | 2 | 0 | 0 |
| To a Gentile at Dock by name Mister Clark | 1 | 2 | 0 |
| | 7 | 18 | 11 |

| | | |
|---|---|---|
| Found by him after his death | | |
| in cash | 1. 14. 0*d*. | |
| and for clothes and other | | |
| items which were sold | | |
| received | 7. 18. 6½*d*. | 8 19 6½*d*. |
| Paid for tombstone | | 1 2 8*d*. |

On the other hand when there was insufficient to pay the debts, the Congregation did not incur further expense. Thus, when one *Samuel HaLevi* died in 1800, his garments fetched 18*s*., four watches made £5 4*s*., a total of £6 2*s*. As the Congregation had laid out £7 0*s*. 7*d*. on his behalf there was no question of commemorating him with a tombstone.[147] Nor did Dikah wife of Moses Isaac the Beadle fare any better when she died in 1815. Her effects were sold and the Congregation took all the money to pay its expenses incurred by it on her behalf.[148]

To be deprived of a tombstone was considered a serious matter. On at least one occasion the Exeter Congregation attempted to force a son to pay his father's debt to the synagogue by refusing permission to erect a tombstone until the debt was paid. Josiah Solomon, the son, appealed to Dr Adler who wrote to the Exeter Congregation in 1855, 'our law demands that the father should not suffer for the son, therefore we are not justified to do such an open insult to an innocent man'.[149] At least one Plymouth Jew living on Congregational relief had a tombstone to mark his grave. He was *Ze'ev ben Naftali* who received 3*s*. 6*d*. per week in 1812 until his death on 7 October 1813.[150] It must have been a comfort to the sick, and especially to the poor, to know that their bodies would be laid to rest with honour and respect.

Members of the Plymouth Congregation often left money to the synagogal officials to recite *kaddish* after their death (particularly if they had no sons), to 'learn' for the benefit of their souls, and to burn a light in their remembrance. Thus, Rachael Benjamin in 1817 left 'one pound to Moses Solomon to say prayers as usual in the school, one pound to Mr Isacher to learn one month. A light to be burnt one month'. Eleazar Emdon wanted a light to be burnt in the synagogue for 12 months.[151] In 1862 Myer Stadthagen left '£3 to my congregation as a memorial to be made every Festival of freewill offerings [i.e. at *yizkor*], and to pay some fit person to say a prayer [i.e. *kaddish*] for me during the year, and a light shall be burned . . .'.[152]

From birth until death, and even after death, a Jew could live, if he or she so chose, within the framework of a Jewish community, which provided the opportunity to maintain the ancient Jewish ways in the new surroundings of Devon and Cornwall.

## CHAPTER EIGHT

# Jewish Philanthropy in the South West

Beginning their classic work on English Poor Law history, Sidney and Beatrice Webb declare:

> Throughout all Christendom the responsibility for the relief of destitution was in the Middle Ages, assumed and accepted, individually and collectively, by the Church. To give alms to all who were in need, to feed the hungry, to succour the widow and the fatherless, to visit the sick, were duties incumbent on every Christian.[1]

From 1597, these functions of the Church were gradually taken over by the Central and Local Government in England.[2] By the last quarter of the eighteenth century the distinguishing feature of the English Poor Law was a

> widespread but haphazard provision for the impotent poor by weekly doles of money, with a persistent belief that it was possible to make a profit out of the labour of those men, women and children who could be set to work.[3]

The latter policy called for the provision of instruments of compulsion, 'which were called Houses of Industry when one of their aspects was emphasized, and Houses of Correction or Bridewells when another side of their function came into view'.[4] With the development of the industrial Revolution it was found that the new capitalist entrepreneurs were so eager for workers to fill their factories that they would even spend money, in the form of premiums given to those charged with looking after the poor, to secure their services.[5]

The Church and State were very differently motivated in their relief of distress. The Church was primarily concerned with the donor and his soul, the poor was the medium of the rich man's salvation. The State was concerned with law and order, as an authoritative Political Economist put it in 1852:

> They [the Poor Laws] are in fact, a bulwark raised by the State to protect its subjects from famine and despair . . . [to] prevent them from being driven to excesses ruinous alike to themselves and to others . . . Without it [the Poor Law] the peace of society could not be preserved for any considerable period.[6]

Jewish philanthropy in London, with the largest concentration of Jews in England, during the eighteenth and nineteenth centuries reflected, in the main, the basic characteristics of general society.[7] These were an acceptance of poverty as a normal state, to be relieved by casual charity dispensed by individuals or endowed bodies; an attempt to humanize conditions for certain categories of person notoriously ill-treated, such as prisoners, slaves and chimney boys; and the relief of special disability or distress by the provision of hospitals and asylums.[8]

In the provincial Jewish community charitable endeavour corresponded more closely to the monastic example of the Middle Ages and to that which was observed generally in villages in the nineteenth century rather than to that which prevailed in the towns. In this connection it has been said that in the towns the rich knew poor streets whereas in the villages they knew poor people.[9] One consequence of this greater intimacy was that all the poor in the village were noticed, whilst in the towns many deserving cases were overlooked.[10] For much the same reason it was easier to differentiate in the village between the deserving and undeserving poor, so the deserving and local poor apparently received, therefore, more help in provincial Congregations than the undeserving and the peripatetic.

Judging by contemporary practice, the wealthier Jews living in Devon and Cornwall in the eighteenth and nineteenth centuries undoubtedly gave donations to individuals who solicited help but, not surprisingly, no records of this type of charity have survived. On the other hand, there is detailed evidence of both the income and expenditure for charitable purposes by the Congregations of the South West and the Jewish societies associated with them.

In much the same way as local parishes distributed charity to the general poor, so, too, did local Jewish communities throughout England give aid to the Jewish poor.[11] Amongst the local resident poor were aged officials of the Plymouth Congregation who were granted a pension after long service. Cantor Benjamin Levy, after having served the Plymouth Congregation for more than 40 years, was granted a pension in 1814 of 'six guineas per month so long as he remains cantor here'.[12] He retained the title of his office and his pension until his death in 1829, aged nearly 100 years.[13] However, when the Congregation's income dropped in the lean 1820s, his pension was cut first to £40 and then to £30 per annum.[14] Widows of officials were also granted a pension. In 1800, the widow of Moses the beadle was granted two shillings per week and free accommodation 'so long as she lives in Plymouth'.[15] Jews, both male and female, who had been long resident in Plymouth, received weekly relief. In 1803, for example, the wife of *Aaron ben Isaac* was given five shillings a week, and in 1812, the same sum was given to a Betsy Isaac, and 3*s*. 6*d*. per week to Benjamin Naftali.[16] In 1816, the Congregation budgeted on an annual basis for its stipendiary poor, and yearly sums of £15 for Lazarus Joseph, and £10 each for Mistress Davis, *Yehiel ben Naftali Hart* and *Zelvelcher* were allocated.[17] An active pensioner was expected to pay her way and Miss Benjamin was 'to receive £6 per half annum and she must continue to make wax candles as previously.[18] Even when the Congregation was financially hard-pressed in 1827 it nevertheless gave 4*s*. every week to 'Mistress Moses, widow of Judah, and to Mistress Abrahams, widow of Mordecai'.[19]

Nor was help given only to the aged and widows. In 1804, one *Solomon ben Hayyim* was in some trouble with the law and the Congregation gave his wife fourteen shillings each week.[20] Similarly, in 1817, when *Judah ben Hayyim Mannheim* (Solomon's brother?) was in trouble 'for forging on the Greenwich Hospital Navy Agent', and his parents were distressed for rent arrears of nine pounds, the Congregation arranged a collection in the Anglo-Jewish community, paid for his defence, and then gave his parents twelve pounds, which was the balance of the monies collected.[21]

The desperate straits to which those living on the border line of poverty could be brought when work failed, or sickness struck down the breadwinner, are disclosed in a pitiful letter of appeal written by Aaron Nathan in December 1827 to the President of the Plymouth Congregation:

> . . . I now have to Inform you beaing on of the helders that I am Drove to the last Extramity, without a farthin in the world having disposed of Everything I could make money of, so as my wife and Sevon Children Should not starve.[22]

His letter to the President and Treasurer[23] 'aroused their great compassion for his distress' and the Congregation made him a gift of three pounds to be doled out over twelve weeks.[24] The help was continued on a lower scale from August 1828 until May 1829 at three shillings a week.[25] To put these sums into perspective it may be observed that when an Ann Usher died at Wallsend aged 102 years, she had received parochial relief of £157 13*s*. in the previous 30 years, i.e. about two shillings a week. This was regarded as a very large sum *in toto*.[26] Towards the end of the century the amounts given by the Plymouth Congregation to regular pensioners had hardly changed. In 1883, two maiden ladies, Miss Bellem and Miss Levy each got 2*s*. 6*d*. a week and two men got 5*s*. per week.[27] These charitable disbursements came from a special charity fund in the hands of the Treasurer.

The cash came from offerings made on Sabbath and Festivals,[28] also from charity boxes in the synagogue. There were four of these in the Plymouth synagogue, one let into the South wall known as the Perpetual Box; a second at the back of the bimah, known as the Weekly Box; and two at the two pillars at the entrance, known as the Poor Boxes. The contents of the first box were probably used to pay for the oil in the perpetual lamp, the second box probably received small but regular amounts, whilst the other two were probably used by worshippers who dropped in odd coins during the reading of the Torah on Mondays and Thursdays each week, and the small offerings of those called to the Torah on those days. The income from these in 1888 was 14*s*., 7*s*. and £2 3*s*. 4*d*. respectively.[29] If the specific charity income was insufficient to meet charitable needs at any particular time, then the Plymouth Congregation voted funds for this purpose from its general purse. In 1808, for example, it allocated £25 to be given to the local Jewish poor for a six-month period, besides out-of-pocket donations to the casual poor.[30]

Apart from the regular amounts to the resident poor, eleemosynary aid was given to the comparatively large numbers of Jewish poor who passed through the South West. The Plymouth Congregation empowered its Charity Treasurer in 1779 to give not more than two shillings to each poor Jew who asked for help, and more in those cases he felt to be especially deserving, provided the President concurred.[31] Furthermore, the Treasurer could direct a poor man to the home of a member of the Congregation,[32] and that member had either to pay for the man's accommodation at an inn, or give the man two shillings.[33] This latter institution fell into abeyance by the time the next regulations were issued in 1835, there being no mention of it in them, but the warden was empowered to 'dispense such casual relief to poor applicants . . . not to exceed Five Shillings to any one individual within a month'.[34] This last sum could be increased to a guinea with the agreement of the President.[35] A similar situation prevailed in Exeter, though there the warden could give only ten shillings, or with the consent of the Honorary Officers, twenty shillings, in a year.[36]

For how many itinerant poor did the Congregations of the South West cater? The first time a survey of Anglo-Jewish charitable endeavour was taken was in 1841,[37] at a time when very few Jews lived in Cornwall, the Exeter Congregation was much diminished in members and vitality, and only in Plymouth did a Jewish community still function properly. According to the Board of Deputies' survey in 1841, the Plymouth Congregation supported four resident widows, and relieved one hundred and twenty casual or itinerant poor. Even here it is not possible to break down the figure further between the casual poor who were resident in the town and the itinerant. For 1883, there is a record of the exact number of itinerant poor and also their principal place of stay beforehand. In that year, the Plymouth Congregation's Treasurer paid out £18 5*s.* to 73 peripatetic poor, an average of five shillings each, with a maximum of twenty-four shillings and a minimum of two shillings to any one person. Although the amount was large for the comparatively small Congregation it pales into insignificance when compared to the amount distributed by the much larger Congregations. In Liverpool, for example, in the twelve months after April 1868, just over £300 was spent for food, cash payments and assistance in emigration.[38] The importance of the small provincial Congregation was to act as a staging-post enabling poor peripatetic Jews to reach their final destination.

The poor came from a wide variety of towns, mostly from the west side of England; Bristol, Cardiff, Dublin, Exeter, Liverpool, London, Portsmouth, Southampton and Teignmouth. Whilst from overseas Jews came from Berlin, Jerusalem, Philadelphia and North Africa.[39]

Although entitled as rate payers to claim parochial relief,[40] Jewish communities rarely allowed their poor to do so. The Board of Deputies' report in 1841 referred to above, covers eight communities who looked after some 50 resident poor and 2,000 casual poor.[41] Of the 50 resident poor, only one, in Plymouth, received parochial relief. The Portsmouth Congregation, a very similar one to Plymouth's in size and tradition, would not allow its poor to apply for parochial relief. The general feeling was that it was not fair to impose on a society which had given refuge and haven, to do so might lead to a backlash of public feeling, preventing the free ingress of future immigrants.[42] There was also another factor, Jews could not easily live within the workhouse as they could not eat there without infringing the dietary laws, though even this problem could be surmounted, as at Liverpool Jews got parochial relief without entering the workhouse. The Plymouth Congregation had several decades earlier, in 1816, considered the advisability of allowing a poor Jew to stay in the workhouse and decided that in certain circumstances one could. A meeting of the vestry had nonetheless to be called to discuss the case.[43]

The impulse to give charity found expression in the South-West Congregations not only in payments to Congregational funds which were then disbursed as charity but also in the formation of societies whose primary or secondary function was to complement the Congregation's charitable disbursements. In forming these organizations the Jews of Devon and Cornwall generally emulated the patterns of similar English societies in the eighteenth and nineteenth centuries.

The eighteenth century in England, particularly in the early part, was a period of association. People became aware of the advantages of associations which could match larger social needs than the individual philanthropist could meet.[44] The end of that century saw the growth of Friendly Societies, the first of which, the Sunbury Friendly Society, was founded in 1773. This had 48 members at foundation, which later increased to 60. A monthly payment of 1*s*. gave a benefit of seven shillings a week in sickness or £7 on death.[45]

The first known Jewish friendly society in the South West was the *Hebra Kaddishah Meshivat Nefesh*[46] and it was inaugurated in Plymouth on Monday, 12 October 1795, the enrolment of members being completed the following week.[47] There were 37 foundation members of whom 13 left at the end of the first year, the number of members dropped to 18 in 1799 and then gradually rose to a peak of 44 in 1811, rapidly falling away once again to 9 in 1819, and fluctuating thereafter between 9 and 13 until 1830, when the records of the society ceased.[48] The fees to the Meshivat Nefesh were more than double those payable to the Friendly Society at Sunbury. There was an initial entrance fee of 2*s*. 6*d*. plus 1*s*. for a rule book,[49] an annual entrance fee of 2*s*. 6*d*. and two shillings was payable each Jewish calendar month.[50] The society's income fluctuated from year to year according to the size of membership. It started at £60 in 1796, dropped to £28 in 1798, and then slowly climbed back to the £60 level at which it remained for the next ten years. Peak membership and income was in 1810 and 1811, when some 40 members paid in over £90 each year.[51]

The prime purpose of the Plymouth Meshivat Nefesh Society was to provide 'benefits' to its members.[52] They balloted for some eight or ten prizes of thirty shillings each, though fifteen shilling consolation prizes were also available after 1800, and occasionally even five shilling prizes as well.[53] It is clear that to some extent at least the society was a lottery rather than a Friendly Society in the accepted sense of the term. In fact, a closer modern analogy would be 'a football pool syndicate', because appreciable sums from the Society's assets were invested in National Lotteries. In 1800, for example, £3 9*s*. was spent on three-sixteenths of a lottery ticket, and in 1804 two-sixteenths of a ticket were bought.[54] Additionally, a ticket in the 'Shakespeare' lottery was purchased for £3 3*s*. 6*d*. in 1800 and 1804.[55]

The secondary purpose of the society was to give members and their guests a gourmand dinner on the Sunday of *Hanukah* each year. In 1800, 17 members and 19 guests consumed 3 geese, 2 turkeys, 23½ lbs of veal, 4 tongues and 5 bottles of gin, together with the usual accompaniments.[56]

Yet another function of the Society, arising out of the zeal of the members to complement the Congregation's charitable endeavour, was to distribute the interest on the Society's investments to a few Jewish poor. £2 13*s*. was paid out 'to several poor persons as settled by the committee' in 1806, and to five poor women and a man in 1808.[57] The amount given to the poor increased as the Society's investments grew. In 1811, £5 6*s*. was paid out to the poor and £2 was allocated for the relief of Jewish poor in Jerusalem.[58] These charitable disbursements rose to a maximum of £10 11*s*. in 1815, and then sharply declined the following year to £2 18*s*. which was shared among eight people, and then ceased altogether.[59]

It is possible that the Society, or a similar one in Exeter, provided death benefits enabling the family to pay for a funeral and an income during the seven days of mourning when the breadwinner is forbidden to go to work. An account of the effects of Simon Levy, silversmith of Woolcombe near Newton Abbot who died in 1802, showed that they were worth £35, an amount of £11 for funeral expenses was noted as 'being money received from a club for that purpose'.[60]

Just as the Friendly Societies were at double risk (the poor got tired of contributing and the rich of managing)[61] so too were the societies, including the Meshivat Nefesh of Plymouth and its successors which will be described shortly, of the Congregations in the South West. There was a comparatively limited number of those able to organize and even fewer prepared to do so, and for these latter there was ample opportunity and need to occupy all their spare time arranging the affairs of the Congregation. The account book of the Plymouth Meshivat Nefesh Society ends with the year 1830, no other books of the society or references to it have been noted after that date, and so it is likely that it quietly petered out about that year.

There was in Plymouth another Jewish society which apparently fulfilled more closely than the Meshivat Nefesh the function of a Friendly Society. It was the Jewish Brotherly Society, founded in 1823,[62] and disbanded about 1860.[63] This society was apparently originally recruited from the ranks of Plymouth's poorer Jews, as Charles Marks

> one of the late Presidents of the Society in a most moving speech very properly alluded to some who *had* been members and because fortune had prospered them in business had thought it becoming to leave a Society which deserved to be supported.[64]

Once again, the annual dinner was an important part of the activities of this society. The *Jewish Chronicle* carried a report that 30 members had dined on Christmas Day, 1851, which was the 28th anniversary of the society.[65] Toasts were proposed to the British Government 'under which we so peaceably live', The Chief Rabbi, Sir Moses and Lady Montefiore, and the Reader and second Reader of the Congregation—Reverends Stadthagen and Woolf. It was considered that the affairs of this society were in decline, there being only some 30 diners in 1851 and again in 1852, due in part to many being absent from the town and 'several having left to go to foreign parts'.[66]

Towards the middle of the nineteenth century, the Anglo-Jewish community felt it was time to put its charitable house in order.[67] There was much duplication of effort, whilst important social work, particularly the education of poor children, was neglected. 'It is desirable', wrote Henry Faudel in 1844, 'to resort to a plan of centralisation . . .'.[68] Furthermore, the community, largely British born and anglicized, was becoming sensitive to the sight of itinerant Jewish poor knocking at the door for aid. In 1845, for example, the Jews of Liverpool founded the 'Liverpool Society for the Suppression of Mendicancy and the More Effectual Relief of the Deserving Poor' in order to discourage itinerant mendicants.[69] Four years earlier, in 1841, a Jewish 'Society for helping itinerant poor' had been founded in Plymouth. A relic of this society, and apparently the sole surviving trace, is a book still in the possession of the Plymouth Hebrew Congregation. The following is a translation of the title page:

> Indication Book of the Society for Helping Poor Strangers which was established here, the holy congregation of Plymouth, Sunday, 9 *Shevat* in the year: 'He that is gracious unto the poor lendeth unto the Lord, And his deed will He repay unto him.'[70]

The book consists of a number of thickened pages which have been perforated with holes. Horizontally opposite the lines of holes are inscribed in alphabetic order the Hebrew and, later, the English names of regular worshippers in the synagogue. The holes form six vertical columns and above these is written, in Hebrew, 3*d*., 6*d*., 1*s*., 18*d*., 2*s*., half-crown. The purpose of these holes was to act as an *aide memoire* to the Society's Treasurer of the monies offered to the Society during the Reading of the Law on Sabbaths and Festivals. On these days an observant Jew may not write, and to ensure that no offerings were forgotten the Treasurer tied a piece of string, coloured according to the day on which the offering was made, in the appropriate hole. The book was made by Samuel Cohen, as may be seen from his modest signature at the foot of the title page.[71]

The Society appears to have disbanded by 1854, for in that year Samuel Cohen gave the book to the Congregation so that it could be used to record offerings to the new Cemetery Fund. This is evident from the inscription on the title page which reads in translation:

> I made it, as *supra*, and now I give it to the said holy congregation for the service of the Cemetery Improvement Charity in the year: 'May our name be inscribed for life in Thy book.[72]

It may be mentioned that the Jewish community in Exeter also had a Society for Charity to the Poor about 1830, but apart from a brief entry in an account book there is no information about its functions other than its name.[73]

Soon after the Plymouth Society for Helping the Itinerant Poor had disbanded, another took its place. It went by the grandiose title of the 'United Jewish Hand-in-Hand Benevolent Society, Plymouth, Stonehouse and Devonport' and it was founded in 1861.[74] Once again, the main purpose of the Society as set out in its revised rules of 1876 was to obviate the necessity of resorting to private collection, by relieving from the Funds of the Society urgent cases of distress, itinerant or otherwise.[75] These last two words 'or otherwise', refer to a change of policy decided upon in 1873, for until that year only itinerant poor were helped, but in 1873 the society had surplus funds and these were utilized for the benefit of the resident poor as well.[76]

The Hand-in-Hand's income was derived from an annual subscription of 12*s*. per member, donations, particularly those made in the synagogue on Sabbaths and Festivals, and bank interest on its deposit account. The income rose gradually from £12 2*s*. 11*d*. in 1865, to its maximum in 1879 of £19 4*s*. 6*d*. and then fell away to £8 7*s*. 6*d*. by 1888.[77] Apparently, there was a general appreciation by the Jews in Plymouth of the work of the society, as in many years the nonmembers' donations amounted to one-third of the total income.[78]

Over the period for which accounts are extant, 1865–88, some seventy per cent of the society's income was expended on charity disbursements, ten per cent on administrative expenses (a commission was paid to the collector of the membership dues), and twenty per cent on refreshments for the members.[79]

Usually, a half-crown or five shillings was given to the itinerant poor, though larger sums are also recorded in the society's minute book. In 1874, for example, a Mrs Lipshitz was given 15*s*., '5*s*. now, 5*s*. at her confinement, 5*s*. the week after',[80] £1 was given to 'a poor woman and child just arrived in town',[81] and in 1875 a visiting *magid* (itinerant preacher) and cantor were given 15*s*. each.[82]

Another characteristic typical of charitable endeavour in both Gentile and Jewish society in England during the eighteenth and nineteenth centuries was to remove the poor to another parish, so that somebody else would have the responsibility of looking after them.[83] The Exeter Congregation, though hard pressed financially in 1844, found it expedient to give £3 to Miss Catherine Ezekiel 'to enable her to go to friends in Cincinnati'.[84] The Hand-in-Hand Society in Plymouth gave ten shillings to a Judah Levy to get him to London in 1875, and one pound to a Mrs Scheertal to help her return to Germany in 1886.[85] To a large extent it would probably not be wrong to regard all payments to the itinerant poor as subsidies to help them on their way to another place where others would have the responsibility of them.

In England it was 'Elizabethan charity that made the pregnant discovery that if poverty is to be relieved provision must be made not only for those unable or unwilling to work but also for many who are willing to work, but unable to find employment'.[86] European Jews were unfortunately only too familiar with poverty which came through no fault of the poor. The Chmielnicki massacres, expulsion from Prague, the Pale of Settlement or discriminatory legislation were any of them more than sufficient to reduce the wealthiest Jew to a pauper on the mere flourish of a signature. Accordingly, Jews were more disposed to give a coreligionist the capital or means to set him up again, and the poor Jew felt little or no shame in asking his fellow Jew for help. To ask a Gentile for help, however, was a very different matter, as Shemoel Hirsch in the early nineteenth century recalls asking a Quaker for charity:

> When I came to the door I could scarce enter for shame, as it was a different mode of begging from that practised by the Jews. To them I need no ceremony but plainly and plumply go in and say, 'I am a poor Jew', and they understood what was wanted.[87]

Indeed, Shemoel whilst still a young lad of 14 had been set up as a pedlar with stock and tray by the Jews of the Great Synagogue, London.[88] When illness left him penniless, the Jews of Exeter set him up again.[89] When he was robbed in Newcastle so that 'in the twinkling of an eye I was reduced from a gentleman to a beggar'[90] a Newcastle Jew gave him the money to get to Manchester, and there the Jews made a collection for him to set him up once again.[91]

Within a few months he had £25 worth of stock and some pounds in ready cash in his pocket again. In other words he was set up four times within a couple of years by various Jewish communities, who would no doubt have continued to do so, had not his spirit been broken by his ups and downs. In different circumstances but similarly motivated, the

Plymouth Hand-in-Hand Society, in line with the Chinese proverb, 'He who gives his neighbour a fish feeds him for the day, but he who gives him a fishing-rod feeds him for life', in 1879 made a grant of fifteen shillings to one Leon Isaacs to enable him to buy a sewing machine.[92]

When the United Jewish Hand-in-Hand Benevolent Society of Plymouth, Stonehouse, and Devonport died away yet another organization came into existence, this time called the Plymouth Hebrew Board of Guardians. It was founded in 1909, and once again it helped the Plymouth Congregation to supplement its charitable aid to resident and casual poor. Changed circumstances gave it two new purposes: to help obtain licences for pedlars and to speed emigrants in transit on their way to America.[93]

To complete the picture of Jewish charitable societies in operation in Plymouth, three ladies' societies may be mentioned. One was the Jewish Female Amicable Society of Plymouth. All that is known of this society is that it existed, and in 1843 gave to its Honorary Secretary—a man, Mr Reuben Abrams—a silver snuffbox in appreciation of his services to them.[94] Then there was the Ladies' Hebrew Benevolent Society, founded in 1830,[95] and still active in 1868, when Jacob Nathan left a bequest of £100 to it.[96] There are no extant records of its work, and it merged with the Plymouth (Jewish) Ladies Benevolent Society early in the 1930s.[97] This latter society had been founded by Mrs Asher Levy[98] on 20 July 1890, when she convened a meeting to re-form an old Ladies Philanthropic Society which had lapsed.[99] The particular purpose of the Ladies' Benevolent Aid Society was to give a small dowry to any poor Jewish bride.[100]

There is no evidence of disbursements by the Plymouth Jewish charitable societies to any non-Jews, and it is unlikely that there were any. On the other hand, Jews as individuals in the South West would have made their normal contribution as citizens to the Poor Rate,[101] as well as responding to requests for help from time to time. The author, when Rabbi of Plymouth, followed in his predecessors' footsteps and disbursed financial help on behalf of the Congregation to Gentiles as well as Jews who came to ask for it.[102] Then, too, there were particular cases when the Jewish community made special efforts for local charities. There was, for example, a 'benefit concert' type of service in the Exeter Synagogue on behalf of the Royal Devon and Exeter Hospital in 1815.[103] The connection between the Exeter Congregation and the Hospital goes back to at least 1797, and probably earlier, when the 'Warden of the Jews' was granted two votes, one in his private and one in his representative capacity, in the election of a new surgeon.[104] In Plymouth, William Woolf was elected Warden of the Poor in 1855 and continued to serve for the next decade, becoming vice-Governor of the Poor Law Guardians in 1864. Nor was he the only Jew among the town's Guardians, he had Josiah Solomons and Joseph Solomons for company during most of the time of his service,[105] and they were followed by Israel Roseman in Stonehouse in the 1870s.[106] They must assuredly have had some reputation for charitable work in the town to have been elected in the first place.

In Exeter, Alexander Alexander was elected a Guardian of the Exeter Poor Corporation from 1877 until his death in 1887, serving as President for one year in 1882.[107] Then there were the wide ranging bequests of £1,800 to 21 Plymouth and Exeter charitable or social institutions in the will of Jacob Nathan who was described as 'one of the most liberal benefactors of modern Plymouth'.[108] No doubt the epithet was earned by benefactions during his lifetime as well as those which were *post mortem*. In 1887, Harry Bischofswerder, proprietor of the Wheal Helena Mine,[109] built the Jubilee Hall for the use of the Penzance public where, for some years just before the turn of the century, he and his wife annually entertained 350 of the poor of Penzance to New Year's dinner. 'All who served were afterwards the guests of these kind and hospitable people'.[110]

One special category of poor person has always had a special niche in the charitable affections of diaspora Jewry—the Jewish poor of the Holy Land. The Penzance Congregation operated a Jerusalem Fund in 1830, which remitted a few pounds a year until 1865, when the Congregation was virtually defunct.[111] The first bequest in the will of an 1832 Plymouth cholera victim, Meyer Jacob Cohen, whose estate was under £100 was one guinea to be sent to the poor of Jerusalem.[112] An 1854 appeal on behalf of the Famishing Jews of the Holy Land, organized by Chief Rabbi Adler, elicited a substantial response of £51 7*s*. from the Jews of the South West.[113] Besides this special appeal, both the Plymouth Congregation and Leon Solomon in Dawlish made regular remittances to Adler, which he transferred to Palestine.[114] At least one emissary, a Rabbi Nissim from Jerusalem found himself stranded without the means to get back. Rabbi Adler gave him a letter of recommendation to Leon Solomon who presumably made it worthwhile for Nissim to travel from London to Dawlish to get help.[115] In 1875, the Plymouth Congregation must have written to Adler complaining it had not received a proper receipt for £14 18*s*. 9*d*. it had sent to the poor scholars of Jerusalem. Adler in turn wrote to the Rabbis administering the Fund and they wrote back[116] on Friday, 7 *Heshvan* 5635 (= 5 November 1875)[117] explaining that all such moneys were divided between the four holy cities (Jerusalem, Hebron, Safed, Tiberias) and that a receipt signed on behalf of any one of the cities was good for them all.

During the eighteenth and, to a greater degree, the nineteenth centuries the needs of groups with special disabilities were recognized to an increasing extent.[118] Gray, for example, considered that the London

Jewish Hospital for lying-in-women was 'the first model of one special form of hospital which was to play a considerable part in the social economy of the poor'.[119] There was never a sufficient number of Jews in the South West with special disabilities to make the foundation of hospitals or asylums for them a matter of necessity, but local Jews played their part in the financial support of such institutions when they were founded in London.[120] In 1860, for example, 40 Jews in Plymouth and their friends collected nearly £24 in response to an appeal for the Jews' Hospital, London.[121] Similarly, after World War II, the Plymouth Congregation supported the South Wales Jewish Home for the Aged in Cardiff. Elderly Jews from Plymouth in need of care were given preferential entry because of the special relationship.

Plymouth's proximity to the Dartmoor prison at Princeton provided the members of the Congregation with further opportunities to practise loving-kindness. At the end of 1814 there were at Dartmoor 2,340 American prisoners of war.[122] Among these were a number of Jewish soldiers and sailors. The most famous of them was Commodore Uriah P. Levy,[123] who fathered the law abolishing flogging in the U.S. Navy. Another Jewish sailor was one Captain Levi Charles Harby. Whilst this man was a prisoner, a Jewish baker came up daily from Plymouth to sell bread. One day a loaf was offered to Capt. Harby which he refused; the baker, however, insisted. Inside the loaf was a newspaper telling of the battle of New Orleans. This apparently encouraged Harby to escape (?with the help of the baker). He got back to his own navy and served with signal success.[124]

According to the late Dr Alexander Carlebach of Jerusalem, his great-great-great uncle, David Joel, settled in Exeter about 1825 and was a visitor of prisons—

> He once helped a Polish Jew to escape from prison by changing clothes with him, and the next morning the warders found him in the cell in the place of the prisoner.[125]

Many Jewish traders operated at the prison,[126] and it is reasonable to suppose that they helped their coreligionists in one way or another.

The next Jewish prisoners of war to occupy Dartmoor Prison, of whom anything is known, were those brought back from the Crimean War. They were Polish Jews and many were accompanied by wives and children. Revd Stadthagen regularly visited them, and provided spiritual and material comforts for them. In January 1855, Chief Rabbi Adler wrote to the Superintendent at Dartmoor Prison, probably prompted to do so by the Plymouth Congregation, regretting that the Jewish prisoners had to work on their Sabbath.[127] Abraham Joseph, a Plymothian who had moved to London, collected £11 in 1855 from the Jews of London and sent it off to Stadthagen, together with a donation of £6 from Sir Moses Montefiore. This sum enabled 'those captives in a strange land' to celebrate Passover.[128] The amount collected for one week's food points to a group of 30 or 40 Jews. Dr Adler sent further sums to Stadthagen for the benefit of the prisoners in 1855 and 1856.[129]

The Jewish connection with the Prison was maintained when it was used to incarcerate civilian prisoners. Chief Rabbi Adler wrote to various convicts in 1861 to help them with their private problems,[130] as well as to the wardens of the Plymouth Congregation urging them to arrange for religious comforts to be made available to Jewish prisoners.[131] At about the same time he wrote to the Governor of the Prison intimating that *matzot* for Jewish prisoners would be sent from the Great Synagogue, London, and that the Plymouth Congregation's Reader should be permitted to visit the Jewish convicts regularly.[132] Apparently, though, it was not the Governor who made difficulties for Revd Stadthagen to visit the Prison but the Plymouth Congregation.[133] So that when there was only one prisoner, Adler asked the Governor to transfer him to Portsmouth 'where it will be easier to attend to the convict's spiritual needs'.[134] Successive ministers of the Plymouth Congregation continued to visit Dartmoor Prison, as well as the Exeter and Channings Wood prisons, with the co-operation of the Visitation Committee of the United Synagogue, London, until the 1980s.

To sum up, it appears that itinerant poor Jews were helped on their way by the Congregations of the South West, whilst the local indigent and distressed were looked after either by the Congregations or their allied societies.

## CHAPTER NINE

# Jewish Inventors, Writers and Artists in the South West of England

For its size the Jewish community settled in Devon and Cornwall in the late eighteenth century and the first half of the nineteenth century produced a remarkable number of inventors, writers and artists. The late Cecil Roth has expressed the view that in no other area in England did the Jews produce such a flowering of cultural talent at this period.[1]

The inventors generally gave an account of their 'improvements' in their own literary productions. One leading figure was Lazarus Cohen (1763–1834),[2] a member of the Exeter Congregation, who invented an improved reaping machine and exhibited it at the Agricultural Society of Leeds in 1790.[3] Like many an inventor both before and after him, he poured his own money into it and tried to finance its development with

> subscriptions from the opulent of our people [i.e. Jews]. Some said it was more like charity than promoting the Arts and Sciences, others told me they were no farmers. Another wanted me to satisfy him who I was known to, as if any person applying his mind to improvements were likely to be of bad character. So I dropped it![4]

Cohen's solution to inventors' difficulties was that an Institution with a Fund should be founded which would 'enable inventors to carry out their projects, which would redound to the honour of the Jewish nation'.[5] In 1808, he published in Exeter an appeal to his brethren for religious loyalty entitled *Sacred Truths*, and in 1825, *A New System of Astronomy*.[6] This book deals with gravity and its influence on the planetary system and the tides, and the laws governing the winds. In the sixth part he gives interesting and sometimes novel translations of some of the key words used in the first chapter of Genesis.[7] He quotes medieval Jewish commentators such as Ibn Ezra and Yarchi and mentions Buxtorf and David Levy in support of his arguments.[8]

Another Exeter Jew with inventive and scientific interests was Alexander Alexander.[9] He was born in Sheerness in 1805 and came to Exeter in 1826, where he married the daughter of Moses Johnson, whose house and shop he occupied until his death in 1887.[10] In 1833, he was appointed optician to King William IV after dedicating to him his first publication. This he called:

> A treatise on the Nature of Vision, formation of the eye and the causes of imperfect vision. With rules for the application of artificial assistance and observations on the dangers arising from the use of improper glasses.[11]

According to his preface he was optician-in-ordinary to their Royal Highnesses the Duchess of Kent and the Princess Victoria. There were some 200 subscribers to the book including some 70 surgeons in Devon and Cornwall, the local Members of Parliament, clergy, local Society, and two Jews. Alexander invented a ventilating eye-shade for the Earl of Caernarvon who was one of his patients. He mentions it in a further treatise: *Observations on the Preservation of Sight and Hints to Spectacle Wearers*.[12] He also invented a Graphic Mirror.[13]

Another inventor was Israel Joseph Solomon, born in Falmouth about 1800. He invented and patented an improved magnesium flash to economize consumption of magnesium ribbon by interlacing it with a ribbon of zinc or tin, or by electro-deposition.[14]

Perhaps rather more academic were the literary and scientific brothers Henry and Solomon Joseph from Plymouth. After some fossil remains had been found at Oreston near Plymouth in 1859, Henry Joseph (1831–88) wrote an account of them[15] which attracted favourable notice.[16] His brother Solomon (1834–1900) wrote a shipboard diary of his journey to Australia in 1859 which gives much information about the Jewish middle-class emigrant in the second half of the eighteenth century.[17]

But perhaps the most distinguished scholar of Jewish background to come out of the South West was Orlando Haydon Bridgman Hyman. He was the son of Simon Hyman, a German Jew from Devonport, who married Mary Corse (or Cawrse)[18] of Nut Street, Plymouth, in St Andrew's Church, Plymouth on 7 January 1813.[19] They had two sons. The elder, called after his father Simon Hyman, entered the Navy as a midshipman in December 1829. He distinguished himself putting down a minor mutiny in 'a prompt and determined manner', but was killed by a snakebite in Madras Roads.[20] The younger was Orlando, born 14 April

1814. B.R. Haydon, who accepted responsibility for the two when he married their widowed mother, managed to get the boy through school though he was 'imprisoned twice and arrested once' for unpaid school bills.[21] Orlando won a scholarship to Wadham College, Oxford, when he was sixteen, and there suffered great privations 'living on bread and water when not invited out'.[22] He subsequently became Ireland Scholar, Fellow of Wadham from 1835–78, and he was later eulogized as 'offering in his task a type of scholarship which I had never been in contact with before'.[23]

Isaac Gompertz (1773–1836), one of the fifteen children of Barent and Miriam Gompertz who converted to Christianity and/or married out of the Jewish faith, was a minor poet. His poem, *Devon*, was published in Teignmouth in 1825 and an extract from it was incised on his tombstone.[24]

Two Jewish women writers who had lived part of their lives in Devon apparently were influenced by their residence there in both their lives and writings. One was Grace Aguilar who came to the county in 1828 when she was twelve years old with her father, on account of his bad health. She wrote popular expositions of Judaism for Jews, and also defended her faith against external attacks. Beth Zion Abrahams writes of Grace's 'early Devonshire perod' and continues:

> This change of scene had the most marked influence on Grace, bringing her into intimate association with scenes and people[25] that left an indelible impression on her thought and work . . . Deeply religious she was in Devonshire drawn to compare the teachings and spiritual content of her own faith with that of her neighbours.[26]

The other woman was Amy Levy (1861–89). She was a Londoner but went to Devon and Cornwall in 1886 for the sake of her health. In a sense this was a return to her family's English birthplace, as she was a descendant of Alexander Moses, founder of the Falmouth Congregation. Whilst in the South West she wrote sketches, poetry, and articles. Oscar Wilde called her 'that girl of genius' and she contributed to *Woman's World* when he was editor.[27] She eventually committed suicide in 1889.

Even a Jew in a comparatively humble trade could display surprising erudition and put up a spirited defence of his faith. Solomon Ezekiel,[28] born in Exeter in 1786, was in Falmouth about 1812–15,[29] and later moved to Penzance where he was a plumber until his death in 1867.[30] In the early part of the nineteenth century he dissuaded Sir Rose Price from establishing in Penzance a branch of the London Society for Promoting Christianity among the Jews.[31] He was the initiator of the Penzance

Hebrew Society for the Promotion of Religious Knowledge[32] and as the main lecturer was responsible for publishing tracts on the lives of Abraham and Isaac, and on the Jewish Festivals.[33]

The compiler of the first bibliography of Africana acquired his love of books from the secondhand bookshops of Exeter and Bristol. He was Sidney Mendelssohn, born on 31 December 1860,[34] in Exeter, where his father was minister, and who moved to Kimberley with the rest of his family in 1878. He took an intense interest in Africana and built up a magnificent collection which he bequeathed to the Union Parliament together with substantial sums for its maintenance and improvement. He wrote *The Jews of Africa* and *The Jews of Asia*, both published posthumously. His major work, *South African Bibliography*[35] has been described as a monumental work which 'reveals an historical and critical analysis that is not approached in any other Colonial bibliography . . . and is the foundation of a South African culture'.[36]

To complete this account of Jewish literary figures associated with the South West it may be added that the parents of Israel Zangwill, the best-known Anglo-Jewish writer of the nineteenth century, settled in Plymouth where his mother had relatives. Israel happened to be born in London (21 January 1864) whilst his parents were on a visit there but his elder sister and next youngest brother were all born in Plymouth. Eventually the family moved to Bristol, where Israel was educated, and then to London.[37]

There was a strong artistic tradition in Devon, and to a lesser extent in Cornwall, during the latter part of the eighteenth century, perhaps influenced by the success of Sir Joshua Reynolds. Like Sir Joshua, most of the better-known artists made their way to London to seek fame and fortune. The Jews of the South West played a notable part in this tradition, making a distinct contribution to the local and national artistic heritage.

The earliest-known Jewish artists in the West Country were the three sons of Nechaniah Daniel of Bridgwater, Somerset. They were Abraham, who worked as a miniature painter in Plymouth; Joseph, a miniature painter at Bath; and Phineas, miniature painter mainly at Bristol. All of them received instruction from their mother 'a very ingenious woman'.[38] A major problem in discussing the affairs of this family is that the three brothers all had establishments in both Bath and Exeter at various times, and frequently advertised themselves and signed their works simply as Daniel—apparently with a deliberate desire to cash in on one another's clients.[39]

Joseph made Bath his main centre, though his widow Mary, daughter of Alexander Wright,[40] and one son John were in Exeter in 1806, and three 'natural lawful and only other children', Courtney, Joseph, and Alexander were in Bristol in that year.[41]

Abraham practised principally at Plymouth, as a miniature painter, engraver, and jeweller. He first appeared in the Plymouth Congregation's books in 1779 when he promised one guinea (not paid according to a note at the side of the list of donors) to the War Levy.[42] It was in this year, 1779, that Abraham took Samuel Hart as a fourteen-year-old apprentice,[43] at which time he must have been in his middle or late twenties.[44] In 1788, Abraham paid £8 16*s*. 6*d*. to the Plymouth Congregation, it is not clear for what purpose.[45] Abraham is listed in the *Universal British Directory, 1798*,[46] as a miniature painter, as is Samuel Hart.

The only known signed work of Abraham Daniel is a portrait of Rabbi Moses Ephraim of Plymouth.[47] Alfred Rubens quotes the opinion of Graham Reynolds, a leading authority on miniatures, about miniatures which have hitherto been attributed to Abraham Daniel or Daniel of Bath:

> The miniatures which at present are attributed to Daniel of Bath have many clearly marked characteristics, though, as is usually the case, these are easier to recognize than describe. Both the high lights and the shadows on the face of his sitters are rendered with unusual breadth. In particular, the light on the bridge of the nose is broad, and the shadowed parts of the nostrils and the lips are emphatic and given sharp edges. The eyes are large and more open, and the shadow under the top eyelid again is well marked and prominent. The miniaturist draws the hair softly, in large masses and without much detail. There is not usually much work on the background. The miniatures have, compared with others of their time, an unusually glossy appearance, almost as if they were painted in thin oil, which of course they are not. I think this effect is to be attributed partly to the breadth of lighting and partly to the gum in the pigments.
>
> In a few of the miniatures so attributed (e.g. one in Mr. Alan Evans' collection)[48] there are signs of a slightly different technique: more stippling on the face, touches of opaque white on the costume and a more heavily painted background. It might be possible to apply these criteria to a hypothetical distinction between the styles of two men otherwise closely similar, and one might go on to attribute one style to Joseph and the other to Abraham Daniel. But this would be highly speculative in the present state of our knowledge; and the differences are not wider than may be seen at times in the work of the same miniaturist.[49]

Brian and other art dictionaries credit the Daniel miniatures to Abraham Daniel following S. Redgrave's *Dictionary of Artists* (1878). Redgrave probably heard of Abraham from S.A. Hart and attached to him some biographical data relating to Joseph. Hart had by then forgotten Joseph but it does seem from what he had written about him years before and from the obituary notices which appeared after Joseph's death that it was Joseph who was originally regarded as the more eminent of the brothers. The matter is not free from doubt and Basil Lond in his *British Miniaturists* (1929) mentions a miniature of Dr Harrington by Daniel signed 'A.D.' at the back but unfortunately this picture cannot be traced, although Major R.M.O. de la Hey, the then owner of the collection to which it belonged, made a careful search for it in the 1950s. Examples of the Daniel miniatures are not uncommon but until a signed work appears their authorship cannot be settled conclusively nor can one rule out the possibility that the work of the two brothers is indistinguishable.[50]

Abraham made his will on 10 March 1806,[51] and died the following day. He left £50 to Elizabeth Codbury, his mistress, the same sum to one of his 'natural' sons, Edward Elliot Thomas Daniel, £100 to his other 'natural' son, William Daniel, and £20 to the charity of the Jewish Synagogue, Plymouth. The estate was valued at £1,500 and the residue went to his sisters, Rachel who married Solomon Nathan of Plymouth (the mother of Jacob Nathan, the Plymouth Congregation's largest single benefactor), and Rebecca, wife of Isaac Alman of Bristol. He appointed Joseph Joseph and Samuel Hart to be his executors, and John Sweet, Gent., Thomas Williams, surgeon, Samuel Hart, Solomon Isaac, and Manley Hart, silversmiths, all testified that he was on the point of death on the tenth of March and unable to sign the codicil appointing the executors.

The Samuel Hart mentioned above was also an artist. His father, Henry Hart, had arrived in England as a young man of 21 from Bruck, Anspach and settled in Plymouth.[52] Henry was a well-established merchant figuring in the Plymouth Town Rental books from 1777 until 1806, i.e. until shortly before his death in 1808.[53]

Henry apprenticed his son to Abraham Daniel in 1779,[54] but Samuel was an unsuccessful character. He is said to have been frustrated as an artist and, having failed to qualify for a studentship at the Royal Academy, he returned to Plymouth.[55] There he became an unsuccessful merchant, always short of money but just managing to hold his head above water. In 1812, for example, he had not yet paid over the £10 legacy which his father had left the Congregation in 1808, nor £50 which his bachelor brother Menachem had left in 1809,[56] and he owed £25 13*s*. on his own account for unpaid seat rentals and offerings. So he deposited a £100 bond on the Plymouth Market which produced £5 per annum which was taken by the Congregation until all debts were paid.[57] When it became difficult for him to earn a living in Plymouth he moved to London about 1820 where his son Solomon Alexander was beginning to make a name for himself.[58] In London he joined a coterie of artists, of whom the most distinguished, though perhaps least able to cope with

*Illustration 17.*
Samuel Hart in Haydon's *The Mock Election.*

life, was B.R. Haydon. For Haydon to paint a good likeness he had to feel an interest in the character of his sitter.[59] In 1827, Haydon was painting 'The Mock Election'[60] and needed

> a model for the official who swears in the member. He bethought him of Hart Senior, a man of peculiar ugliness, and wrote to his son, Solomon,[61] who was a friend of his to ask if his father would sit. Solomon who was devoted to his father and very sensitive about his appearance refused indignantly. Haydon wrote a long repentant letter and asked old Hart[62] to come to breakfast as a proof of forgiveness. The old man went with heart overflowing. But no sooner had he started than the son felt an uneasy suspicion. He argued that such a betrayal was impossible, yet his uneasiness could not be allayed, and he set out for the prison,[63] where he discovered Haydon just finishing a wonderful likeness of his father swearing in a dandy on a piece of burnt sugar stick.[64]

It has been said that Haydon was married to a Jewess, Mary Hyman,[65] but she was no Jewess, having been baptized at St Neot where she was born on 2 June 1793.[66] The error appears to have emanated from Mary Russel Mitford, who wrote: 'Poor Haydon's wife was a most beautiful woman, just like the Rebecca of "Ivanhoe", a Jewess born and by her first marriage'.[67] Miss Mitford was much taken with the Jews. To one correspondent she wrote: 'Do you know much of the Jews? I have always been interested in the whole race, and my friend, Miss Goldsmith . . .'.[68] To some extent the Jew, Simon Hyman, financed Haydon, as he left £52 10*s*. per annum, the interest of £1,000, to Mary.[69] Samuel Hart had the satisfaction of seeing his son become a fashionable society painter and when he died at the end of 1838, he was probably aware that his devoted son was soon to be elected an RA. For an unsuccessful man, Samuel Hart had perhaps done not so badly, after all.

Before turning to the career of Solomon Hart, which flourished in the second quarter of the nineteenth century, it is as well to look at the Jewish artists of Exeter and to a lesser extent Cornwall, who helped, together with the Daniel brothers and Samuel Hart, to lay the foundation of his success.

One of the first Jews to settle in Exeter was Abraham Ezekiel, who arrived there some time about 1745 and became a successful and respected figure in the town.[70] He was a goldsmith but no work of his is known. He and his wife Sarah[71] had six children, including two sons, Henry Ezekiel[72] and Ezekiel Abraham Ezekiel. The last named was an outstanding man. He was a versatile artist, and not only engraved portraits and *ex-libris* but also painted miniatures, which he successfully took up at the age of 40,[73] whilst continuing the trade of engraver as well as that of silversmith and scientific optician.[74] An advertisement in the *Exeter Flying Post* in 1784 gives a good indication of the wide range of his artistic activities and commercial ability.[75] He took the opportunity to remind the public: 'N.B. The large Perspective view of Bideford, engraved by him from a Drawing by Mr. Jewell, to be had, Price Five Shillings . . .'.[76]

The following account of his artistic output is given by Alfred Rubens:[77]

> Britten records a watch by him dated 1794.[78] While apprenticed to a Jeweller, he produced, self-taught, an etching "View of Bideford" from a drawing by Jewell.[79] The British Museum has four examples of his work—a portrait of Micaijah Towgood (1700–92), dissenting minister of Exeter, engraved in line after Opie,[80] and published in 1787; a stipple engraving of the same portrait published in 1794; a portrait of John Patch, surgeon at Exeter, engraved in line and stipple after Opie, and published in 1789, and a stipple engraving of the portrait of General Stringer Lawrence (1697–1775) by Sir Joshua Reynolds, published in 1795.[81] The Exeter City Library possesses copies of his engravings of a portrait of Thomas Glass, physician at Exeter, published 1788; a portrait of William Holwell and another of Rev. John Marshall, schoolmaster at Exeter, after Keenan, published in 1798 on which he is described as 'engraver, optician and goldsmith. Another engraving by him is entitled 'The Breastplate of the 3rd Exeter Volunteer Corps embodied in 1800'. Fincham records fourteen *ex-libris* signed by Ezekiel. A miniature painting attributed to Ezekiel by the late Basil Long is in the Royal Albert Memorial Museum, Exeter.

When Ezekiel died of dropsy after a long illness on 13 December 1806, the *Exeter Flying Post* printed the following obituary:

> On Saturday last died aged 48, Mr. E.A. Ezekiel, of this city engraver and jeweller. He had long lingered under the complaint of dropsy, and contemplated dissolution with a most religious resignation. He was followed to the grave by many respectable persons, who have for several years past enjoyed the pleasure of his agreeable conversation and the attachment of his unshaken friendship. In the profession of an engraver he possessed a correct taste, and happy facility in making designs to meet the ideas of his employers, and as a workman, he was certainly unequalled out of London. His portraits of several distinguished characters in this City and neighbourhood will always be admired for their faithful execution; they never fail to excite the reward due to his merit, while they renew the presence of the person

> whose likeness he represented with great correctness. In a word there are few men whose loss will be more felt, not only by his immediate friends and connections but by the public at large. A discourse was delivered at the grave, previous to interment, by the chief Priest of the synagogue; who truly and affectingly held up the deceased as a pattern for imitation, both as a good son and brother, a good man and a citizen of the world.[82]

E.A. Ezekiel was regarded as a respectable scholar and linguist. Even as late as 1830, the *Exeter Journal* included his name among 'Persons of Eminence, Genius and Public Notoriety, Natives of Exeter'. A miniature portrait of him was in the Anglo-Jewish Historical Exhibition, 1887, but it has apparently disappeared.[83] His sisters carried on the business for some years but without Ezekiel's flair it faded away and Catherine had to be assisted to emigrate to her relatives in Cincinnati.[84]

Another Exeter Jew, though one also with Portsmouth connections, who engaged in artistic work was Moses Mordecai. He has no less than twnty *ex-libris* to his credit besides two in which I. Levi of Portsmouth collaborated.[85] He worked about the middle of the eighteenth century, and for part of the time at least, he was in London. His trade card, a charming little plate which calls special attention to his skill as an heraldic engraver, reads:

> Engraving
> in Seals, Stamps,
> Plate, Copper Plates
> and Pewter
> by M. Mordecai
> No. 55
> Houndsditch near
> Bishopsgate Street
> Arms neatly painted on Vellum

Mordecai also worked as a goldsmith in Exeter and entered his mark at the Exeter Assay Office in 1788.[86] He is recorded in the *Exeter Directory*, 1792, as one of the 'Principal Traders' of that city,[87] his jeweller's shop being in Fore Street. In 1803, acting on behalf of the small local community, he took up the lease of the Jewish cemetery in Magdalen Street renewing it in 1807.[88] A family pedigree dated 1799, illustrated with signs of the Zodiac written and painted by him, was exhibited at the Anglo-Jewish Historical Exhibition.[89] Moses Mordecai died in 1809 and left a most interesting will which was referred to earlier.[90]

There was a minor Jewish artist in Falmouth in the first quarter of the eighteenth century, Simon Solomon, son of Bella and Israel Solomon, an uncle of Joseph Israel Solomon. In Israel Solomon's *Records of my Family* he describes his uncle as

> by trade a painter, who possessed artistic qualities which could not expand at that time in the town of Falmouth. His lifelike panel paintings of fish, and also a painting for a large round table, the subject of which was from the History of Joseph and his brethren, and his transparencies when national illuminations took place, were the admiration of the inhabitants of Falmouth.[91]

He was married to Kitty Solomon, a granddaughter of the founder of the Falmouth Congregation. He was a sickly man and died in 1825 'much respected and lamented'.[92]

The most distinguished Jewish painter to come out of the South West was Solomon Alexander Hart, who worked chiefly during the Victorian era.[93] He was born in Plymouth in 1806, sent to a school in Exeter for a year when he was seven years old, but returned home to a school run by a Unitarian minister.[94]

Solomon Hart wanted to be apprenticed to an engraver, but his impoverished father could not afford the apprentice fee. Another engraver offered to take him 'on these hard terms: seven years service from seven in the morning to seven in the evening without any premium. This being a case of cruelty to animals[95] . . . I went to the British Museum to study in 1821'.[96] Alfred Rubens gives the following account of his artistic career:[97]

> At the age of fifteen he was studying at the British Museum, and was admitted as a student at the Royal Academy in 1823. He was only able to pursue his studies in the evenings, as during the day he earned his livelihood by colouring old prints and making copies of Old Masters in miniature on ivory. At this period he also painted some miniature portraits. In 1826, at the early age of twenty, he exhibited at the Royal Academy a portrait of his father, "Mr. Samuel Hart, Professor of the Hebrew Language", and thereafter he exhibited frequently, including "Study of the late Samuel Hart" (1839); "Israelites" (1840); "Scene in a Polish Synagogue" (1841); "Duke of Sussex" (1841); "Barroe Helbert Ellis, Esq. of the Hon. East India Co's. Civil Service" (1844); "Simchat Torah, Leghorn Synagogue" (1850); "Alderman Salomons, M.P." (1852); "Sir Anthony de Rothschild for the Committee Room of the Jews' Hospital"; "The Rt. Hon. David Salomons, Lord Mayor of London" (1856); "The Rev. Dr. Adler, Chief Rabbi, painted for the Vestry Room of the Great Synagogue" (1857); "Rev. A. L. Green" (1858); "The Eve of the Sabbath" (1868); "Sir Moses Montefiore, to be

placed in Town Hall of Ramsgate" (1869); "Proposal of the Jews to Ferdinand and Isabella" (1870); "Menassah ben Israel before Oliver Cromwell" (1872).

Hart also did some work as an engraver, and with his father's assistance, was able to instruct his brother, Marx, in the rudiments of wood engraving. In 1830 he exhibited at the Society of British Artists (Suffolk Street) his "Polish Synagogue; Elevation of the Law", which was bought by Robert Vernon for £70, and subsequently bequeathed to the National Gallery. Success now was rapid; Lady Montefiore paid 150 guineas for one of his pictures, and in 1839 he was elected a full member of the Royal Academy.

At this stage in his career Hart, like many another, lost the friendship of B.R. Haydon. Hart generously acknowledged his debt to Haydon having 'received much good advice from him' but it was virtually impossible for any successful man to remain on friendly terms with Haydon.[98] Rubens continues the story of Hart's career after his election as a full member of the Royal Academy:

About 1840, the Duke of Sussex selected Hart to paint his portrait for the Jews' Hospital. Hart attended on the Duke at Kensington Palace, and was surprised to find that the Duke knew all about him. 'You forget', said the Duke, 'the peculiarity which distinguishes my family. We collect a quantity of information and facts concerning persons and their affairs which we never forget. I know when you lived in Newcastle Street, Strand, over the milk shop where you struggled all day to get bread for certain members of your family whom you supported, and when you could only afford time in the evenings to pursue your studies at the R.A.' The sittings were protracted, as there were continual interruptions. The Duchess would call the Duke away for a game of billiards; the Duke of Cambridge would call to examine the picture, and remark, "Very like, very like", and the sitter made matters difficult by smoking continuously. Hart was impressed by his enormous stock of tobacco, cigars and pipes. However, the picture was eventually finished, and the Duchess congratulated Hart on it and commissioned him to paint a copy of the head. She said that she thought he had a difficult subject in a corpulent man, but had avoided coarseness and had made him look like a gentleman. Hart subsequently painted a portrait of Rabbi Isaac Levi, which was presented to the Rabbi by the Duke of Sussex. Between 1854 and 1863 Hart was professor of painting at the Royal Academy, and, in 1864, he was appointed librarian.

He died a bachelor at 36 Fitzroy Square, London, on 11 June 1881, and left the bulk of his fortune to his sister-in-law, Mrs Margaret Hart of Baltimore and then to his nephews and nieces who all lived in America.[99] But he did not forget the synagogue in the town from which he had departed 66 years earlier to seek his fortune—he left the interest on £1,000 to the Plymouth Congregation.[100]

As might have been expected in the light of the intense artistic tradition in the Jewish communities of the South West in the latter part of the eighteenth and the early nineteenth centuries the Plymouth and Exeter Congregations possess some beautiful ritual silver.[101]

The Exeter synagogue possesses four silver pointers, two of which are very elegant. One was given to the Congregation in 1812 by *Simcha Isaac ben Zvi SGL* and has the London silver mark of 1810–11, whilst the other is hallmarked with a crown over a 'V' and with five castles. Of the other two, one has an Exeter hallmark with the maker's initials 'G.F.' and the second has no marks at all. It also owns a small but aesthetically designed breastplate with a London silver mark of 1869–70, which was given by Solomon Elsner in memory of his first wife, Rosina, who died in 1861.[102] The crowning glory of the silver used to decorate a Scroll of the Torah is the pair of bells which often surmount it. The Exeter Congregation has two sets of such bells, both most beautiful. One pair was given by seventeen ladies of the Congregation in 1821,[103] the other was acquired in 1813.[104] It may also be mentioned that an Exeter silversmith, probably Jewish, was responsible for a fine pointer inscribed '*Israel ben Naphtali Hirsch*, Truro, 1836', which came from Falmouth and is now in the Jewish Museum, London.

The Plymouth Congregation also has a fine collection of ritual silver. It has a number of pointers including:

(1) A solid piece, ten inches long, with the London Assay mark of 1745. It belonged to Joseph Sherrenbeck.[105] The ring at the top is a watch bow, hallmarked Exeter 1803 with maker's initials 'S.L.'[106]

(2) A silver pointer, 9 inches long and very elegant. It belonged to Abraham Joseph in 1765 and is not hallmarked.

(3) A pointer, 11 inches long with a ball at the top and middle. No hallmarks. Some simple shapes, stars, circles, commas, trefoils, are cut out of it, giving it an oriental appearance. The inscription implies that it was made by Judah the son of Abraham Ralph of Barnstaple, in 1782.

(4) A pointer with the same names which are inscribed on the silver shield of 1784.[107]

(5) A very fine pointer 11 inches long with the London Assay mark of 1813–1814, with the maker's initials 'S.A.' There is a well written inscription which, translated, reads, 'The gift of Samuel Hart, the pericope of "Get thee out", 1814'.[108]

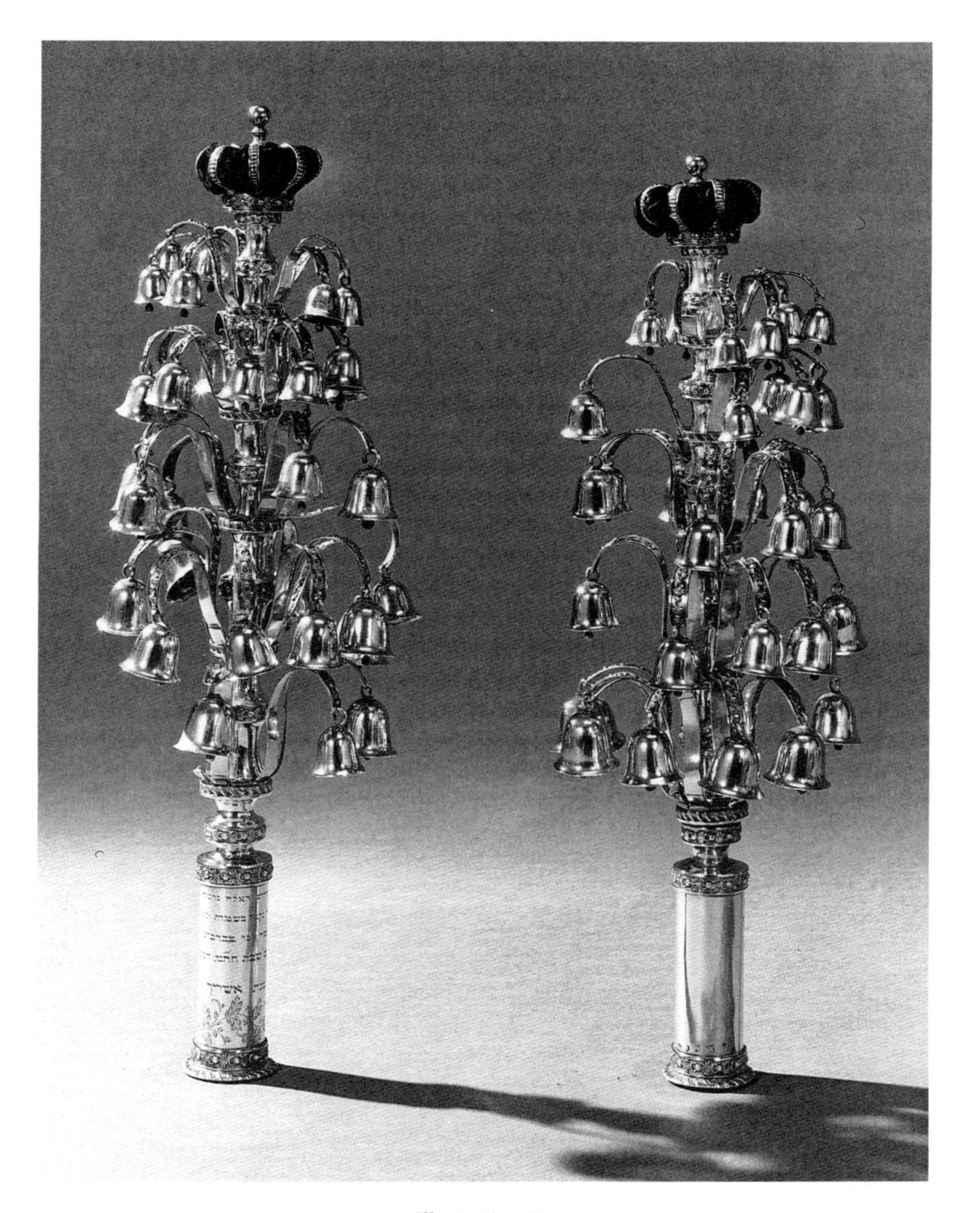

*Illustration 18*
Pair of silver Torah bells, Exeter, 1813.

A fine pair of eighteenth-century bells are currently in use in the Plymouth synagogue. They have a London hallmark of 1783–4 with the maker's initials 'I.R.' There are no inscriptions on them. The Congregation has no breastplates earlier than the late nineteenth century, though one of these is an interesting piece, apparently cut out from masonic regalia. Amongst its other pieces of ritual silver are a *kiddush* cup[109] given by Joseph Joseph in 1775 when he was only nine years old, and hallmarked London 1755–6; a beautiful spice box in trefoil shape with exceptionally fine filigree work, given by Mrs B. Moss to the Dock *Minyan* Room about 1825 (the date has been obliterated) and when that closed down, handed over to the Plymouth Synagogue in 1844; and a magnificent ewer and basin for the priests to wash their hands before the ceremony of *duchaning* (blessing the congregation at Festival day services). These were given by Mrs Levi Barent Cohen and her sons (who were priests) in 1807 to the Congregation in appreciation to it for burying her son and their brother who died in Madeira and whose body was brought back to Plymouth. In such circumstances, corpses were preserved in a cask of brandy. In this case, the body could have been shipped on to London but that would have entailed further delay in the burial, and Jewish law requires burial as speedily as possible.

## CHAPTER TEN

# The Acculturation and Assimilation of the Jews of Devon and Cornwall

The immigrant Jew, whether he was from Central Europe in the eighteenth century or from Eastern Europe in the nineteenth century, when he first arrived in the South West, was ethno-centric.[1] That is to say, he had a favourable evaluation of his own group's system culture and values which set him apart from the English society into which he arrived. It appears, for there is no evidence to the contrary, that in every case a process of acculturation took place.[2] In other words, the immigrant began to assume in many respects the cultures and the values of his host society but nevertheless retained his identity as a Jew. In some or many cases, it is not possible to quantify,[3] the changes brought about by the process of acculturation led to assimilation, the disappearance of ethnic identity, so that the Jew and his or her descendants became totally absorbed and submerged into Gentile society. In this chapter an attempt will be made to describe the way in which a minority, in this case the Jews of the South West, by a breakdown in its distinctiveness became incorporated in the course of time into the system of social relations which constituted the wider society around it. In order to do so successfully it will be necessary to study the change in cultural characteristics of the Jewish minority in response to those of the surrounding majority, in other words the process of acculturation and assimilation.

What were the characteristics which made the immigrant Jew in the eighteenth and nineteenth centuries stand out amid his 'English' contemporaries?[4]

The men were identifiable by their hirsute appearance, the married women by their covered hair; the male newcomers stood out on account of their foreign or typically Jewish-style dress; immigrant Jews could be recognized by their broken English, and also by their typically Jewish names; the Jews tended to engage in a limited range of occupations; and finally, and most basic, though not always as immediately apparent as some of the other characteristics just mentioned, there were the Jews' religious observances, particularly the Sabbath, the dietary laws, and synagogue attendance. These distinguishing characteristics and how they were gradually softened, particularly by the children and grandchildren of the immigrants, until they all but disappeared, as well as the process of acculturation, will now be considered in detail.

When Jewish men first arrived in England from Germany or eastern Europe in the eighteenth and early nineteenth centuries they were easily identifiable by their hirsute appearance. The biblical command, 'Ye shall not round the corners of your heads, neither shalt thou mar the corners of thy beard'[5] was commonly interpreted in Jewish circles to mean that beards were to be left untrimmed and sidelocks were left to grow in a characteristic style.

One of the first stages on the road to assimilation was to attend to the hair. Sidelocks were abandoned; beards, when worn, were trimmed according to the taste of the day—more often than not they were shaved off entirely. The portraits of Jewish males in the South West in the early part of the nineteenth century portray clean-shaven men. Portraits of Abraham Joseph I (died 1794),[6] even the very learned Rabbi and Philosopher Moses Ephraim (died 1815),[7] J. Abraham of Bath and Cheltenham in 1829[8] (formerly of Exeter),[9] Henry Ezekiel of Exeter (died 1831),[10] the sons of Betsy Levy of Totnes,[11] Jacob Solomon (left Exeter 1830),[12] and Moses Solomon (died Plymouth, 1838),[13] all picture clean-shaven men dressed in the height of fashion. Indeed, were it not for the provenance of the portraits it would hardly be possible to identify the sitters as Jews. Shaving off the beard was a deliberate act, and at least one Polish Jew from Cornwall, Phillip Samuel, after removing his for fear it would arouse prejudice against him, afterwards regretted the act as a sign of religious weakness.[14]

The same applied later in the century when it again became customary for men to wear beards. Once again the Jewish males dressed fashionably and trimmed their full beards, in the style of the times. There is a delightful pen portrait of a Jew educated beyond his ability, and consequently a failure in life, who had been supported by his family. He was Elias, brother-in-law of Alexander Moses of Falmouth, born in London but settled in Falmouth. About 1824

> Elias was a well instructed Englishman, and very free in speech on political subjects. I remember him; his dress was then ample, the coat in Louis XIV fashion, waistcoat the same, breeches with buckles fastening at the knee, long wool stockings with shoes, and heavy,

> large white metal buckles. His hair, with a long quantity behind the neck, tied with a large black ribbon in a knot, and white necktie, very ample, folding around the throat and half covering up the chin.[15]

Another distinguishing mark was the wig covering the Jewess' hair. According to Jewish law, unmarried girls may display their hair but after marriage this is not allowed. In oriental countries Jewesses covered their heads with a kerchief, but in eastern Europe they often wore wigs, known in Yiddish as a *sheitel*, a name often used pejoratively, either with or without an additional covering.[16]

There is strong evidence that married Jewish women in the South West, or at least the more wealthy who could afford to have their portraits painted, abandoned the wig and wore their natural hair. More than twenty extant portraits of eighteenth- and nineteenth-century married Jewesses in Devon and Cornwall depict them with uncovered hair. Only of Betsy Jacobs of Totnes (1759–1836) is there a portrait which shows her wearing a lace cap completely covering her hair.[17] Moreover, only one reference to a Plymouth Jewess exchanging her hair for a wig on marriage has been noted, and that reference implies that to do so was exceptional. The allusion is in a postscript to a letter from Abraham Joseph II to his aunt. He was about to wed his first cousin Rosa and wrote to her mother in 1856: 'I think if the fashionable ladies were to see [how] well my dear Rosa looks in the new headress [they] would all lose their old ones.'[18]

Coupled with his unkempt appearance, the Jewish male newcomer was also clearly recognizable by the foreign style of his clothes. Shemoel Hirsch described the clothes he wore when he arrived at Gravesend in 1821 as a 14-year-old lad:

> A pair of German boots to the knees and a large tassel to each of the tops, a pair of small clothes, black silk waistcoat, lead nankeen longcoat down to my heels any laps or opening behind, and a hat a small crown and wide brim.

Although he had no luggage with him, and at that age could hardly have been adorned with a beard, his appearance was sufficiently distinctive to attract a crowd of 'hundreds of boys' who followed him, mocking and jeering, 'as I went to Duke's Place, [City of London] where the Jews are to be found'.[19]

In 1825 Wolfe Moses, 'a young Jew of foreign extraction' was persecuted by a gang of boys who had gathered to watch the launch of a ship. An ironmonger

> bore testimony to the outrageous conduct of the boys to the friendless Jew. One lad threw him in the water, knocked him down and ill-

treated him, so that in desperation he drew a penknife and wounded Hitchcock in the thigh.[20]

Most Jewish men from eastern Europe wore a *Kaftan,* a long flowing coat once part of the national Polish costume, but from the eighteenth century restricted to Jewish use, possibly because of some natural tendency amongst Jews, particularly *Hasidim,* to conservatism in dress. Some kept their style of dress because their livelihood was to some extent dependent on a Jewish appearance. Thus it was obviously to the advantage of the old clothes man, a predominantly Jewish occupation in the eighteenth century, to look the part. He walked the streets with a pile of hats perched upon his head and capacious garments hung from his shoulders, over which a bag was slung. Equally recognizable was the Jewish pedlar with his broad-brimmed hat and pedlar's tray with its assorted bric-a-brac.[21]

Contemporary prints all portray a similar type of appearance and it should not be thought that this was merely a caricature. That the type was familiar in Devon is clear from an account by a French privateer captain who was imprisoned at Millbay Prison, Plymouth, in 1807. He escaped from the prison 'disguised as an old Jew man with his bag over his shoulder.[22] The disguise was too successful: a young lad called Corbière, dressed as a girl, also escaped with him and together they went to Plymouth Dock Theatre. Some American sailors seeing what they thought was a nice looking girl in the company of a bearded old Jew considered them, or her, fair game, a fight broke out and it was soon discovered that the girl was no girl, and the Jew no Jew![23]

As soon as the immigrant Jew's clothes needed replacing, he would be involved in the process of anglicization, because his new clothes would almost certainly be of the English pattern. The outfit acquired for a young man, the nephew of a comfortable Jewish merchant in Exeter, who arrived from Leghorn in 1732 is listed in Ottelenghe's *An Answer*:

| | £ | *s.* | *d.* |
|---|---|---|---|
| To a Blue Coat | 2 | 0 | 0 |
| To a Hat, Wigg, Silver Snuff-Box and shoes and stockings | 1 | 10 | 0 |
| To a Broad Cloth Sute | 6 | 0 | 0 |
| To a new Hat, Wigg, Shoes, Stockings | 2 | 10 | 0 |
| To 3 new Shirts and 2 old ones | 1 | 10 | 0 |
| To a pair of Plus Breeches, and a Camblet Waistcoat | 1 | 0 | 0 |

He must have looked very elegant![24]

The portraits of Solomon Joseph (left Plymouth 1859),[25] Alderman Eleazer Emdon (1841–1900),[26] Henry Joseph (died 1888),[27] Solomon Solomon (born 1833),[28] and Abraham Joseph II (died 1868),[29] all depict benign-looking patriarchal figures dressed in the typical clothes of their period. So much so that without foreknowledge of the name of the person portrayed it would be very difficult to identify the sitter as either Gentile or Jew. A very good illustration of this difficulty is a photograph published in *Devonia* which was taken on the Barbican, Plymouth, about 1840.[30] It shows some 22 Plymouth notables including 'Mr. Cornbloom great frequenter of the Barbican, a Jew' yet it is not possible to pick out Mr Cornbloom. By this time, however, Mr Cornbloom was already a well assimilated and non-observant Jew to the extent that the billiard hall he owned was open, at least from 1829, on the Jewish Sabbath,[31] and the Plymouth Barbican was then 'the usual rendezvous for Plymouth men after church service on Sunday mornings'.[32]

Before leaving the topic of dress, it may be added that a glance at the portraits of the Jewesses living in Devon and Cornwall in the eighteenth century is enough to see that they wore the fashionable dress of their time.[33] Some of the smaller pieces of textile used in the Plymouth Synagogue emanating from the early nineteenth century may well have been the little iron aprons or dresses of wealthy Jewish women, and donated by them for use in the Synagogue.[34]

Even after some years had passed and the immigrant was clean-shaven and dressed like everyone else, his foreign origin was betrayed as soon as he began to speak. Often he pronounced 'w' as 'v', 'b' as 'p', and 'th' as 'd', as the following lines of the old clothes man indicate:

> She sold me some pargains and gave me some meat,
> Vich though it was *trypha* [treifah] I couldn't but eat,
> Den to give her a kiss, dears, I thought it no sin . . .[35]

Manasseh Lopes, according to scurrilous handbills which circulated in Plymouth in 1805, is supposed to have spoken in the same way:

> One day says his wife to her dear little Mosey,
> 'I think for some time quite dead looks your nosey!'[36]
> Says Moses, 'Vy sure the vomans not jealous—
> Pray haven't you had quite enough of your fellows!'
> 'My dear', says his wife, 'I humbly beg pardon'.
> Says he, 'You forget ven you valk'd in the GARDEN
> . . .[37]

Further evidence on the way in which an immigrant spoke, this time a Moroccan Jew, is to be found in the rather charming account of his cross-country peddling activities recounted by a Turkish rhubarb seller to Mayhew about 1820:

> When I go across de countree of England, I never live in no lodging houses—always in de public, because you see I do business dere; de missus perhaps dere buy my spices of me. I lodge once in Taunton, at a house where a woman keep a lodging house for de Jewish people wat go about wid de gold tings—'Jewellery'.[38]

There was not a great deal which the immigrant Jew could do about his foreign accent, but the home born Jew undoubtedly spoke in the same way as the Gentile members of his social class. Even a not so well-educated Jew, born in England, seems to have spoken a normal English, as the following letter with its phonetic spelling indicates:

> Mr. Alexander Samuels
>
> Sir,
>
> On the 3rd of August last I wrote to the former officers of the Congeratation of my Circumstances and that I was sorry to inform them that I stood in the gratest of want of beaing assisted by them, to which I never had any reply—Mr Harris on the last Holladays Handed my wife a half Sovering Stating that Mr. N. Joseph Sent that—I no not if that Sum was from is [= his] privite Purse: or if it was from the Congaratations money—be that as it will I now have to Inform you beaing on [= one] of the helders that I am Drove to the last Extramity, without a farthin in the world having disposed of Everything I could make money of, so as my wife and Sevon Children Should not starve. I am now forced to apply to you to me call [= to meet with *Kohol*, the governing body of the Congregation] and Congaratation in my behalf so as to Grant Somthing In moderatation so as to help my family—and to send me a trifle before the meet [= they meet; or, the meeting] for I can assure you I have not a thing to fetch me a farthin.
>
> Your Ansur as Soon as Posable will oblige
>
> Your obdt and Humble S$^{t.}$
> Aron Nathan
> 17 Pearl Street,
> Stonehouse.
> Thursday December 1827.[39]

The following letter of application by Abraham Franco to become a member of the Plymouth Congregation in 1829 with its phonetic spelling adds weight to the argument that even semi-literate but English-born Jews spoke the normal English of their class:

> To Mr M. Mordecai, Chief Elder and Gentelmen of the Hebrew Congregation of Plymouth.
>
> With Humble respect I take the Liberty to introduce a Petition hope you would not think it a trouble of tacking it in your kind consideration. I am strongly persuaded that you would not deny a humble Prayer on the same subject as I onws [= once] tocke the Liberty to a former Congregation of this Town.
>
> It is I my wife and Familie Particular wish of becoming Members to the Hebrew Society as I formerly enjoyed with my for Fathers. I hope Gentl: you would not deny me that Blessing which I and my Familie for so many years past have been deprivt of. Gentl: I tacke the liberty to inform you that a Gentl: in this Town Promist me His kind assistans with your kind Approbation and assistans.
>
> I remain Gentl:
>
> Your Humble Servant
> A. Franco[40]

Of course, even a well-spoken Jew might adapt his speech to suit his needs. Samuel Coleridge met his match when he remonstrated with a Jew for crying 'Ogh Clo':

> The Jew stopped, and looking very gravely at me, said in a clear and even fine accent, 'Sir, I can say "old clothes" as well as you can: but if you had to say so ten times a minute, for an hour together you would say *Ogh Clo* as I do now', and so he marched off. I was so confounded with the justice of his retort, that I followed and gave him a shilling, the only one I had.[41]

Further examples of some weird phonetic spellings could be multiplied from the Minutes of the Exeter Congregation which were kept in English over the period 1823 to 1869, though, it may be noted, 'th' is never represented by 'd' in these minutes.

The adult immigrant generally wrote, and presumably spoke Yiddish, his mother tongue, when communicating with his fellow Jews. Sooner or later the vast majority must have picked up a smattering of English words, at least sufficient to converse with customers on prices, names of articles, and to use the everyday nouns and verbs of kitchen English. Nonetheless, the immigrants felt more at ease using Yiddish, particularly when transacting synagogal affairs.

It was probably a feeling that it was somehow more 'proper' to use Yiddish which prompted Moses Mordecai of Exeter to write his will in Yiddish in 1808,[42] and Moses Jacob of Redruth to sign his will with his Hebrew name in 1807.[43] Samuel Hart, born in Plymouth in 1755, penned a letter from London in Hebrew (and not the perhaps more natural Yiddish) to the Plymouth Congregation's officers in 1824.[44] Perhaps he used Hebrew because he wanted to impress his readers both with his knowledge of Hebrew as well as his beautiful calligraphy, as he was trying to earn a living as a Hebrew teacher at this period.[45] The minutes of the Plymouth Congregation continued to be written in Yiddish until 1834, by which time more than half of the adult males of the Congregation had been born in England, and if the minutes were written in Yiddish then it may be supposed that the proceedings were to some extent conducted in that language, or were intelligible to most of those who attended the meeting.[46]

Nonetheless, by the turn of the first quarter of the nineteenth century there was a strong movement towards anglicizing the conduct of Congregational affairs in Exeter and Plymouth, and probably in Falmouth and Penzance. This was accomplished by dropping Yiddish and using English instead for the rules, minutes, and other official documents. The preamble to the Exeter Congregation's Regulations in 1823 succinctly makes the point:

> The vestry members of the Congregation of Exeter taking into their consideration the state of their regulations are unanimously of opinion they require translating into English . . .

Obviously, Yiddish had ceased to be the daily language of most vestry members of the Exeter Congregation by 1823, even though this cannot be shown for certain until 1838, when the spare pages after the regulations were utilized as a 'Waste'[47] Minute Book. The record was made during the course of the Congregational meetings, and was made in English, indicating that the proceedings probably took place in English.[48]

It does not occasion any surprise to find the Exeter Jews anxious to translate their regulations into English in 1823, as the more important members of the Congregation including the three chosen to do the translation were all Exeter born. The three translators were Henry Ezekiel, Eleazar Lazarus and Simon Levy, born in 1773, 1789 and 1791 respectively.[49] Henry Ezekiel's familiarity with English and capacity to express himself may be gauged from the following dedicatory preface to a Hebrew Almanack which he intended to publish in 1817:

> Revd and Learned Sir,
>
> In dedicating these few pages to you I trust I shall not be thought presumptuous, my only motive being that this book might be found useful particularly to those of our Community that reside and travel beyond Seas and who cannot afford to purchase an annual *Luach* [Calendar]. It was by chance I met with this little Book printed entirely in English, entitled *A Portuguese Jew's Calendar* . . .[50]

By 1838, it would appear that some, if not most, of the members of the Exeter Congregation were unable to read Yiddish or Hebrew easily. Writing to the Chief Rabbi in 1838, M.L. Green and A. Alexander added a postscript:

> It will particularly oblige our Congregation for your answer to be sent us in English.[51]

Anglicization of the Plymouth Jews' speech took place about the same time as in Exeter. It is noteworthy that Aaron Nathan wrote in English to the officers of the Congregation to ask for help in 1827.[52] Similarly, an H. Ralph writes in 1824 in English asking for the use of the Congregation's house, and in good English at that:

> Sir
>
> Understanding that the House at present Occupied by Mrs Myers is about to be vacated and that the meeting respecting the disposal of it to another person will be held on the coming Sunday I have presumed to offer myself as a person to fill that situation when your interest on my behalf will always be considered an everlasting favour. My circumstances in life being well known no one would doubt my being an eligible person to fill the place hoping you will use your best influence for me on that day.
>
> I remain your
> most obt servant
> H. Ralph.[53]

Even earlier, the accounts and minutes of the Meshivat Nefesh Society founded in 1795 were beautifully written up in impeccable English. Only one example of this need be given. Opposite the name of Angel Emanuel is a note: 'May his name remain a monument of his virtues. This dear good friend went to the West Indies and died of a fever prevalent there 24/1/1797'.[54]

By 1835, the Plymouth Congregation's rules and minutes were written in English and even announcements in the Synagogue had to be made in English.[55] Moreover, the minute book in which the Congregation's proceedings were recorded in Yiddish abruptly breaks off in 1834, and a new book, which cannot now be traced, was begun which almost certainly was kept in English.

Even in the most conservative area of Jewish life, that associated with death, English in the form of tombstone inscriptions that are now extant first appeared in Plymouth in 1825,[56] being invariably used after the 1850s. The earliest surviving English inscription in the Falmouth Jewish cemetery is dated 1831,[57] in Penzance is 1841,[58] and in Exeter is as early as 1810 and invariably on all stones thereafter.[59]

Even when a Jew had anglicized his appearance and dress, and spoke impeccable English, he could still often be identified by his name. Names such as Aarons, Abrahams, Benjamin, Cohen, Ezekiel, Hyman, Isaacs, Israel, Jacob, Joseph, Lazarus, Levy, Marcus, Mordecai, Moses, Myer, Nathan, Samuel, Solomon, Woolf, used either as forenames or surnames indicated to the man in the street that the bearer was Jewish. Similarly, names ending in -owsky or -ovitch, and other foreign sounding names were frequently identified as being 'Jewish'. Often a Jew would anglicize his name so that he would not be the victim of discrimination, imagined or real, or merely as an attempt to integrate and merge into the wider society around him. The anglicization of names was effected in the South West of England in a variety of ways:

(a) Where a name had a meaning in its original language it was often translated into English. Thus Isaac Karmey, synagogue teacher in Plymouth 1857, became Isaac Stone;[60] Elimelech Hichtenfeld became King Field.[61] When xenophobic passions ran high a Jew with a foreign sounding name might keep it in private life, but trade under its translation. Such a one was Ephraim Holcenberg who traded as 'Woodhills'.[62]

(b) Often a well-known name, either locally or nationally, was adopted because of its similarity to the original. In this way Raussman became Roseman, Durckheim became Dirk and the praenomen Lemmle became Lemon. Jews whose patronymic was Jonah became first Jonas, then Jones or Johnson.[63] A Jewish family in Plymouth seems to have started off life as Katzenellenbogen, modified this whilst still abroad to Katenellson, and arrived in England with the patriotic name of Nelson.[64]

(c) Surnames were modified by aphesis. In this way the synagogue cantor Abrahams became a well known operatic star Braham and a Plymouth Aaron became Aron.

(d) Lack of familiarity with English spelling on the part of the immigrant, or inability to understand his poor pronunciation on the part of the listener, often led to a name appearing in several different forms. Thus Burstein in 1851 was Bernstein in 1858 and Burenstein in 1866. A certain man joined the Exeter Hebrew Congregation in 1830 and the folio for his account is headed '—Luvis'. In 1833 his folio was headed '—Lewis', which the following year is amplified to 'J. Lewis'. In 1837 he metamorphoses into 'Lewis Schultz' and in that name he leased a shop in Fore Street Hill in 1838. In the censuses of 1851 and 1861 he is called 'Lewis Schultz' but in 1861 he also signed a document as 'Lewis Salz'. His name finally achieved a stabilized form on the tombstone which marks his last resting place and there it appears as 'Louis Shultz'.[65] This type of confusion is typical of Jewish immigrants' names.

Forenames were anglicized even more quickly than surnames. Israel Silverstone, who came to Exeter from Poland about 1830, called his first children by traditional Jewish names—Bella, Sara, Rebecca, Isaac—but after ten years of residence in England come Clara, Maurice, John, Selina and Fanny. The children of Jackson Marks who were born in Poland were called Myer and Leah, those born in Plymouth were called Mathilda, Henry and Julia. The East European immigrants were under some pressure to anglicize their names. One little girl at Manchester Jews' School was told by her Jewish headmistress, 'You can't go through life with a name like *Taube*'. So against her parents' wishes she changed it to Tilly.[66] Eleven-year-old Mordecai Kushelevitch quickly changed his nickname, which in Pikeln, Lithuania, was Mara, to Mark, and his surname to Woolfson, his uncle's surname, when he arrived in Devonport in 1934. This was done, he later recalled, because Mark Woolfson was easier to pronounce than Mordecai Kushelevitch, less 'foreign', and, for personal reasons, made it easier to identify him with his Devonport family.

The following short list of Hebrew names illustrates the way in which the changes took place in Devon and Cornwall, as well as generally in England:

| *Hebrew name* | *English equivalent or derivative* |
|---|---|
| Alexander | Sender |
| Aryeh (usually used with Judah, q.v.) | Lyon, Lion |
| Asher | Lemon or Angel (from the hypocoristic Lemmle or Anshel) |
| Avigdor | Figdor and Victor |
| Baruch | Barrow and Barnett, Benedict |
| Benjamin (often used with Ze'ev, q.v.) | Wolf, Woolf[67] |

| *Hebrew name* | *English equivalent or derivative* |
| --- | --- |
| Bilah | Betsy |
| Eliezer | Lazarus, then Lawrence |
| Elijah | Elias |
| Gershon | George |
| Hayyim | Hyman, then Harry |
| Hirsch | Harris, then Harry[68] |
| Isaiah | Josiah |
| Issacher (often used with Dov = a bear)[69] | Barnett, Barent and later Bernard |
| Jonah | Jonas, Jones and possibly Johnson |
| Judah | Lion,[70] Lippa (hypocoristic), Lewis, Louis. Lewis led back to Levi in eighteenth century. Lion led to Lyonell then Lionel |
| Meir | Myer |
| Menachem | Emanuel |
| Michael | Mitchell (as a surname) |
| Mordecai | Mark, Marcus |
| Moses | More, Morris, Maurice |
| Simcha | Bunam or Joyful (translation)[71] |
| Ze'ev | Wolf (translation) then William |
| Zvi | See Hirsch, above. |

In Jewish religious documents, such as the wedding contract or bill of divorce, as well as when called to the Torah, a Jew is named by his own Hebrew name and then described as *ben* (= son of), or a woman by *bat* (= daughter of), and then follows the father's Hebrew name. The very use of English names instead of the Jewish one in the context of synagogal life is itself a very strong indication of the process of acculturation. Until 1810, whenever a man's name is mentioned in the minutes of the Plymouth Congregation he is invariably referred to as *X ben Y*. In 1810, an 'M. Johnson' is mentioned. In 1812, two more men, 'Abraham Levy' and 'Lion Levy' are referred to; in 1816, of nine men present at a meeting, the names of five are given in their Hebrew form and four in their English form and the following year only three are in the Hebrew form and five in the English.[72]

The most basic, though not always the most immediately noticeable, factor differentiating the Jew from his fellow citizens was his religion.

The day to day observances of Judaism necessarily restrict free social intercourse between Jew and Gentile, so that even in a State where there is freedom of action, office holding and the like for all, regardless of religious affiliation, the fully observant Jew cannot but help live in his own 'ghetto'. The Jewish laws relating to permitted foods, observance of the Sabbath, Synagogal attendance and marriage, oblige observant Jews to live within walking distance of the synagogue, and to eschew certain occupations and leisure activities.

Considering these observances, perhaps the factor which most limits social intercourse is adherence to the dietary laws. According to a strict interpretation of these, a Jew cannot easily eat at the house of a non-Jew. It is therefore not surprising to find that Jewish salesmen travelling in Devon in the eighteenth century made special provision for themselves. Some inns kept special utensils in a locked cupboard which Jewish pedlars used. The Jews used to write their Jewish names in Hebrew with chalk across the face of the saucepans when they had finished with them, so that those who came afterwards would know by whom the pan had last been used.[73] Mention has already been made of the Moroccan Jew, a Turkish rhubarb seller, who told Mayhew that about 1820, he stayed at a lodging house in Taunton which catered particularly for Jews.[74] Gentile customers of Jewish talley-men have wonderingly told the author that when their suppliers brought their wares they would accept an offer of a cup of tea, but always without milk and in a glass!

A growing laxity in the observance of the dietary laws on the part of many members of the South-West Congregations is discernible as the nineteenth century progressed. In 1821, it was still most unusual for a Plymouth Jew to eat *treifah* food, and one who was wrongly accused of so doing reacted most strenuously.[75] In the following four decades standards dropped drastically, both in the communal butcher shops in Plymouth and Exeter and in the homes of members. In 1860, for example, Nathaniel Hart's servant bought *treifah* fat at the Plymouth kosher butcher, but Revd Stadthagen informed the President who declared Hart's household *treifah*. Hart confessed that he took a *treifah* loin, but offered the rather lame excuse that it was on medical advice. Dr Adler was appraised of this incident and wrote to Stadthagen, taking him to task for allowing *treifah* meat to go out from the shop at all: 'how do you conduct the butcher shop?'[76] At no period does there seem to have been a Jewish kosher butcher shop in Plymouth. In the 1930s, two Gentile butchers kept kosher departments, obtaining the meat from London. By 1960 there was only one shop,and that only opened its kosher department (in a garage) on a Thursday morning, the minister acting as a *shomer* and porging the meat. It closed in 1964, and those who wanted kosher meat had it delivered by rail or parcel post.

Another feature of the Jewish religion which tended to set its adherents apart was Sabbath observance.[77] Two aspects of this were evident to the general public; that Jews did not trade on Saturday and that they did not tend their fires on this day.

Until the early nineteenth century most workers were paid late on Saturday, so that the main shopping period of the week was after 7 p.m. on a Saturday night. It was therefore no great hardship for Jews to keep their shops closed from sunset on Friday to sunset on Saturday, and indeed this was the norm for them.[78] John Solomon, advertising bespoke and ready-made clothes in Exeter in 1844, declares, 'No business transacted Friday evening until Saturday evening'.[79] Local tradition in Redruth more than a century after his passing relates that Emanuel Cohen, jeweller, could be seen waiting for sunset on Friday and Saturday evening to close and open his shop.[80] Aaron Levy of Plymouth, watchmaker and jeweller to the Queen, advertises as late as 1853 that he is closed on the Jewish Sabbath. Some shops were open on Sundays, though this was stopped from time to time,[81] and Jewish shopkeepers no doubt made up for some of their loss of trade by trading on the Christian Sabbath. If they were open on both Saturday and Sunday they stood the risk of prosecution, as happened to Abraham Jackson of Shields, who was fined five shillings and had to spend four hours in the stocks for observing neither Sabbath.[82] From 1855, there were a number of attempts to introduce Parliamentary legislation to control Sunday trading, and the London Committee of Deputies of the British Jews[83] sought and largely secured exemption for Jews who kept their shops closed on Saturday.[84] In Plymouth, pressure was brought to bear on Jews who traded on Sunday. The *Western Morning News* reported that 'on account of the annoyance caused by Israelitish shopkeepers and others persisting in keeping their places open on Sundays . . . Mr. Edwards, the able superintendent of police . . . induced them to close their shops'.[85]

As the hours of labour grew shorter, workers were paid earlier in the day, and it became increasingly difficult for Jewish shopkeepers, whose customers were mainly weekly wage earners, to earn a living from a Sunday to Friday week. In 1853, Revd Hoffnung had been accused of condoning Sabbath desecrators who gave him handsome presents.[86] In 1854, the Exeter Congregation, already hard pushed to find a quorum of ten adult male Jews for Sabbath services, wrote to the Chief Rabbi asking if a man who traded on the Sabbath remained eligible to be counted to the quorum.[87] In 1856, Dr Adler wrote to Revd Mendelssohn at Exeter that 'in my name you should check public desecration of the Sabbath'.[88]

The Plymouth Congregation had written to the Chief Rabbi as early as 1839 about Nahum Cornbloom who kept a billiard hall open on Saturday. Rabbi Adler in that case recommended the Congregation to enforce its sanctions against him if he persisted, particularly as he was acting contrary to the 'law of the land'.[89] Naturally, if the men were in their shops on a Saturday morning, they could not simultaneously be in the synagogue for the main service of the week. By 1851, attendance at the morning service on the census Sabbath in the Plymouth synagogue was sparse, only 45 men, women, and children out of a total of 205 souls.[90] There were probably not more than 18 or 20 men who regularly attended the main service of the week. On the same Sabbath, the Exeter synagogue is supposed to have had 44 worshippers, whilst in the Falmouth and Penzance synagogues there were 10 and 16 worshippers respectively.[91]

The difficulties of keeping the Sabbath whilst engaging in retail trade undoubtedly prompted a number of men to become salesmen or enter the wholesale trade where they could arrange their times to suit their religious conscience.

Another aspect of Jewish Sabbath observance which was widely noticed was the law which forbade a Jew to light a fire on the Sabbath day. A Jewish family would often employ a Gentile to tend its fires and put on, or turn off, lights. So common was this practice that there was a popular nineteenth century term for such a person—a Jews' Poker.[92] Jacob Fredman of Plymouth needed to have his fire attended to and called in a passer by, who, with some amusement, helped him. It was the then Commander-in-Chief at Mount Wise.[93]

Of course, there were Jewish religious observances kept by the Jews of the South West which were not obvious to their Gentile neighbours but which nonetheless served to keep the Jews in an invisible ghetto. There were, for example, the laws relating to intimate marital relations culminating in the monthly use by the married women of the *mikveh* (ritual bath). Few Gentiles could have been aware of the existence of the institution, but it nevertheless played its part in preserving the identity of Jewish marital life and tied the observant Jewish family to the vicinity of the synagogue.

At some time during the early part of the nineteenth century most Jewish women throughout England gradually stopped going to the *mikveh*. In 1838, the Exeter Congregation attempted to prevent its women from using any bath other than its own ritualarium by imposing a fine of one guinea on those who bathed elsewhere.[94] By the time Adler took over the reins as Chief Rabbi, barely half of the English Congregations who replied to his questionnaire in 1845 had a *mikveh* at all, the others declaring that their members could use the sea, Montpelier baths, and public or private baths.[95] The Exeter Congregation's reply on the point is worth quoting in full as it well illustrates the attitude of mind of a community, representative of most of Anglo-Jewry, which had abandoned a religious precept but which lacked the courage to say so outright to the Chief Rabbi:

> **Is there a Mikveh?** One was formed on the same building as the *Shool* at a cost of not less than £80, but from the necessity of being built on the second floor and the apparatus to heat the water being

> above that again and the difficulty of obtaining a supply of water and the injury it produced to the premises we were reluctantly impelled to abandon its use within the last eighteen months and consequently the Public Baths are now resorted to where there is a bath constructed which on investigation is found to be within two inches of the prescribed rule for size as being Kosher. But we regret to add that on account of a trifling extra expense it is not generally used.[96]

A similar situation obtained in Plymouth when the Congregation's *mikveh* supervisor died. Adler wrote to the President asking for a female to be appointed without delay, 'especially now when every excuse is made not to use it'.[97]

When a Jew marries out of his faith, a procedure known in the Jewish community as 'intermarriage', it becomes evident that he cannot be leading a fully observant Jewish life. In the first place such a marriage is forbidden by Jewish law. Then, children born of a non-Jewess are regarded as Gentiles. Furthermore, the extensive home ceremonials of Sabbaths and Festivals could hardly be celebrated by a non-Jewish spouse. It is not possible to estimate the number of marriages between Jews and Gentiles, i.e. common law marriages, or marriages celebrated in church or registry office in the South West in the period under study. Such marriages were sufficiently common in the eighteenth century in the Plymouth Congregation to provoke a rule against them in 1779.[98] In the latter half of the nineteenth century such marriages may have accounted for up to a third of all marriages in which Jews were involved. The extent of intermarriage by a leading Jewish family in Cornwall is typified by the children of Lemon Hart. He had one son, David, who married a Cornish girl, and four daughters, two of whom married Gentiles.[99] In most cases, Jews and Jewesses who married non-Jews became lost to the Jewish community and few traces are to be found.[100] Some cases of intermarriage noted in the South West of England can be quoted to exemplify them all. Manasseh Masseh Lopes married Charlotte only child of John Yeats of Monmouth at Horton near Windsor on 19 October 1795.[101] According to W.G. Hoskins, Lopes formally abandoned his Judaism only in 1802,[102] and a church marriage involving a Jew and a Christian was technically possible, as it has never been clearly ruled that a church marriage is illegal even if both parties are unbaptized; *a fortiori* a mixed marriage would be legal.[103] Another Jew, Simon Hyman, married Mary Cawrse (or Corse) in St Andrew's Church, Plymouth, on 7 January 1813:[104]

> at 16 she was such an exquisite girl and so persecuted by men of fortune, that her friends hurried her into marriage with a man old enough to be her father, because he was rich; he was ruined and died.[105]

Even though a man who lived with a non-Jewish mistress was debarred from certain synagogal rites, he did not necessarily lose all contact with his family and the Jewish community. Abraham Daniel, noted miniature painter of Bath and Plymouth,[106] bequeathed £50 to Elizabeth Codbury 'with whom I lived', £50 to his natural son Edward Elliot Thomas Daniel, £100 to his other natural son William Daniel, and then, surprisingly perhaps, £20 to the charity of the Jewish synagogue at Plymouth, and the remainder to his sisters, Rachel,wife of Solomon Nathan of Plymouth, and Rebecca, wife of Isaac Alman of Bristol.[107] Abraham Franco, in spite of being married to a Gentile wife, strongly desired to be accepted as a member of the Plymouth Congregation together with his wife.[108] We have seen that Samuel Ralph was married more than once in church, and yet had Jewish burial.[109] Isaac Gompertz, the minor poet, married Florence Wattier in Church in 1818 and baptized all his children at birth; yet he was buried in the Jewish cemetery at Exeter. Moses Ximenes, a frequent visitor to Baruch Emanuel Lousada at Peak House, Sidmouth, was baptized in 1802. Yet he remained on good terms with the Sephardi authorities. The Lousada's did not abandon their ancestral faith, the fact that they remained friendly with Ximenes and other apostates is a pointer, perhaps, to their own commitment to Judaism.[110] Nonetheless, the children of such men were entirely lost to the Jewish community, and they themselves contributed but little to its spiritual and economic welfare.

There were a number of converts to Judaism in the South West. Paradoxically, they provide still further evidence of a weakening of Jewish religious bonds, rather than betokening a virile, vital community attracting converts because of its strength. Most of the converts were women who wanted to marry a Jew. The conversion of women to Judaism is physically easier than for men, as males are required to be circumcised. There was, for example, one '*Miriam daughter of Abraham our father*'[111] in Exeter in 1832.[112] She was the wife of Jonah Solomon[113] and mother of Baruch Jonas and of whom the wardens of the Exeter Congregation wrote to Chief Rabbi Hirschel in 1838 that her immersion for the purpose of conversion had taken place in Exeter under their supervision about 1810.[114] According to Jewish law, as Baruch had been born before his mother's conversion, he was not a Jew, and his conversion took place on 7 October 1811.[115] In 1813, the family of another Exeter Jew were converted to Judaism:

> The mistress of *Menahem Mendel ben Ze'ev Wolf* of Exeter immersed on Tuesday 21 *Ellul* 5574 [= 6 September 1814] and her name is Rebecca and her daughter, about four years old, immersed on the same day and her name is Rachel, and another aged two years and her name is Leah.[116]

The earliest convert to Judaism noted in Plymouth was one Abraham the Proselyte whose son Isaac was born in Thursday, 10 February 1785 and circumcised a week later.[117] In 1823, two *orchim*, i.e. either travellers or residents in Plymouth without seats in the Synagogue, were called to the Torah. One was Ze'ev son of Abraham our father and the other is referred to as 'd°'[118] The 'ditto', however, refers more probably to him being an *orach* than a proselyte. The children of Abraham Franco and his Gentile mistress who were orphaned by the cholera epidemic of 1832 were converted by the London Beth Din on Tuesday, 6 November 1832.[119] Then there was a woman who was converted before her child was born. She was Leah, wife of Judah Pinner, whose son Joseph was born on 20 January 1866.[120] A contemporary of hers was a Martha who was sent to Holland to be immersed and accepted as a proselyte in 1872.[121] On her return to Plymouth she married one *Zvi ben Isaac* on Sunday, 9 *Adar I* 5632 (= 18 February 1872) but there is no record of the marriage in the Plymouth Congregation's Marriage Registers; so presumably this was purely a religious ceremony, a civil one having preceded it. The first marriage with civil effect conducted under the auspices of the Plymouth Congregation between a Jew and a convert was that of Jacob Nathan Brock who married Eva Lavinia Kinsey Atkins in 1890.[122]

Much the same picture of assimilation and intermarriage applied to the second generation of the East European Jewish immigrants to Plymouth who began to intermarry with Gentiles after the 1914–18 war. The trend continued until the Second World War when it was estimated that about a third of the Plymouth Jewish community took Gentile spouses. Once again, even where the spouses became Jewish—which was by no means always the case—the trend denoted a weakening of Jewish religious and social attachment. Sometimes the offspring of such marriages remained loyal members of the Jewish community. All too often the children opted out of the Jewish faith, occasionally with feelings of antipathy to it on account of some real or imagined slight received by them or their parents. The high degree of intermarriage was directly responsible for the attenuation of the Jewish community in Plymouth after the Second World War. Many of those who left the town then, did so to give their children a wider Jewish social life in the hope that they would marry Jews.

With most of the Anglo-Jewish community bent on anglicizing itself in the early part of the nineteenth century, it was no wonder that some Jews went all the way along the road and became Christians. Some of the eighteenth-century converts to Christianity have already been mentioned.[123] Among others, there was an Isaac Polak of Penryn, 'a Jewish Priest', who in 15 February 1760 married Mary Stoughton, a widow, of the same town.[124]

In the early part of the nineteenth century there was every encouragement for Jews to become Christians, for that was the period when the London Society for Promoting Christianity among the Jews was becoming increasingly active, opening schools for Jewish children and distributing tracts amongst the adults. The London Society's increased success was directly due to an improvement in its financial standing and social status in the years immediately following 1811.

The story of this change in fortune, which had far reaching effects on Anglo-Jewry, prompting the Jewish Day School movement as a countermeasure,[125] had its roots in Devon. It started in 1795, when a Miss Jane Parminter bought some land two miles from Exmouth and built a house there resembling the Church of San Viale at Ravenna in a strange circular design which she named Á la Ronde.[126] The house stood in many acres of parkland and nearby were some fine oak trees said to be 400 years old. She also built a chapel and manse, with a miniature almshouse and equally miniature school. The manse was to be occupied by a married but childless ordained minister and the almshouse by four industrious women of good character with some independent means who would continue to spin or crochet to help support themselves, 'but in the case of a *Jewess who should have previously embraced Christianity* becoming a candidate she should be preferred to all others'. The school was for six girls of poor and indigent parents and again 'the children of *Jewish* parents should in all cases be preferred'.

Miss Parminter's interest in Jews was occasioned by Zionist feelings, with the proviso that the eventual re-settlement of Jews in Palestine would be preceded by their conversion to Christianity.

Jane Parminter died in 1811, and shortly after her death a Mr Lewis Way (who had inherited what was then an enormous fortune of £300,000 under circumstances which a writer of fiction would hardly have dared to invent) was riding with a friend along the road which leads from Exeter to Exmouth. Way was suddenly struck by the sight of Á la Ronde and asked about the strange dwelling. His friend gave him full particulars adding that Miss Parminter had recently died and her will contained an interesting codicil. With reference to the group of oaks, so he said, she had decreed:

> These oaks shall remain standing, and the hand of man shall not be raised against them until the Jews are converted to Christianity and are restored to the Land of Promise.

The recital of this curious injunction, which was in strict accordance with Miss Parminter's known interest in Jews, had a profound effect on Lewis Way. Impressionable, impulsive and deeply religious, Way was obsessed with the novel tale. It seemed to him that the finger of Providence had pointed out to him a cause to which he could devote his life and fortune which had come to him, so to say, from Heaven itself—he would dedicate his life to the conversion of the Jews. The oaks of Á la Ronde at Exmouth were directly responsible for a flood of pamphlets, sermons, charitable bequests, and verse:

> List to the voice of the aged trees,
> Pass them not heedless by;
> I hear in the sound of the moaning breeze
> The earnest and heartfelt cry
> Of her who willed that these trees should stand
> Till the Jews should return to their Fatherland.

Way went to the London Society for Promoting Christianity amongst the Jews, which was then struggling with an overdraft of £14,000, and promised £10,000 and his further aid. With Lewis Way's money giving it status the Society employed missionaries, printed tracts, arranged for schooling for Jewish children, provided jobs and subsidies, and accepted donations from churches up and down the land. In the ten year period ended 1825 these donations averaged £32,000 per annum, and they produced some results. The Society made converts, perhaps a dozen or two dozen a year for some years, until there were very few Jews left in England prepared to convert to Christianity, and the Society was obliged to go overseas to find converts. There was, however, a certain irony in the fabrication of such an edifice on the foundation of the romance of the oaks of Á la Ronde—the report of the codicil was *ben trovato*.

Quantitatively, the crop of converts in the nineteenth century from the Jews of Devon and Cornwall was very slight indeed, qualitatively it was not much better, though it included one M.S. Alexander, whose conversion as the discussion below will show, was to have a profound, if incidental, effect on the Roman Catholic revival in England in the latter part of the nineteenth century.[127]

Alexander's conversion was almost certainly based upon religious considerations, but wealthy Sephardim like Lopes and Ximenes did not forsake their faith on doctrinal grounds. To be sure, their loyalties to Judaism and its observances were but tenuous before their conversion, social ambition motivated them, and they largely attained their desires.[128] Family squabbling seems to have prompted Joseph Ottolenghe's conversion in Exeter in 1734.[129]

Apart from the few converts in the eighteenth century already mentioned, the other known Jewish converts to Christianity in the South

West mostly converted in the period 1820–30, the heyday of the English Missionary societies in England. It has been suggested that the missionary societies were responsible, in part, at least, 'for blunting religious prejudice by deploring persecution [of the Jews] and promoting (in admittedly insidious charities) their temporal and spiritual welfare'.[130]

The most important convert to Christianity in Devon was undoubtedly that of Michael Solomon Alexander. He was born in Schoenlanke,[131] Germany, in 1799. From 1815 until 1819 he taught in a German *cheder*. Informed that he could have a situation in England if he learnt *shechitah*, he studied the subject and became proficient. In 1820 he came to London but was disappointed in his hopes and instead went as a private tutor to a Jewish family in the country, perhaps in Nottingham. It was apparently whilst within this man's home that his interest in Christianity was aroused. He eventually became the *shochet* and second cantor to the Plymouth Congregation, an office which his Christian friends enthusiastically described as an 'Inspector of Meat which is an honourable office and bestowed only on Priests of unblemished reputation'[132] but which his erstwhile Jewish friends were quick to point out was a very minor office indeed—the truth, as so often, lying somewhere between the extremes.[133] In Plymouth, Alexander took in pupils teaching them Hebrew and German.

One of his pupils was Revd B. Golding of Stonehouse who in discussions awoke Alexander's latent interest in Christianity, with the result that after some hesitation and correspondence with the Chief Rabbi, Alexander was suspended from his duties. The announcement of his intending baptism caused intense public interest, and more than a thousand people came to witness the event at St Andrew's Church, Plymouth, on 22 June 1825.[134] The Revd John Hatchard, vicar of St Andrew's, published the sermon he preached on the occasion, and Alexander added an appendix in the nature of an *Apologia pro vita mea*.[135] On 9 November 1825, Mrs Alexander[136] followed her husband's example, but this time the baptism was at Allhallows Church, Exeter, perhaps to spare additional embarrassment to her family in Plymouth.[137] Alexander did not lose any time in getting to work. He converted an Exeter Jew who was baptized at Bristol in 1825, and probably had a hand in the conversion of another Exeter Jew who was baptized on Christmas Day, 1825.[138]

Alexander was promised an appointment as missionary to Poland,[139] but he eventually became a home missionary of the Society for the Propagation of Christianity amongst the Jews and Professor of Hebrew at Kings College, London.[140] When a United Protestant Bishopric was established in Jerusalem by the Church of England and the German Lutheran Church, Alexander was chosen for the position. The appointment was wryly noted in the Jewish press:

> The Bishop of Jerusalem is to get £1,200 per annum, better pay than his 20 shillings a week as a slaughterer of animals to the Congregation at Nottingham or Plymouth.[141]

The establishment of the Bishopric in co-operation with the Lutheran Church was one of the prime factors which led John Henry Newman to embrace Roman Catholicism, as he, with many others of the High Church party, did not recognize Lutheran orders.[142]

In Exeter, in 1820, a converted Jew called Hyman Isaacs published a tract attempting to persuade his coreligionists to join him in his new faith. His rather superficial theology published as *A solemn and affectionate address to the Jews clearly demonstrating . . . that Jesus . . . is the only true Messiah*[143] could hardly have converted the more sophisticated Jewish families of Exeter. There is some reason to believe that Hyman Isaacs later renounced his new faith and changed his name to Hyman Levy, because an Exeter Jew, in 1825, called Hyman Levy Davis Isaac changed his name at about that time on account of its similarity to 'Hyman Levy the Penitent'.[144] Unless there was a coincidental similarity of names, it would appear that Hyman Levy was the former Hyman Isaacs. The penitence may have been short lived, because in 1832, when his sons Colin, aged 20, a hatter, and George Christian, aged 18, a shoemaker, escorted a drunk blacksmith from Totnes out of Exeter and stole his watch, he was referred to as 'Isaac the converted Jew and Christian Missionary'.[145]

Such criminal 'sprigs of Judaizing Christians' were all too common and posed a considerable problem to the missionary societies. A spokesman for the London Society for Promoting Christianity among the Jews attempting to allay subscribers' anxieties that all the Jews 'who have professed Christianity have turned out to be hypocrites' admitted, 'we should not deny that deceit is an awful feature in the Jewish character' and could only explain it by blaming Christians for their ill-treatment of the Jews through the centuries.[146] Lewis Way himself was imposed on by 16 young Jews in his home at Stansted who professed conversion, and were duly shaved and baptized, when a rumour reached them that Way was bankrupt. The next day all the converts decamped, stripping the house of all they could lay their hands on, including their host's silver spoons. Macaulay commemorated the event:

> Each, says the Proverb, has his taste. 'Tis true
> Marsh loves a controversy, Coates a play,
> Bennet a felon, Lewis Way a Jew,
> The Jew the silver spoons of Lewis Way.[147]

Other Jewish converts to Christianity in the South West, who did not come from the twilight world of the criminal class, include:

(a) Solomon Gompertz (1806–83), brother of Isaac Gompertz the poet, was baptized in 1823 at Otterden, Devon. He later attended Peterhouse College, Cambridge, and became a clergyman.[148]

(b) The four children of Jacob and Elizabeth Levy who were all baptized in St Mary's Church, Truro, on 22 February 1822.[149]

(c) The five children of Abraham Simmons, a hawker, and his wife, née Jane Barker, who were baptized in St Andrew's Church, Plymouth, on 23 March 1825.[150] According to a contemporary newspaper report Abraham Simmons himself was 'shortly to be baptized'.[151] Jane Barker herself was probably not a Jewess, so her children would not have been Jewish in the eyes of Jewish law.

(d) Harriet, 'the nineteen year old daughter of the late Moses Hyman, the Jewish Priest at Falmouth who was publicly baptized at the Wesleyan Methodist Chapel, Falmouth' in 1830.[152]

In this last case, the girl's mother and friends remonstrated with her so sternly, that Harriet ran to the Mayor for protection. When he summoned the mother to appear before him, Mrs Hyman with other members of the Falmouth Congregation agreed to take the girl back if she stopped going to Church, but without such an undertaking, Mrs Hyman would not let the girl come home as she had other children at home whose spiritual welfare could have been threatened. Harriet was to have been married within a few weeks of the dispute, but the match was broken off on account of her interest in Christianity. This interest was consummated on Sunday, 7 November 1830 when she entered the Church.[153]

Generally speaking the Jewish community, both in London and the Provinces, avoided direct confrontation with the Christian public on the subject of conversion. The reluctance of Anglo-Jewry to have anything to do with 'mad' Lord George Gordon who became a Jew is well known.[154] In the latter part of the nineteenth century, English Gentiles wishing to convert to Judaism were sent to Holland to be initiated there, rather than in England.[155] From Plymouth, for example, a certain *Zvi ben Isaac* sent his Gentile wife-to-be to Holland in 1872 to be converted there so that he could marry her.[156] The last thing that a Jewish community such as that of Plymouth wanted to do was to engage in a public disputation on the respective merits of Christianity and Judaism. Once, however, about 1835, a debate was virtually forced upon the Jews of Plymouth by a Joseph Wolff. This colourful personality started life as a Polish Jew, became a Roman Catholic, then a Protestant, and finished as the rector of a Somerset parish married to Lady Georgiana Walpole. Before he became rector, he was employed as a missionary. Let a Plymouth Jew, Lewis Hyman, tell the story of Wolff's attempt to force a religious debate:

> [About 1835] gentlemen of the genus of Wolff were as plentiful as blackberries, and generally visited the provincial towns on Passover or Festivals, with their saddlebags stuffed with tracts, patted by parsons, and fondled by dowagers . . . Then Joseph Wolff, a star of the first magnitude visited Plymouth. He placarded the walls defying the Jews to mortal combat on the evidences of Christianity. Our community, ever timid,[157] declined until strongly urged, a meeting was held on a Sunday in Passover. A number of Jews and Christians were present, among others Dr. Cookworthy, a celebrated physician, Chairman; Rev. Hatchward, Vicar of St Andrews; Mr. Newton, one of the founders of the Sect of Plymouth Brethren, etc. After a number of arguments on both sides, Abraham Joseph, one of the disputants, produced the *Asiatic Journal* of 1833 [in which Wolff had written a long account of various visions in one of which The Devil, Mohammed, Abraham, Isaac and Jacob, the Apostles with Jesus, and St Paul all appeared to him and gave him a crown]. Wolff was dumbfounded. Asked by the Chairman if the statement was correct, he did not deny it. Dr. Cookworthy, after hearing the account read by Abraham Joseph immediately vacated the Chair, saying, 'Gentlemen, I am done—the meeting is dissolved'. This unlucky contretemps frustrated Wolff's grand battle at Plymouth, also at Exeter where he had announced to lecture. After this he visited more genial climes.[158]

It is noteworthy that Abraham Joseph relied on an attack on the personality of Wolff rather than discuss the relative merits of Christianity and Judaism. On the other hand, there were a number of literary polemics against the activites of the 'conversionists' in the early part of the nineteenth century,[159] but these were generally addressed to the Jews themselves. One such work emanated from Exeter in 1808 from the pen of Lazarus Cohen[160] entitled:

> *Torat Emet*: Sacred Truths addressed to the Children of Israel residing in the British Empire, containing strictures on the book entitled *The New Sanhedrin* tending to show that Jews can gain nothing by altering their present belief.

In his book he warns his fellow Jews of the work of the conversionists and particularly against a statement put out by the Sanhedrin founded by Napoleon that intermarriage was permitted.[161] A similar work by Hart Symonds, issued from London in 1823 and entitled *The Arguments of Faith*, was originally euphemistically addressed to Sophists and Epicureans. After two years' residence in Penzance, the author was emboldened to address his work to the London Society for Promoting Christianity amongst the Jews and attempting to refute their arguments.[162] The title of a polemic work by Selig Newman, a Jew who spent some years in Plymouth, speaks for itself:

> *The Challenge Accepted: A dialogue between a Jew and a Christian. The former answering a challenge thrown out by the latter respecting the accomplishment of the prophecies predictive of the advent of Jesus.*[163]

Other polemical works were patronized by the Jews of the South West. S.I. Cohen's *Elements of Faith,*[164] for example, had eight subscribers in Falmouth alone compared to nine in the rest of the provinces out of 267 in all. M. Sailman's *The Mystery Unfolded,*[165] a sweeping exposure of the means employed by the London Society for Promoting Christianity amongst the Jews to obtain converts, had nine Jewish subscribers—and five non-Jewish ones including an archdeacon!—in the South West out of 337 subscribers in the rest of England and overseas.

Some Jewish parents in Plymouth, as elsewhere, sought to discourage their children from leaving their faith by a provision in their will disinheriting any who did so.[166] Samuel Hyman, a Plymouth pawnbroker who died in 1839, left an instruction in his will that children of his who married out of the faith shoud be disinherited.[167] As early as 1818 Emanuel Levy of Exeter envisaged the possibility that a beneficiary of his will might 'renounce or cease to believe in the rites and ceremonies of the Jewish faith . . . or marry a person not professing that faith', and declared that such a person should forfeit any benefit under the will.[168] At the end of the nineteenth century, when less observant Jewish merchants began to keep their shops open on Sabbaths and Festivals, one staunchly orthodox member of the Plymouth Congregation made his sons promise that, after his death, they would keep the shop closed on the second days of Festivals. These second days, being Rabbinic in origin, are not kept in Israel, they were abolished by Reform Judaism, and in laxly observant circles even when Jewish merchants closed their businesses on the first Festival day they operated them on the second. The sons gave their word and the founder of the business died content. Not in his wildest dreams did he imagine that the sons would open the shop on Sabbaths and the first days of Festivals, and as he had only extracted a promise to close on the second days, they opened up on all the other days, including the Day of Atonement, with a good conscience![169]

What then were the causes of the progressive assimilation of the Jews in the South West of England, and for that matter the rest of England, in the nineteenth century? Probably, they were very similar to those which brought about a weakening of the ethnic and religious identities of other immigrant minorities, such as the Poles and Germans in America, or the Huguenots and Irish in England. There were, doubtless, a number of factors each of which contributed its share: the social pressure of any majority upon a minority to make it conform—particularly in the case of children educated in State schools; a weakening of parental authority in the lives of children better educated and more able to take advantage of new opportunities for economic and social progress; a hope that by adjusting himself, chameleon-like, to his contemporary society, he would go unrecognized, and hence be less subjected to the possibility of indignity, or be the victim of social discrimination in educational, job or business opportunities; a desire to enjoy civil rights legally denied to him as a Jew; and perhaps at the root of it all, an uneasy feeling, often hardly admitted to the conscious self, that his culture and way of life was inferior to the culture and *mores* of the wider society around him.

This last point is to some considerable extent dependent on education. A person who is well educated in his own language, discipline, history and general culture is more likely to be proud of it and loath to abandon it. Apparently, in the early part of the nineteenth century when there was a far-reaching and good teaching of Judaism, which includes a knowledge of Hebrew—the language of the Jews' prayers, the Bible and post-Biblical literature and history, as well as laws and customs with their rationale—firmly committed Jews, in the main, were produced. When standards of Jewish education dropped, there was a fall in the standards of Jewish commitment. But this to some extent begs the question. Did the drop in Jewish educational standards lead to a decline in Jewish commitment, or did an assimilation-bent community allow its educational system to deteriorate? There can be no clear-cut answer to this question for the nineteenth century any more than there is for a similar dilemma in the late twentieth. All that can here be attempted is to describe the educational system in the South West of England for imparting Judaism to children, and its ultimate breakdown.

In the late eighteenth century, as well as in the nineteenth, all the Congregations of the South West made arrangements for their children to be taught. The Plymouth Congregation appointed Moses Isaac[170] in 1781 to be the communal teacher at a salary of £42 per annum. The charity fund paid £10 of this sum and the rest was paid by the vestry members. A term's notice was necessary before withdrawing a child.[171] At a meeting in 1797, it was decided that 12 parents had to contribute annually £13 19*s*. for 23 children, probably all boys, and every seatholder paid twopence weekly to make up the deficit in the teacher's wages.[172] In 1806, *Simeon ben Nathan* was appointed as teacher at £50 per annum on condition that *cheder* (school for Jewish studies) was held each day 9.00 a.m. to 12 noon and 1.00 p.m. to 4.00 p.m., Sabbaths and Festivals excepted.[173] He remained some years because in 1811 the Congregation allocated him £15 per annum for teaching poor children (and two guineas for each extra poor child). He had to learn[174] with these poor children either from 10.00 a.m. to 1.00 p.m., or from 2.00 p.m. to 5.00 p.m.[175] It thus seems that the children of the poor only got half of the hours of instruction that the rich children received.

The children's theoretical studies were given practical application when it was decided in 1811 that seats should be provided on the west wall of the Plymouth synagogue for the children, and that their teacher had to remain with them during the times of prayer.[176]

The Exeter Congregation also had its *cheder*, and the communal factotum had to teach in it. Again, the wealthier parents paid for their children's tuition, but there were recurrent difficulties over the children of the poor. A note in the Minute Book of the Exeter Congregation in 1851 tells its own story:

> Hoffnung to have 10/6d. for each poor child taught—not seconded.
>
> Hoffnung to have 1 gn. for each poor child taught—not seconded.
>
> Hoffnung to teach poor children gratuitously—carried.[177]

In the light of the resolution which was finally carried it is not surprising to find that the Exeter Congregation became disenchanted with Revd Hoffnung's efforts and soon after advertised for a young man able to teach.[178] The Congregation got a new man, Revd Albu, but hardly profited by the change. This is apparent from the minutes of the Congregation in 1853 that contain proposals

> that Mr Albu shall be compelled to teach the children at the houses of those parents who require him to do so.—Not passed.
>
> that a committe be formed for the purpose of providing a teacher and to submit such rules to the board as shall be beneficial to the teaching of the children.—Unanimous.[179]

In its early days, at least until the dawn of the nineteenth century, there is some reason to assume that this educational system functioned well. Samuel Hart was educated in Plymouth in the decade 1760–70, his Hebrew letter written when he was seventy years old is the work of a scholar.[180] Abraham Joseph II, privately tutored in Plymouth from about 1804 to 1815,[181] in later life displayed an intimate acquaintance with the Hebrew language and Jewish philosophy and lore, as well as possessing a fine Jewish library.[182] There seems to have been a Hebrew-speaking circle which met in Plymouth in 1844 every Friday night[183] to converse in

Hebrew and study Hebrew texts. This is apparent from a letter[184] written in a consciously 'modern Hebrew' style and sent from London by one *Abraham ben David* to Henry Solomons at Plymouth.[185]

As the nineteenth century progressed the hours of *cheder* were seriously curtailed in Plymouth, for whereas in 1806 children went to *cheder* for six hours a day,[186] by 1822 it had been cut down to two hours a day from 9.00 a.m. to 11.00 a.m., Sunday, Monday, Tuesday and Wednesday. Furthermore, the special teacher had been dispensed with, and instead the *shochet* 'to hold school in the Community's house to teach children Hebrew in so far as they are able to learn'.[187] It sounds, and it probably was, a poor and haphazard Jewish education which the children received.

This impression is confirmed by the state of educational affairs which Chief Rabbi Adler found when he began his ministry in England. 'I purpose', he declared in his installation sermon in 1845, 'to superintend your establishments for education',[188] and he did. Without delay he sent out a questionnaire to every community in England and the replies indicate that, with the exception of Birmingham, there was no properly organized Jewish education outside of London and not much inside it, either.[189]

Although there are replies from the Exeter, Falmouth and Plymouth Congregations, the more detailed reply of the Bristol Congregation to the question 'What Schools?' throws light on a situation which was probably typical of all the communities of the West and South West of England. It replied,

> None. The more affluent children are sent to Hebrew Boarding Schools and others are taught by the *Shammas* and *chazan*, as far as relates to Hebrew. And as to other Branches of Education, by English Schools or Teachers in the locality.

The Exeter Congregation in its reply stated that the *cheder* met in the vestry room on Sundays and Tuesdays from 4.00 p.m. to 7.00 p.m. Even this attenuated instruction was only attended by ten boys and one girl out of 20 boys and 13 girls in the Congregation. The Falmouth Congregation merely reported that Mr Rintel taught Hebrew and German, whilst the Penzance Congregation did not reply at all. The Plymouth Congregation informed the Chief Rabbi that Mr Woolf taught Hebrew reading and translation to a varying number of pupils and 'regret the defective state of education and pulpit instruction'.

For the next twenty-five years Adler attempted to remedy a situation which, left unattended, could only lead to the breakdown of Jewish life in all but the largest communities. His solution, like that of Solomon Hirschell twenty years earlier, was to set up Jewish day schools in each community, where secular and sacred subjects would both be taught. He repeatedly wrote to provincial Congregations urging them to establish day schools and offering financial help as well as advice on syllabus and text books.[190] Without Jewish day schools the situation became desperate. In Exeter, by 1862, most of the Jewish children had virtually no knowledge of their religion. Myers Solomon of Exeter wrote to the *Jewish Chronicle* in 1862 that he travelled the country five days out of seven to seek a livelihood but on the Sabbath he called the children together. He asked a boy, nearly 13 years old (the magic age of *bar mitzvah*!) what his belief was as a Jew. The boy replied that he believed in the Father, Son, and Holy Ghost![191]

Dr Adler's efforts to establish Jewish day schools bore fruit in Plymouth, for he induced Jacob Nathan, wealthy bachelor and ready supporter of all his charitable appeals, to make himself responsible for founding such a school in Plymouth. In the last eighteen months before his death in May 1867, Nathan busied himself collecting monies for this project. By his will, he left £3,000 for this purpose;[192] and a school, the Jacob Nathan School, was opened in 1869. In 1874, there were 24 pupils, 11 boys and 13 girls, who were taught Hebrew and Jewish subjects by the minister of the Congregation, Revd Rosenbaum, and secular studies by a Mr Williams, assisted by a Miss Mitchell.[193]

The school, as a day school, did not last very long, and it probably closed down in the early 1890s. No records of the school appear to have survived, there are no registers, no record of the governors, nor any minute of its daily activities. In the absence of such records it is, of course, difficult to ascertain how effective was the education or what part the school played in ameliorating the effects of assimilation. One significant pointer, perhaps, to its influence is that at least two of its former pupils, J.B. Goodman (who, in his later years, acted as a lay-reader for the Congregation on the High Festivals) and B.H. Emdon, both played a beneficial and prominent part in the affairs of the Plymouth Congregation in the mid-twentieth century.

The name of the school, however, was kept alive, to ensure that its trust income was not lost. Accordingly, the new Religion Classes which the Congregation started were still called the Jacob Nathan School, a fruitful source of later confusion.

By chance, a few pages of minutes of the school *cheder* committee covering the period October 1895 until March 1898 have survived.[194] The topics revealed by the minutes are strangely reminiscent of those dealt with over the previous two centuries!

> The Vestry be recommended that the management of the school be vested in the hands of the school committee.

Every child (of 5 and under 13)[195] requiring private lessons from Dr Berliner[196] shall pay a weekly school fee of 6d.

Non Members children requiring lessons from Dr Berliner must have sanction from the school committee.

A school should be held at Devonport, and a clause to that effect be inserted in Rev Dr Berliner's agreement.

Mr Titleboam proposed that the *Shabas* school be open to all Jewish children.

In 1896, the school hours in Plymouth were: Sunday 10.30 a.m.–12.30 p.m.; Monday, Tuesday, Wednesday 5.00 p.m.–7.00 p.m.; Saturday 2.00 p.m.–3.00 p.m. Whilst at 63 George Street, Devonport they were: Sunday 2.30 p.m.–4.00 p.m.; and Thursday 5.15 p.m.–7.00 p.m.

There were 24 pupils in the Plymouth *cheder* and 8 at Devonport, making 32 children in all receiving Congregational Jewish education, including some 10 girls. The number dropped rapidly. By December 1897 the number of pupils had fallen to 18 in Plymouth (and 6 of these joined in 1897), arranged in four classes, and 5 in Devonport. Complaints of late coming, absenteeism, late arrival on the part of Dr Berlin, and poor teaching are frequent. The committee was so dissatisfied that it proposed to examine the children monthly, on a scheme of studies set up by Dr Berlin, E. Plaskowsky and E. Orgel. The number of children attending continued to fall, and in 1906 there were only 15.[197]

By the 1930s, the hours of tuition had somewhat lengthened, though whether tuition had improved it is difficult to say. Mr Lionel Aloof remembers *cheder* then being each weekday night from 5 p.m. until 8 p.m. and on Sundays from 10 a.m. to 1 p.m. The teachers were good but the discipline was poor, as some of 'the scions of the top hat members were very *chutzpadick* [= cheeky] towards the teachers'![198] Coincidental with the assimilatory movement there was also a distinct process of acculturation amongst the Jews in the South West during the nineteenth century. The basis of this process was a good secular education, but there is an almost complete silence on where this was acquired in the first half of the nineteenth century. Apparently, the only Jew who refers to his secular education in the South West in this period is Solomon Alexander Hart. He relates that in 1813

> at the age of seven I was sent to school at Exeter. There I remained a short time. With regard to this school I may say that the aphorism 'Spare the rod and spoil the child' was not neglected in my case. I returned to Plymouth in 1814. Being an Israelite, I was debarred from entering Dr. Bidlake's Grammar School which was restricted to Churchmen. Consequently I was placed with Rev. Israel Worsley, a Unitarian Minister. With him I remained for the best part of five years. He

> was a most excellent man, but took greater interest in the composition of his sermons than in the classes of his schoolroom.
>
> I dare say, however, that my own shortcomings were more due to my own want of attention than to his. The lessons in French were always given on Saturday when I was absent. I detested arithmetic. My sums were done for me by a boy named Martin, a son of a brewer in Plymouth.
>
> I liked Ovid on account of the picturesqueness of the stories. One of my Master's sons avenged himself because I would not enter his class. He afterwards compared notes good-humouredly, and I certainly did not express gratitude for the frequent infliction of the cane.[199]

It is not clear why Hart was sent to Exeter for a year. Perhaps it was occasioned by a domestic matter such as sickness or the death of his mother.[200] Presumably Jewish boys (and perhaps girls, too) started to go to non-denominational schools at about this time. In the latter part of the nineteenth century when school attendance became compulsory, the process of acculturation was no doubt accelerated when the children played the games of childhood and made friends with their non-Jewish school fellows.[201]

The career of Solomon Hart represents a link in the process of social acceptance by English society of the middle-class Jew. 'He takes his place with the first Jewish baronet, first Jewish barrister, first Jewish MP, first Jewish Master of the Rolls, and first Jewish Senior Wrangler' was the view taken of him after his death.[202] He modified the popular prejudice that Jews are concerned only with money.'[203] It was not only that as a Jew he had become an RA and even Librarian of the Royal Academy, but also because throughout his career he had chosen Jewish subjects, and dignified ones, to present to the English public.

There were many others in the South West who trod a similar path to Hart's, even if they were not quite so famous. In a previous chapter on the leading artistic and other talented Jews in the South West the names of many others have been mentioned whose lives and careers exemplified the acculturation and eventual assimilation of the Jewish communities in Devon and Cornwall.

There is some evidence to indicate the educational levels attained by Jews in the South West in the nineteenth century. The signatures in the marriage registers of the South West seem to suggest that the Jews of Devon and Cornwall were more literate than their Gentile neighbours and their coreligionists in London. In the general population of England in 1837, 33 per cent of grooms and 49 per cent of brides signed the marriage register with a cross.[204] The figures for Cornwall in 1840 were very similar—33 per cent of the grooms and 54 per cent of the brides.[205]

In the Bevis Marks Synagogue, London, the percentage of illiterate grooms in the period 1837–43 was 6.5, though for the previous 20 years it had been 12 or 13 per cent.[206] In the first one hundred marriages recorded in the Plymouth Congregation's Marriage Register from 1837 until 1891, one groom and two brides signed with a mark (a circle and not a cross, on theological grounds) and three grooms signed with their Hebrew signature. The sample is comparatively small, but is taken over a long period. One of the illiterate brides, Rachel Bowman, came from Hull, and 'formed a clandestine union' with a Sam Alexander of Plymouth whom she married in 1851.[207] Sam and Rachel were both born in Poland.[208] The other bride married the illiterate groom in 1854,[209] but they do not figure elsewhere in the Congregational records. The grooms who appended Hebrew signatures were Abraham Burstein (Bernstein) in 1851,[210] who was presumably a recent immigrant, as he does not appear in the 1851 census; Israel Rousman (who signs *Isser ben Leib*) and who married Rachel Freeman in 1870;[211] and a Marks Levy of London who also married in 1870.[212] There were no signatures by a mark in the marriage registers of any of the other Congregations in the South West, but again the numbers of marriages recorded in them are too small to draw any conclusions in this regard. It should also be borne in mind that the Jews of the South West in the nineteenth century formed part of the urban population which was generally better educated than the rural.

Another way of acculturation was to take up the responsibilities, if not the legal obligations, of citizenship. In 1798, there was a call for 'loyal citizens to train themselves to arms' in order to repulse any invasion. Jews in Exeter and Plymouth were admitted to and enrolled in the Volunteer Companies. Jacob Levi of Portsmouth wrote to the Governor at Portsmouth asking to be enrolled as a Volunteer, adding that Jews had been enrolled as such 'at Bristol, Dover, Plymouth, Exeter, Liverpool, Gosport and many other places'.[213] The Governor sent Levi's letter to Pitt who passed it on to Henry Dundas on 1 June 1798.[214] The Governor wrote: 'I proposed to them the formation of a company entirely of their own sect . . . but that they declined'.[215] The Jews were anxious to do their duty as citizens and not as Jews. A few weeks later a Falmouth Jew made some sort of proposal, perhaps to pay for a Volunteer Company of Jews, to the Earl of Mount Edgecumbe who passed it on to Henry Dundas with a covering letter.

> I received some days ago the enclosed letter from Mr. Symons, Falmouth. I am not acquainted with him nor does his offer come with any recommendation but as his proposal is a very handsome one . . . I submit it to you.[216]

Unfortunately, Symons' letter was sent on to the Admiralty and no trace of it can now be found. There is no record in the official sources of Jews, other than Lemon Hart, serving as officers in the Fencibles, Militia, or Volunteers in either Devon or Cornwall in the period of 1797 to 1804, so those who were admitted must have been in the ranks.[217] Additional evidence regarding the part some Jews played in the defence forces of Cornwall at this period is to be found in some notes prepared by Geoffrey H. White, a former editor of *The Complete Peerage*, and a descendant of Lemon Hart of Cornwall.[218] Of Lemon Hart he wrote:

> When a French invasion was threatened . . . Hart raised a company of volunteers in Cornwall. These were styled the Ludgvan Volunteers, and Mr. Hart was appointed their captain by—.[219] He also received a letter of thanks from ?Duke of Portland. I saw the letter more than 50 years ago but have forgotten details.

The Ludgvan Pioneers were raised late in 1798, with Captain Lemon Hart in command for a short while. Pioneers were small corps designed to assist in such work as blocking roads, building defences and constructing batteries. He left the Pioneers for a short while when it became the Ludgvan and Marazion Volunteers and became a first lieutenant in the Mounts Bay Fuzileers, a two company corps raised in March 1797. In 1800 Hart returned to the Ludgvan and Marazion Volunteers.[220] The Exeter Militia List of 1803 has been published.[221] There were 7,320 males in Exeter of all ages according to the 1801 census and the names of 3,102 males aged between 17 and 55 years appear in the Militia List. Of these possibly 17 were Jews. Some of the constables or officers who were required by the *Amended Act for the Defence and the Security of the Realm, 1803*, to return the lists added additional biographical information. Two of the Jews were listed as opticians, there were two silversmiths, three travellers, one pedlar (Israel Stone, if he was Jewish), and a reader in the synagogue. No occupation was listed for five of the men but they were described as 'Jew'. The reader, Moses Levy, was discharged, a Samuel Polock (if he and two other men called Polock were Jews) was already in the Volunteers, three men were 'infirm', and six 'were willing to serve'.

Only an assimilated Jew could serve of his own free will in the Navy. Such a one was Samuel Nathan of Teignmouth, son of a centenarian Polish Jew who died in Exeter. Samuel served in the Teignmouth lifeboat crew when young, and subsequently under Lord Raglan and was in the battles of Balaclava, Inkerman, and Sebastopol. He was twice decorated for bravery, once by the English and once by the Turks.[222]

Some Jews served in the Navy in an involuntary capacity. The Press gangs of the early nineteenth century did not differentiate between Christian and Jew,[223] indeed Jewish pedlars and especially tailors were at extra risk: being often bandy legged on account of their trade they were taken for sailors.[224] The name of Ordinary Seaman John Levy who was flogged round the fleet at Sheerness for desertion in 1802 suggests that he might have been a Jew, whilst the name of Ordinary Seaman Jacob Cohen of the 44-gun frigate HMS *Sibylle*, who received on 17 June 1809 from Emanuel Hart of Plymouth £2 19*s*. 6*d*., being his share of prize money in the capture of the *Espiegle*, is even more suggestive.[225] It is said that B.A. Simmons, the *shochet* at Penzance, was impressed and fought at Trafalgar, losing a finger in the battle.[226]

Jewish Plymothians played their parts in both the First and Second World Wars. Two Rolls of Honour hang in the Plymouth synagogue, testifying to the pride of the Congregation in the civic service of its members. The names on the Rolls are virtually a roll-call of the membership of the Plymouth congregation:

*H.M. FORCES*
*1914–1918*

Abrahams, James S.
Abrahams, John
Abrahams, Joseph
Bash, Aaron
Bence, E.
Brand, Cecil
Brand, Sydney
Brock, Mark
Cohen, M.M.
Franks, E.
Fredman, David
Fredman, J.I.
Fredman, Israel
Fredman, Louis
Goodman, H.
Joseph, I.
Lazarus, Sim
Levy, Victor
Milner, I.
Morris, M.
Pinkofsky, M.
Sanger, M.
Silk, Barnet D.
Spark, Joseph
Weinberg, David
Woolfstein, M.
Woolfstein, S.

*H.M. FORCES*
*1939–1945*

Aloof, L.
Aloof, P.
Angel, E.
Angel, L.
Angel, M.
Begleman, W.
Bence, H.
Bromberg, D.
Caplan, H.R.
Cohen, B.
Cohen, G.
Cohen, H.H.
Cohen, I.
Cohen, L.
Cohen, P.
Cohen, W.I.
Dubovie, I.
Emdon, R.B.
Fredman, D.
Fredman, L.B.
Goldberg, S.
Gordon, M.E.
Greenberg, B.
Greenburgh, H.
Hack, L.
Hurwhitt, M.
Jordan, D.A.
Kaphan, N.
Lee, L.D.
Lee, P.
Lewis, R.I.
Marks, S.
Melichan, H.H.
Milner, H.H.
Peck, B.
Peck, H.
Richman, A.
Richman, H.
Richman, J.
Richman, N.
Richman, S.
Robins, G.
Robins, R.
Roland, A.
Roseman, I.
Roseman, I.
Rutman, M.
Salsberg, H.
Sanger, H.
Sanger, S.
Silver, I.
Simons, H.
Solomon, M.D.
Spiers, L.D.
Telfer, R.L.
Wiseman, A.

*KILLED IN ACTION*
Nissam, M.
Silverstone, Jack

*KILLED ON ACTIVE SERVICE*
Simons, R.

*SPECIAL CONSTABULARY*
Brand, A.
Brock, E.
Erlich, A.
Gordon, S.
Jacobs, M.
Lazarus, G.
Robins, E.
Roseman, Meyer
Solomons, M.

Women included in the 1939 Roll are Pte L. Angel and Sgt I. Silver in the Army Transport Service, Sgt Barbara Cohen and Aircraftwoman L. D. Lee in the Women's Auxiliary Air Force, and Driver Sylvia Goldberg in the First Aid Nursing Yeomanry.

The Rolls of Honour do not tell the full story. There is an army barracks at Crownhill, outside Plymouth. Here was stationed during and after the First World War, elements of the 38th, 39th and 40th Battalion of the Royal Fusiliers. Throughout that war, leading Zionist militants such as Trumpeldor and Jabotinsky had campaigned to be allowed to establish a Jewish regiment which would fight alongside the Allies. At first, the British authorities were less than lukewarm about the idea, but they were unable to resist the pressure. This came from the public and press who saw young able-bodied Jewish civilians who, because they were Russian-born, or had been born in a part of Poland which regularly changed hands between the Russians, Germans and Poles, were disqualified from service in the British Army. On 23 August 1917, the *London Gazette* officially announced the formation of a Jewish regiment, the 38th Battalion of the Royal Fusiliers. It was later joined by the 39th Battalion, which was made up largely of American Jewish volunteers, and after that by the 40th Battalion which was formed from Palestinian Jewish volunteers. The regiment served in the Middle East where it fought against the Turks with distinction, and is known as the Jewish Legion. David Ben-Gurion, first Prime Minister of Israel, served in it. In a letter to his wife at the end of July 1918 he writes,

> The camp we are staying in is not far from the beautiful port of Plymouth . . . called Egg Buckland . . . I was intoxicated by the charming scene . . . green mountains and valleys covered with silk, fertile fields . . . The Sabbath is observed here, and on that day we are let off all training, apart from marching to the synagogue together with all the officers, headed by the colonel.[227]

A Jewish family called Kauffman had been resident in Plymouth since 1851. Henry Kauffman, who may well have belonged to this family, joined the 38th Battalion. On 30 December 1917 he married Annie Goodman in the Plymouth synagogue.[228] Another wartime romance was that between a Jewish soldier, Harry Israel Woolf of the South African Infantry, stationed at the Raglan Barracks in Stonehouse, who married Isabella, daughter of Joseph Cohen of Plymouth.[229] A less happy ending came to a young Westminster lad whose remains are marked by a tombstone in the Congregation's Gifford Place cemetery:

In loving memory of
John Lithman
late of the Judeans, 38/40 Royal Fusiliers
died Jany 8th 1919/5679
aged 16 yrs & 9 months
Erected by his sorrowing parents
M.H.D.S.R.I.P.

The Hebrew inscription refers to him simply as 'of the Jewish Army'.[230] Another war grave of a Jew serving in the Judeans was that of eighteen-year-old Myer Nyman of Swansea who adopted the military name of Michael Burns and who died on 2 February 1919.[231] Yet another war grave was that of Stoker 1st class Harry Phillips of HMS *Vivid* who died on 2 April 1918 aged 29.[232]

Ranks were omitted on the World War I tablet, but included on that of World War II. In the Second World War there were 14 officers, the highest ranking being Colonel Telfer, a professional soldier. Major G. Robins received the MBE for work which he did beyond the call of duty, and Sgt R.B. Emdon died as a result of injuries received at Dunkirk. Only those who were actually members of the Plymouth Congregation at the time the tablets were written are commemorated. Thus the names of 23-year-old Sgt Morris Solomon of the Royal Australian Air Force, buried in Gifford Place,[233] and Isabella Mary Vassie, a member of the Women's Auxiliary Air Force, of Wolsely Road, Devonport, who died aged 20 on 11 October 1942 and was buried at the non-Jewish cemetery at Weston Mill were not recorded.[234] The names of Mark Woolfson of Devonport, who moved with his family from Devonport to Bournemouth on his demobilization, or David L. Maxwell, a Canadian volunteer pilot or Maurice Overs, a petty-officer in the Marines, who became members after the table was written, were therefore omitted, even though they had done their military service in Plymouth.[235] The last two also provided Second World War local romances. David Maxwell marrying Gussie Holcenberg and Maurice Overs marrying Ruth Bloom.[236] Perhaps because he had already left Plymouth, the name of Morry Smith was also not included. He had a barber shop *cum* fruiterer's in Union Street, Plymouth. He joined the army and became middleweight boxing champion of the British Expeditionary Force.[237]

It was not only in a military capacity that the Jews of the South West and the rest of Anglo-Jewry desired to do their civic duty, but they also played a full part in civil life. Geoffrey Alderman has discussed the Jewish dimension in British political life.[238] For the first hundred years after the re-settlement, Jews in England were content to keep a low profile. In 1753 Jewish merchants prevailed upon the Whig government to give its blessing to the *Jewish Naturalization Act*. This Act enabled foreign-born Jewish merchants to become naturalized, thereby saving themselves costly 'alien duties' on their imports, without being obliged to receive the Sacrament and take the Protestant Oaths of Allegiance. There was a wave of country-wide agitation against the Act. 'Jew' and 'Whig' became synonymous terms in the political rhetoric of 1753–4.[239] The Exeter City Council passed a resolution soundly condemning the Act. The eighteenth century, alas, was still an age of religious bigotry and the Act was repealed. From about 1830, the Jews of England began to take an active interest in politics at local and national level. A number of Jews who had abandoned their faith helped to pave the way for their still-committed brethren. Sir Menasseh Masseh Lopes was returned as MP for Romney in 1802, becoming Sheriff of Devon in 1810; in 1814 his nephew, Ralph Franco, became MP for Westbury; in 1818 Ralph Bernal was elected at Lincoln; and in 1819 David Ricardo, the political economist, became MP for Portarlington.[240]

At this period it seems that insofar as Jews displayed any bias it was towards the Conservative interest, but by the middle of the nineteenth century the Jewish vote inclined to the Liberals.[241] At Plymouth in 1857, for example, 'one of the Liberal candidates, James White, was quick to point out, that as a citizen of London, he had always given his vote to Rothschild and his support to the removal of Jewish disabilities'. To make assurance doubly sure he announced that he had given ten pounds 'for distribution among the poor Hebrews'![242] With one exception, William Woolf, the Jewish voters of Plymouth voted for White and his running mate.[243] Enfranchisement had been in the air since the 1820s; the emancipation of Catholics took place in April 1829. The Reform Bill was passed in 1832: the only substantial body of male English citizens left without Parliamentary vote and excluded from Parliament was the Jews. For them, too, relief was in sight. In 1835, an Act which incidentally relieved voters from the necessity of taking any oaths threw the franchise open *de jure* as well as *de facto* to professing Jews.[244] To secure the franchise and the right to municipal office probably meant more to the wider Jewish community than the right to sit in Parliament, a right granted only in 1858. True, the right to sit in Parliament affected all English Jews, but in practice only a handful of the wealthiest, aristocratic families could hope for a seat, whereas municipal office was open, especially in the provinces, to the middle-class Jew, related to many of and known to all his coreligionist townsmen.

Just how great was the change in the attitude of the English middle-class between 1825 and 1845 is exemplified by the election of Solomon Ezekiel[245] to the Penzance Board of Highways in 1845. The *Voice of Jacob* in its report sums up the change:

> On 25 March [1845] a burgess eulogized the Jewish character for sobriety, loyalty and general good conduct. He proposed Mr. Solomon Ezekiel who was elected unanimously. There has been a great change of late, for 20 years ago such a proceeding would almost have produced a riot.[246]

In Plymouth, or rather Devonport, the climate had changed somewhat earlier, for Phineas Levy[247] was one of the 75 Commissioners elected following a writ of *mandamus* in 1829.[248] After 1830, election of Jews to municipal office became almost commonplace in the provinces. A Jew was elected as a councillor in Southampton in 1838 and at Portsmouth in 1841; in 1839 as a founder member of the Corporation of Birmingham. Sir David Salomons served as Sheriff of Kent in 1839, whilst E. Lousada was Sheriff of Devon in 1842–3.[249] Some seven years later in 1836 Charles Marks[250] was elected an Assessor of Plymouth.[251] After these two came a succession of Jews taking part in civic affairs: the office of Guardian of the Poor for Plymouth was filled by William Woolf, as well as Josiah and Joseph Solomons in the 1850s; Eliezer Emdon was a Poor Law Commissioner in the 1860s, whilst Israel Roseman acted in that capacity in Stonehouse in the 1870s.[252] In 1862, Woolf became the first Jew to serve on the Plymouth Town Council and continued to do so at least until 1879.[253] Concurrently with Woolf a father and son played their part in civic life. The father began his rather late in life when he had already completed his biblical span of three score years and ten. He was Abraham Emdon[254] who was elected to the Town Council in Devonport in 1869 and became a member of the General Purposes Committee for Morice Ward in 1870.[255] He died in 1872, and his seat was taken by his son, Eliezer, who was elected unopposed.[256] Eliezer's service as a Poor Law Commissioner in the 1860s has been mentioned and he continued in public office for the next 40 years. In 1896, he was elected an Alderman for Ford Ward and in 1897 for Keyham. He was proposed as Mayor but declined the honour on account of his wife's poor state of health.[257] Other local politicians include Aaron E. Lyons[258] who was elected to the Stonehouse Board of Health in 1881, becoming chairman of the Urban District Council in 1890, leaving in that year for London to study for the bar,[259] and Myer Fredman, who was elected to Clowance Ward in 1893, becoming Mayor of Devonport in 1911, serving on the Devonport Council for 34 years until his death in 1927.

This tradition of public service in Plymouth was continued in the twentieth century: Lionel Jacobs who had a furniture shop in the Octagon was a Conservative Councillor for Millbay Ward in 1903 and for Valletort Ward in 1920; Eleazar Orgel stood for the Council in 1905 and lost by some 30 votes; Harry or Hyman Nelson, jeweller of Union Street, was nominated to the Board of Guardians in 1906; S. Robins was elected to the Council in 1936; Mrs Hester Robins was a Liberal Councillor for St Aubyn's Ward in 1929 and chaired the Public Health Committee. She was awarded the OBE in 1934 for her public service. At her death in 1936, her life and work were marked by a memorial service in the Plymouth synagogue; both Ernest Brock and his wife L.A., but better known as Cissie Brock, were active councillors for many years. A memorial service for Mrs Brock was widely attended; A. Fredman was elected to the Board of Guardians; I. Fredman was a councillor for St John's Ward, Devonport; Dr M.E. Gordon was a Conservative councillor for Nelson Ward in 1949; and Arthur Goldberg,[260] after a lifetime of public service, was elected Lord Mayor of Plymouth in 1961. It is apposite to mention here that Isidore Joseph,[261] who as a young man left Plymouth for Torquay in order not to open a business in competition with his family and who later returned to Plymouth, was elected Mayor of Torquay.

In the light of the foregoing it may well be said that the Jewish community in Plymouth built up a splendid tradition of public service throughout the nineteenth and twentieth centuries.

Curiously, there does not appear to have been any such tradition in civic service on the part of Jews in Exeter. There appears to have been only Alexander Alexander who was a Guardian of the Poor, and who acted as vice-President of the Exeter and Chairman of the St Paul's Ward Liberal Association, and he did not aspire to any office by public election.[262] It may be that as a Cathedral city Exeter was more resistant than Plymouth to accept Jews in a representative capacity.

In the latter part of the nineteenth century, professing Jews began to stand for Parliamentary constituencies in the West Country as well as in Devon and Cornwall.

Frederick Goldsmid, son of Isaac Lyon Goldsmid who had long campaigned for Jewish emancipation, was elected as a Liberal MP for Honiton in 1864, served only one year before his death, and was

*Illustration 19*

Arthur Goldberg—born Plymouth, 1908; educated at Plymouth College; Plymouth solicitor; Lord Mayor of Plymouth, 1961; died Plymouth, 1982.

succeeded by his son Julian who eventually became Deputy Speaker of the House.[263] A London Jew, Israel Abrahams, tried to gain a seat in Devizes in 1863. The *Western Morning News*, which adopted a consistently anti-Semitic stance,[264] opposed Abrahams, primarily because he was a Jew.[265] Abrahams lost, though not necessarily because of his religion. When a Jew, one Myer Jacobs became Mayor of Taunton in 1877, the local vicar wrote to him suggesting that if he were not a Christian he ought to resign.[266] By the turn of the century, the citizens of Plymouth, however, were prepared to vote for a candidate on his own merits and for the political views he represented. The religious faith of Sir S.F. Mendel, of London, whose grain ships *Nina* and *Rosina Mendle* called regularly at Plymouth, did not prejudice his chance when he contested Plymouth in the Liberal interest unsuccessfully in 1895, won it in 1898, and lost it

*Illustration 20*

The opening of the Holcenberg Collection, Plymouth City Library, 1963. (Rear row, l to r: Ald. I. Joseph, Ald. N.W. Lamb, Rabbi B. Susser, Mr D.L. Maxwell. / Front row, l to r: Ald. A. Goldberg (Lord Mayor), Mrs I. Joseph, Mrs A. Goldberg (Lady Mayoress), Mrs D.L. Maxwell, The Earl of Mt Edgecumbe.)

again in 1900.[267] When Mendl stood in 1898, Marcus Adler, Actuary of the Alliance Assurance Company and a staunch Conservative, wrote to Myer Fredman, the most prominent Jew in Plymouth, saying that he had been asked by leading members of the London community to enlist the support of Plymouth Jews for Mr Guest, the Conservative candidate. Marcus Adler's father had been Chief Rabbi and his brother Herman was then the Chief Rabbi. So his appeal was interpreted as a 'quasi-pastoral letter' and caused a local sensation. Not only did it seem that the Chief Rabbi was favouring the Conservatives but it also seemed that the Adler's preferred a Gentile to a Jew. Herman Adler wrote to *The Times* protesting that he had always kept aloof from party politics—which was not true—and telegraphed the Plymouth Liberals that 'My brother wrote without my authority'. As this was a purely secular matter Marcus needed no permission. 'In any case, Herman was exceedingly careful not to repudiate the letter, and Marcus never issued a disclaimer or a withdrawal.'[268]

Sir Julias Vogel, later to be Prime Minister of New Zealand, unsuccessfully fought the seat at Penryn and Falmouth in the 1880s, spending some £5,000 in the campaign.[269] In the general election of 1900 some Jewish Conservative candidates were viciously abused by the local Liberal press. The *Cornish Echo*, for example, attacked Nathaniel L. Cohen, a candidate for Penryn and Falmouth, as 'a stock exchange operator and a Jew'. He lost the election by twenty votes.[270] Arthur Strauss, the son of a Mayence Jew who had Cornish mining interests, was elected MP for Camborne in 1895, though he lost his seat in the general election of 1900.[271] Generally speaking, the Jews of the South West voted for Liberal candidates and Jews stood in the Liberal cause. This may be seen as quite natural as the Liberal party had put itself forward, rightly or wrongly, as the party which had supported Jewish aspirations for enfranchisement, and had sponsored the first six Jewish MPs.[272] For example, Mr Morrison, Liberal MP, in 1861 thanked the Jews of Plymouth, who, except for two, had voted for him *en masse*. He made the point explicitly, 'none have worked harder than our brethren of the Jewish persuasion (loud cheers) who, with the Roman Catholics, have voted for me and have shown their appreciation of what has been done in favour of them to promote the cause of religious freedom'.[273] There were, however, three nineteenth-century councillors in Plymouth, William Woolf, Lionel Jacobs, and Eliezer Emdon, who represented the Conservative cause.

In these ways, then, the Jews of the South West in the course of the eighteenth and nineteenth centuries, gradually lost their distinctiveness in dress and speech, broadened their education, widened their range of economic pursuits, extended their role in civic affairs and modified their religious practices until they were hardly, if at all, distinguishable from their fellow citizens in whose midst they lived. Through conversion in the early part of the nineteenth century, through acculturation and eventual assimilation, usually after intermarriage, through emigration to avoid this fate for themselves or their children and to seek better economic opportunities, the Congregations of the South West dwindled away until they died out in Cornwall and barely survive in Devon.[274]

# EPILOGUE

The foregoing pages have discussed the rise and subsequent decline of the Jewish communities in Devon and Cornwall, as well as various aspects of the social, religious and economic life of their individual members. By and large, it cannot be said that the Jews made any outstanding contribution to the life of the South West in any single sector, with the possible exception of the artistic in the first quarter of the nineteenth century. They did not introduce new industries as the Huguenots did elsewhere, nor new methods in old ones like the Jews of London in the furniture, catering and tailoring trades. But in view of their comparatively small numbers this is hardly surprising. Nevertheless, in their lives and vocations, in their religion and its practices, they added to the life of the two counties, particularly in the two leading cities of Devon.

Visible traces of the rich eighteenth- and nineteenth-century Jewish presence in the South-West are now few. Apart from the beautiful synagogues at Plymouth and Exeter which are still in use, as well as the disused ones in Falmouth and Penzance, and the cemeteries of the five Congregations, the Jews of the modern period have left behind them little more than the medieval Jewry of Exeter. There are some place names with tragic connotations—Jew's Woods in Plymstock,[1] and Jew's Lane at Herland Cross in Cornwall;[2] a few street names—Jews' Court[3] and Synagogue Place in Exeter;[4] some locations, for reasons unknown, vaguely identified with Jews—a Hebrew Brook at Kenegie, Lifton,[5] and The Jew's House at Polperro;[6] a few expressions in folk speech—Jews' fish,[7] a Jew beetle,[8] it's cold enough to shave a Jew,[9] *Makom Lamed* (= 'L. Place', i.e. London);[10] a Jewish section in the Plymouth City Library;[11] and some charitable bequests that seem to have been swallowed up in the anonymity of the Welfare State:[12] these are the main reminders of the once flourishing Jewish communities.

As the Falmouth and Penzance Congregations have disintegrated and died, and the Exeter Congregation just maintains its synagogue and cemetery, so too the Plymouth Congregation, in the course of the latter part of the twentieth century has slowly declined. Regular weekday services were abandoned after the Second World War. By the end of the 1970s, the Congregation had reduced its officials from a minister, *chazan* and *shammas* to a part-time minister, and when he left in 1981, he was not replaced. Instead, the Congregation relies on visiting officiants using its accommodation as a holiday flat. By the end of the 1980s it is often difficult to obtain a minyan on Friday nights and Sabbath mornings, as well as on the second days of Festivals. If the decline continues, then perhaps its beautiful synagogue, too, will be opened once a year for services only at the High Festivals. Still, prediction is exceedingly difficult in social and economic life. Perhaps, there may yet be an influx of Jews into the South West. The University of Exeter has a number of Jewish academics and students, and these could be a valuable aid to a renewed Congregational life. Torquay, as well as other suitable spots, may well thrive as a retirement centre of the South West. Along with others, Jews may retire there in greater numbers, or settle there in order to take advantage of new economic opportunities.

In their heyday, at the end of the Napoleonic Wars, there were between 700 and 1,000 Jews in Devon and Cornwall representing some 4 per cent of all Anglo-Jewry; in 1989 there were barely 150 Jews in the two counties and the percentage had dropped to negligible proportions. The sun seems to have settled on the organized Jewries in Cornwall, their future in Devon is very uncertain.

Many of the descendants of the Jewish settlers in the South West, like many of the Flemings, Huguenots, Irish, Scots and other nineteenth and twentieth century European immigrants to England, have been absorbed into the wider reaches of English society and have lost most, if not all, of their national and religious characteristics. Others are to be found scattered over the world and still firmly attached to their faith and people. Gumpert Michael Emdin journeyed from Emden and encamped in Amsterdam. And he journeyed from Amsterdam and he encamped in Plymouth. And his descendants journeyed from Plymouth and encamped in South Africa. And their descendants journeyed from South Africa and encamped in Canada. And in the ninth generation a young man[13] journeyed from Canada and came home to study the Torah in Jerusalem. These pages are the story of those Jewish settlers who made their home in Devon and Cornwall—and their tribute.

# NOTES TO PAGES 1–25

## *Chapter One*

1. I Kings, vii, 14 and I Kings, xvi, 31.
2. M. Margoliouth, *History of the Jews in Great Britain* (1851), I, 9–15 (afterwards quoted as Margoliouth, *Jews in Great Britain*).
3. J. Bannister, 'Jews in Cornwall', *Journal of the Royal Institution of Cornwall*, II (1867), 324 (afterwards quoted as Bannister, 'Jews in Cornwall'), usefully summarizes all the arguments.
4. S. Applebaum, 'Were there Jews in Roman Britain?' *Transactions of the Jewish Historical Society of England* (afterwards quoted as *TJHSE*), XVII (1950), (afterwards quoted as Applebaum, *Roman Britain*), p. 189.
5. Margoliouth, *Jews in Britain*, I, 23. For 300 Ancient British expressions which are also Hebrew homonyms and synonyms, see H. Rowlands, *Mona Antiqua Restaurata* quoted in T.S. Duncombe, *The Jews of England* (1866), p. 25, where he mentions some 30 examples; and M. Margoliouth, *Vestiges of the Historic Anglo-Hebrews in East Anglia* (1870), p. 14, and p. 65 where he quotes eight phrases.
6. E.N. Adler, *History of the Jews of London* (Philadelphia, 1930), p. 1.
7. White's *Devonshire Directory* (1850) (afterwards quoted as White, *Devon Direct.*) p. 41. The houses near Chudleigh (Ordnance Survey (1960) SX 87/839,765) called Jews Houses should not be confused with these smelting ovens. They take their name from their proximity to Jew's Bridge, for which see above, p. 24.
8. A.K. Hamilton-Jenkin, *The Cornish Miner* (1962), p. 68f. The term may have originally come into use during the medieval period, (see above, p. 22).
9. W.C. Borlase, *Antiquities of Cornwall* (1769), p. 163. See also T. Hogg, *Manual of Mineralogy* (1828), p. 74, and *Journal of the Royal Institution of Cornwall*, IV (1871), 227.
10. *Gent. Mag.* XCVI (1826), 125.
11. The name Marazion is itself suggestive of Hebraic origin, meaning either 'sight of Zion' or 'bitterness of Zion'. For Market-Jew, see Additional Note 1, p. 335.
12. *JC*, 1 June 1860.
13. C. Roth, *The Rise of Provincial Jewry* (1950) (afterwards quoted as Roth, *Provincial Jewry*), p. 91, but cf. C. Roth, 'Jews' houses', *Antiquity*, XXV (1951), 98, 66–68 where he discusses place names associated with 'Jew', such as Jews' Tower at Winchester, Jews' Mount at Oxford, Villejuif near Paris or Judenberg in Germany. He points out that often there is no proven Jewish community at that place. He suggests that when the origin of a large structure was unknown, it would be ascribed to the Jews, 'loosely corresponding to the

term "Cyclopaean" in vogue today—or yesterday—to describe massive structures of great or even mysterious antiquity.' But a Jew's House is only 3 feet high and could hardly be called even a large, let alone gigantic, structure.

14. Applebaum, *Roman Britain*, p. 205.
15. *Devon and Cornwall Notes and Queries*, 3rd ser. XI, p. 456.
16. Applebaum, *Roman Britain*, p. 190.
17. Ibid.
18. Lady Aileen Fox, *Roman Exeter* (Exeter, 1952), Plate X b.
19. Cf. P. Romanoff, 'Jewish symbols on Ancient Coins', *Jewish Quarterly Review*, XXXIII (1943), pp. 14, 15.
20. *Shin* is used as an abbreviation for *Shadai* (= Almighty) and is used on the phylactery of the head.
21. Applebaum, *Roman Britain*, pp. 197–9, quotes all the literary references.
22. See Mark, xv, 42f.
23. *Jewish Encyclopaedia* (New York, 1901) (afterwards quoted as *Jew. Encycl.*) s.v. JOSEPH OF ARIMATHAEA.
24. For the persistence of the legend see C.C. Dobson, *Did our Lord visit Britain?* (Glastonbury, 1936); the seventh edition of this work was reprinted for the third time in 1959. And also Brendan Lehane, 'Did Christ come to Britain?', *Weekend Telegraph*, 116, 16 December 1966.
25. C. Roth, *A History of the Jews in England* (3rd edn, Oxford, 1964) (afterwards quoted as Roth, *Jews in England*), p. 2.
26. M. Margoliouth, *Vestiges of the Historic Anglo-Hebrews in East Anglia* (1870), p. 14.
27. Bannister, 'Jews in Cornwall', p. 335.
28. M. Margoliouth, loc. cit. V. Newall, 'The Jews of Cornwall in Local Tradition', Jewish Historical Society of England, *Miscellanies* (afterwards quoted as *MJHSE*), XI (1979), 119–21, collects most of the evidence relating to Jews in ancient Cornwall. All her material is to be found in B. Susser, 'The Jews of Devon and Cornwall from the middle ages until the twentieth century', unpbd. PhD diss. University of Exeter, 1977. She doubts if there was any major connection.
29. Roth, *Jews in England*, pp. 5, 6.
30. Ibid. p. 6.
31. Ibid. p. 11.
32. Starrs, charters and records of court cases are the main sources for the history of the medieval Anglo-Jewry, and therefore accounts of Jewish life in England are heavily coloured by financial transactions. Evidence is now coming to light of medieval intellectual activity in England, similar to that of the Jews of Northern France. See Ephraim Auerbach, *'Mitoratam shel Hachmei Anglia milifnei Hagirush'*, *Sefer Hayovel Tiferet Yisrael* (1966), pp. 1–56, and C. Roth, 'The Intellectual Activities of Medieval English Jewry', *The British Academy Supplemental Papers*, VIII, (Oxford, 1949).
33. Roth, *Jews in England*, p. 12.
34. Ibid., p. 13, but there is no evidence that the medieval Exeter Jewry ever had its own cemetery (M. Adler, 'The Medieval Jews of Exeter', *Report and Transactions of the Devonshire Association*, LXIII (1931), (afterwards quoted as Adler, 'Medieval Jews'), p. 222). See, however, the suggestion on 'Jew's Bridge', above, p. 25.
35. W.G. Hoskins, *Devon* (1954) (afterwards quoted as Hoskins, *Devon*), p. 131.
36. Ibid. p. 135.

37. *Victoria History of Cornwall* (1906), p. 525; *British Mining* (2nd ed. 1887) 92, 918 (quoted in *TJHSE*, XXIX (1982), 23).
38. Ibid. p. 540.
39. I.e. *Dieu-le-saut*, May God save him—the French translation of the Hebrew name Isaiah.
40. 1 mark = 13*s*. 4*d*.
41. Adler, 'Medieval Jews', 223. A misprint in J. Jacobs, *Jews of Angevin England* (1893) (afterwards quoted as Jacobs, *Jews of Angevin England*), p. 73, makes Deulesalt pay the fine for his 'boys' instead of 'bonds'.
42. A gold mark was eight ounces of gold.
43. Adler, 'Medieval Jews', p. 222.
44. J. Katz, *Exclusiveness and Tolerance* (Oxford, 1961) (afterwards quoted as Katz, *Exclusiveness*), p. 57.
45. Roth, *Jews in England*, p. 282.
46. *Calendar of the Plea Rolls of the Exchequer of the Jews*, ed. J.M. Rigg (1905) (afterwards quoted as Rigg, *Plea Rolls*), I, 242. See also above, p. 16.
47. Roth, *Jews in England*, p. 22, says that only Winchester Jewry escaped molestation; Adler, 'Medieval Jews', p. 223, asserts that none of the West Country Jewish communities was involved.
48. They were London, Lincoln, Norwich, Winchester, Canterbury, Oxford, and perhaps Bristol (Roth, *Jews in England*, p. 29).
49. Roth, *Jews in England*, pp. 29, 30.
50. Ibid. p. 91.
51. Ibid. p. 29.
52. H.G. Richardson, *The English Jewry under Angevin Kings* (1960) (afterwards quoted as Richardson, *English Jewry*), pp. 18, 19.
53. V.D. Lipman, *The Jews of Medieval Norwich* (1967), p. 73. Cf. Rigg, *Plea Rolls*, I, 148, for an exceptional case where a depositor had *free* access to the chest, and also op. cit. p. 271.
54. Rigg, *Plea Rolls*, I, 82.
55. Ibid.
56. V.D. Lipman, 'The Roth "Hake" Manuscript', *Remember the Days*, ed. J.M. Shaftesley (1966), p. 55.
57. Rigg, *Plea Rolls*, II, 193, 200, 218, 258.
58. Ibid. I, 132.
59. *TJHSE*, VII, (1911), 44.
60. Roth, *Jews in England*, p. 27.
61. Amiot was active in Exeter for some years. In 1204 he lent £5 to Sir Henry de la Pomeroy (for whom see J. Prince, *Worthies of Devon* (1810), p. 645), for which the King exacted a tax of a bezant (2*s*.) for each £1. In 1211, he tenanted a house in the High Street belonging to one Godeknight (Adler, 'Medieval Jews', p. 224).
62. *MJHSE*, I, lxvii: however, Richardson, *English Jewry*, p. 164 suggests that the financial records of the period are incomplete.
63. 'This is clear from the glosses of the English *Tosafists* and from the fact that Richard of Devizes makes a French Jew recommend a lad not to go northward in England, because he will find none speaking Romance' (Jacobs, *Jews of Angevin England*, p. 338). See also *The Sefer Ha-Shoham*, ed. B. Kar (1947), p. 11.
64. Richardson, *English Jewry*, p. 93.
65. Ibid. p. 93, but cf. A.E. Poole, *From Domesday Book to Magna Carta* (Oxford,

1955), p. 2, who writes that 'by the latter part of the twelfth century it was well nigh impossible to tell whether a man was Norman or English'.

66. *TJHSE*, II (1895), 91, and Adler, 'Medieval Jews', *passim*.
67. Roth, *Jews in England*, p. 34.
68. Adler, 'Medieval Jews', p. 227 erroneously gives the date as 1216, see Roth, *Jews in England*, p. 35.
69. Adler, 'Medieval Jews', p. 227 gives the figure of 60,000 marks.
70. See above, p. 9 n. e.
71. Adler, 'Medieval Jews', p. 227. Personal descriptions were not uncommon, e.g. Moses *cum naso* (with a nose), Manasseh *grassus* (big), Benedict *longus* (tall), Deudoné *cum pedibus tortis* (with lame feet) (H.P. Stokes, *Studies in Anglo-Jewish History* (Edinburgh, 1913), p. 65).
72. Jacobs, *Jews of Angevin England*, p. 239. In 1204, they lent 6 marks to Robert fil Ascelin.
73. In 1205, he lent £5 to John Sep (Jacobs, *Jews of Angevin England*, p. 240). This Deulecresse was perhaps the progenitor of an Exeter family of *Cohenim* whose titular name became their surname. In 1266, there was a Deulecresse le Chapelyn of Exeter (Adler, 'Medieval Jews', p. 231), probably identical with Deulecresse le Prestre of Exeter who is mentioned in 1277 (Adler, 'Medieval Jews', p. 233). It is not unlikely that the 1205 Deulecresse was the grandfather of his 1270–6 namesake.
74. Together they lent Eustace son of Albert £8 in 1205 (Jacobs, *Jews of Angevin England*, p. 240). Yveliny also occurs in Exeter records as a Christian name (Adler, 'Medieval Jews', p. 224).
75. Roth, *Jews in England*, p. 36, quoting *Chronicle of Lanercost*, p. 7.
76. Roth, *Jews in England*, p. 52.
77. Ibid. p. 37.
78. Ibid. p. 40.
79. Rigg, *Plea Rolls*, I, 18.
80. She married Alexander II, King of Scotland, at the age of eleven.
81. He would be called 'of Gloucester' after he had left that town to reside in another. If he moved from Exeter, he would most likely then be known as Jacob of Exeter. Moses of Exeter, below, known by that name whilst still in Exeter, was probably so called to distinguish him from another Moses in a different town.
82. Adler, 'Medieval Jews', p. 227.
83. Jewish women frequently acted as financiers in the medieval period and later periods (Roth, *Jews in England*, p. 115).
84. Adler, 'Medieval Jews', p. 228.
85. Ibid. p. 223.
86. The following table, after Roth, *Jews in England*, p. 271, summarizes the royal exactions and illustrates the rapid rise in the rate of taxation and the resultant impoverishment of the community in the second half of the century.

| *Year* | *Amount in marks* |
|---|---:|
| 1221–30 | 33,000 |
| 1231–40 | 63,000 |
| 1241–50 | 117,000 |
| 1251–60 | 72,000 |
| 1261–70 | 6,000 |

Richardson says that various writers have greatly exaggerated the tallages exacted by Henry III (Richardson, *English Jewry*, p. 214).

87. Although the city and castle of Exeter were given by the King to his brother Richard, Earl of Cornwall, in 1231, the Jews of Exeter remained the sole property of the Crown, and continued to pay their taxes to the royal treasury (Adler, 'Medieval Jews', p. 228).
88. Roth, *Jews in England*, p. 45.
89. Adler, 'Medieval Jews', p. 228.
90. Rigg, *Plea Rolls*, I, 75.
91. Roth, *Jews in England*, p. 47. Richard protected 'his' Jews and taxed their resources with discretion.
92. Adler, 'Medieval Jews', p. 229. In 1255, Bonenfant was the pledge for another royal tallage, as well as for £5 5*s*. 0*d*. tax due from another Exeter Jew, Isaac son of Abraham.
93. Richardson, *English Jewry*, p. 172 and Roth, *Jews in England*, p. 49.
94. Rigg, *Plea Rolls*, I, 151, where Belia is his widow in 1267. He was still alive in 1263 when he had business dealings with Bonenfant (Rigg, *Plea Rolls*, I, 182).
95. At his decease the value of his chattels, within and without the *archa*, was 40*s*. Tercia, his widow, paid one mark fine to take up her inheritance.
96. Adler, 'Medieval Jews', p. 230. A rigorous search of Exeter's *archa* by the Sheriff revealed no assets.
97. Published in London, 1925, for the History of Exeter Research Group of the University College of the South West.
98. Paragraph 65, p. 37.
99. But cf. Rigg, *Plea Rolls*, II, 194 where 12 Christians and 8 Jews make inquest on a disputed charter. For the permissibility of Jews making Christians take a Christological oath in spite of a prohibition against doing so (*Sanhedrin*, 63b), see Katz, *Exclusiveness* p. 35. Some Jews relinquished their claims rather than take any oath, see I. Elfenbein, *Teshuvot Rashi* (New York, 1943), p. 327.
100. Roth, *Jews in England*, pp. 59–62.
101. Adler, 'Medieval Jews', p. 231.
102. Rigg, *Plea Rolls*, I, 132. They would not go of their own accord, the Sheriff had to send them. In addition, the Jews probably had to pay a tax for leaving Exeter, even temporarily, to go to London (cf. *Calendar of the Plea Rolls of the Exchequer of the Jews*, ed. H.G. Richardson (1972), IV, 147, item 81).
103. C. Roth, *The Jews of Medieval Oxford* (Oxford, 1951), p. 28.
104. Roth, *Jews in England*, pp. 63–4.
105. Probably the present Newton Abbot. Adler, 'Medieval Jews', p. 231, n. 9, says that one of his clients was a Paulinus of Newton. See above, p. 7.
106. Rigg, *Plea Rolls*, I, 242. William de La Leye, probably the same as this William le Layte, was the guardian of Hugh Fychet who in 1274 claimed that Copin and Jacob Crespin had fraudulently placed an £80 bond in the *archa* (Rigg, *Plea Rolls*, II, 140). The assault may mark the beginning of the dispute.
107. Adler, 'Medieval Jews', p. 231.
108. Rigg, *Plea Rolls*, IV, 49.
109. Ibid. 103, 107.
110. Roth, *Jews in England*, p. 70.
111. Ibid. p. 272.
112. Cf. H. Jenkinson, *Calendar of the Plea Rolls*, (1929), III, 200, 'good, dry pure clean and better wheat without evil moisture'.

113. R.R. Mundill, 'Anglo-Jewry under Edward I', *TJHSE*, XXXI (1990), 1–21, strongly supports the view that the transactions were genuine, and that the Jews had become credit-agents trading in futures.
114. Richard of Devizes (ed. Howlett, p. 437) writing in 1192, 'Exeter feeds men and beasts with the same corn'.
115. Hoskins, *Devon*, p. 107.
116. Roth, *Jews in England*, p. 115.
117. *TJHSE*, II (1894), 85–105.
118. Rigg, *Plea Rolls*, IV, 98.
119. Roth, *Jews in England*, p. 73.
120. See *TJHSE*, I (1893), 15–24. A Jew's property escheated to the King when he or she was converted to Christianity. The ensuing poverty proved a bar to conversion of Jews. Neophytes were therefore given '. . . a home and a safe refuge for their whole lives with sufficient sustenance without servile work . . .' (C. Trice Martin, 'The Domus Conversorum', *TJHSE*, I (1893), 15, quoting Mathew Paris).
121. Roth, *Jews in England*, p. 74. For Christian moneylenders and their methods of evading canon law on usury, see Abrahams, *Starrs and Charters*, ii, civ–cviii.
122. See Richardson, *English Jewry*, pp. 217–23 for a discussion of the subject.
123. Roth, *Jews in England*, p. 75. A perusal of the receipts of perquisites at the Tower from the Jews of London, 1277–8, indicates some dozens of different excuses for 'fleecing' the Jews (Rigg, *Plea Rolls*, IV, 173–94).
124. Adler, 'Medieval Jews', p. 234.
125. *Close Rolls*, 1284, p. 278. See also below, n. 155 and above, p. 16.
126. It is surprising that though his house was confiscated at his death his property was not forfeited; 35 of his bonds valued at £357 were in the *archa* when it was opened in 1290. His clients included Sir Robert le Denys, Richard Bullock, the goldsmith, who was one of the Christian chirographers, two priests, Roger de Moleyns and Arnulf of Hunecroft, and numerous other residents of Devon and Somerset (Adler, 'Medieval Jews', p. 237).
127. In 1238, Ursell of Exeter fled after a Jewish enquiry into coin-clipping (Richardson, *English Jewry*, p. 221, n. 1).
128. When the *archa* was opened in 1290, 24 tallies were found recording loans ranging from 2*s*. to £3 13*s*. 3*d*.; the few that were dated were from 1286–9.
129. J. Parkes, *The Conflict of the Church and the Synagogue* (Philadelphia, 1961), p. vii, and Katz, *Exclusiveness*, p. xi.
130. Roth, *Jews in England*, pp. 76–7.
131. H. Graetz, *History of the Jews* (Philadelphia, 1956) (afterwards quoted as Graetz, *History of the Jews*), iii, 509. Graetz asserts that 'the great misery of the Jews during the Middle Ages began with Pope Innocent III and the fourth Lateran Council'.
132. The ground had been prepared by a series of tracts agains the Jews: *Dialogus inter Christianum et Judaeum de fide catholica* (anonymous, between 1123–48). Bartholomew, Bishop of Exeter's *Dialogus contra Judaeos ad corrigendum et perficiendum destinatus* (1180–4); Peter of Blois, *Invectiva contra perfidiam Judaeorum* (*c*. 1204) (*TJHSE* XVII (1951), 230). Bartholomew's *Dialogus* may have been prompted by the settlement of Jews in Exeter at that period (above, p. 5).
133. Roth, *Jews in England*, pp. 54, 77–8.
134. Ibid. p. 40.

135. Graetz, *History of the Jews*, iii, p. 516.
136. Roth, *Jews in England*, pp. 119–21. Katz, *Exclusiveness*, p. 9, however, asserts that occasional infringements did provoke public admonitions.
137. *Jew. Encycl.* s.v. BLOOD ACCUSATION. Hardly a decade has passed since 1144 without the 'Blood Libel' being raised somewhere. Riots following publication of Ritual Murder charges occurred in Russia, in Tashkent in 1961, and Margelan in 1962 (*JC*, 25 January 1963). See also *Pravda*, 5 May 1993 for recent changes.
138. Roth, *Jews in England*, pp. 13, 18, 21.
139. Ibid. p. 45.
140. 'Little St Hugh of Lincoln.'
141. Roth, *Jews in England*, pp. 55, 56, 78.
142. Ibid. p. 21. Other instances were: in 1234, at Norwich (p. 53); in 1274, London Jews coerced a woman convert to go overseas so that she might return to her ancestral faith; in 1290, at Oxford (p. 83).
143. Ibid. p. 271. It was deemed specially meritorious to assist converts, especially those who had apostatized under duress, to escape from Christianity (*Sefer Hasidim* (Berlin edn. 1891), pp. 200, 201, 209).
144. Ibid. p. 41.
145. In *TJHSE*, VI (1908), 255ff., various dates are given for the incident from 1260–75.
146. Richardson, *English Jewry*, p. 32, and Roth, *Jews in England*, p. 83. For a discussion of the motives leading to apostasy see Katz, *Exclusiveness*, pp. 74–6.
147. M. Adler minimized the total (*Jews of Medieval England* (London, 1939) (afterwards quoted as Adler, *Medieval England*), p. 32).
148. Richardson, *English Jewry*, p. 29, corrects Adler's view that the number was insignificant. On the other hand, Katz, *Exclusiveness*, p. 68, strongly suggests that throughout Europe there were only isolated instances of voluntary conversion.
149. Nicholas le Jew at St Winnow, Cornwall, in 1321, must surely have been a convert (*Bibliotheca Cornubiensis*, p. 914).
150. Perhaps Leo of Bourg, chirographer about 1266.
151. Rigg, *Plea Rolls*, I, 132, 265.
152. Adler, *Medieval England*, p. 309. It is not impossible that Henry and Richard are the baptismal names of Samuel and Solomon.
153. Adler, *Medieval England*, p. 351. She was one of the 28 women who resided in the Domus from 1280. Alice and Claricia were still there in 1308 (*TJHSE*, IV (1899), 54).
154. *TJHSE*, IV (1899), 26.
155. See above, n. 125.
156. She left the Domus in 1309, went back to Exeter, married and had two children, Richard and Katherine. In 1327, she left the children in Exeter and returned to the Domus where they joined her in 1333. Richard was granted a pension of one and a half pence a day even when he left the Domus (*TJHSE*, IV (1899), 26).
157. If Henry and Richard are the same as Solomon and Samuel then the percentage is reduced to about 4.5 per cent.
158. Richardson, *English Jewry*, p. 28.
159. He was Bishop of Exeter from 1280–91 (G. Oliver, *Lives of the Bishops of Exeter* (Exeter, 1861), p. 48.

160. Roth, *Jews in England*, p. 77.
161. In 1286 a wealthy Jewish financier of Hereford had invited his Christian friends to his daughter's wedding which was celebrated with 'displays of silk and cloth of gold, horsemanship and playing sports and minstrelsy' (W.W. Capes, *Registrum R. de Swinfield*, pp. 120–1).
162. Roth, *Jews in England*, p. 78.
163. *TJHSE*, II (1894), 91.
164. *TJHSE*, II (1894), 91, no. 39. V.D. Lipman estimates the Jewish population of Norwich at the time of the expulsion to have been about fifty or sixty (V.D. Lipman, *The Jews of Medieval Norwich* (1967), p. 38). If the number of bondholders and the value of their bonds are any criteria then the medieval Jewish population of Exeter was about 100 at the time of the expulsion in 1290.
165. Roth, *Jews in England*, p. 84.
166. Ibid. p. 72.
167. Richardson, *English Jewry*, p. 225.
168. Ibid.
169. In 1190, the Jews had provided one-seventh of the royal income, by 1290 the proportion had dropped to one one-hundredth (Roth, *Jews in England*, p. 84).
170. Roth, *Jews in England*, p. 85.
171. For the main routes of the Expulsion, see Map 2.
172. Aaron of Cornwall together with Moses Rod were arrested at Uxbridge in 1244 on a charge of stealing a horse (Rigg, *Plea Rolls*, I, 98). He may be identical with Aaron of Caerleon (? Carlyon Bay, 3 miles from St Austell).
173. Quoted by Jacobs, *The Jews of Angevin England*, p. 186.
174. Ibid. p. 188.
175. Quoted by Arthur Bluett, *Cornish Magazine*, ed. A.T. Quiller-Couch (Truro, 1899), ii, 274.
176. Ibid. p. 269.
177. *Journal of the Royal Institution of Cornwall*, XVII (1907), 320.
178. See the descriptive card relating to this object at the Royal Institution of Cornwall, Truro.
179. I am indebted to Mr H. Douch, Curator of the Royal Institution of Cornwall, for this suggestion. He also weighed and measured the figure.
180. *VJ*, 13 March 1845.
181. Though the more usual name for England was *Iyyei HaYam* (see *TJHSE*, XVII (1952), 74, n. 2).
182. *TJHSE*, IV (1899), 215.
183. See above, n. 51.
184. Ordnance Survey, (1960), SX 87/839,765.
185. *Episcopal Registers of Exeter: Edmund Stafford* (1886), p. 223.
186. H.R. Watkins, *Totnes Priory and Medieval Town* (Torquay, 1914), p. 348.
187. *The Place Names of Devon*, eds J.E.B. Gover, A. Marver, O.F.M. Stenton (Cambridge, 1932), p. 469.
188. C. Henderson, *Old Devon Bridges* (Exeter, 1938), p. 47.
189. Exeter City Council Act Book, XIV (1753), 198.
190. C. Henderson, *Old Devon Bridges* (Exeter, 1938), p. 47.
191. White, *Devon Direct*. p. 470.
192. See Roth, *Jews in England*, p. 103, n. 1.
193. Cf. the cross at Oxford built with Jews' money, 1275, (Richardson, *Plea Rolls*, IV, 76).

194. The medieval Jewish cemetery at York is in a place called Jewbury. Perhaps Jewysbrugge in Devon was adapted from a name similar to Jewbury.
195. Compare, for example, the origin of the name Jew's Lane, above, p. 25. There is an old house in Polperro, Cornwall, which is called 'The Jew's House' but it got its name as late as 1922 (letter to author from Mr F. Nettleinghame, 14 June 1963).
196. See Roth, *Jews in England*, pp. 132–48, 158–66.
197. *TJHSE*, IV (1899), 35. For medieval Jews in isolated places see C. Roth, 'The ordinary Jew in the Middle Ages', *Gleanings* (New York, 1967), p. 25.
198. *TJHSE*, IV (1899), 87, 100; XXIX (1982), 9–21. See also M.B. Donald, *Elizabethan Copper* (1955), pp. 90, 299, 300, 343.
199. Item 54 in the *Catalogue of Exhibition of Anglo-Jewish Art and History* (1956), p. 17. Mr M.H. Gans of Amsterdam kindly sent me photocopies of his material on Palache.

# NOTES TO PAGES 26–67

## *Chapter Two*

1. Marranos were crypto-Jews who remained in Spain after 1492, and in Portugal after 1497, when professing Jews were banished.
2. Roth, *Jews in England*, p. 166.
3. Ibid. p. 173. The term Sephardi is used to describe Jews emanating from Spain and Portugal.
4. Ibid. p. 193.
5. Ibid. p. 199, n. 1.
6. For a detailed discussion of Anglo-Jewish statistics see V.D. Lipman, 'A survey of Anglo-Jewry in 1851', *TJHSE*, XVII (1953), 171–88. Jews tend to settle in ports. According to the *Jew. Year Bk*, 1967, pp. 188–90, about half the world's 13 million Jews live in ports. Most of England's Jews live in 50 ports or coastal towns, the remainder, about 80,000, are scattered in another 50 or so inland centres.
7. A.M. Hyamson, *The Sephardim of England* (1951) (afterwards quoted as Hyamson, *Sephardim of England*), p. 330.
8. Roth, *Jews in England*, pp. 269–70.
9. The decennial census in England has never enumerated religious denomination. Even the 1851 census which made a survey of religious places of worship, only numbered worshippers who actually attended a place of worship on 28–9 March, as well as the average attendance on a Sabbath morning for the previous six months, but did not attempt to number the adherents of any religious body.
10. Hyamson, *Sephardim of England*, p. 4, and Roth, *Jews in England*, p. 137.
11. Count Gondomar's letters from London were published as *Documentos ineditos para la Historia de España* (Madrid, 1936). Mr Edgar Samuel kindly supplied this reference.
12. Roth, *Jews in England*, p. 149.
13. See L.I. Newman, *Jewish Influence on Christian Reform Movements* (New York, 1925), pp. 1–3, for the use of 'Judaizing' and its synonymous forms in this sense.
14. R.N. Worth, *Calendar of the Plymouth Municipal Records* (Plymouth, 1893), p. 158. The correct entry in the Widey Court Book reads: 'Itm payd Edward Arnold in full discharge of all demands due from Jno. Lawrenson Hebrew high Jerman who was mayntayned att the charity of the Town of Plymouth att the Univ'sity—ij$^{li}$ xviij$^{s}$'.

15. H.R. Coulthard, *The Story of an Ancient Parish, Breage with Germol* (Penzance, 1913), p. 151. The source of this story appears to be S. Rundle's 'Cornubiana' in the *Transactions of the Penzance Natural History Society*, 1885/6 and quoted in M.A. Courtney, *Cornish Feasts and Folklore* (Penzance, 1890), p. 93.
16. Similarly, the murder of a Jew in the woods at Plymstock, outside Plymouth, led to them being called Jew's Woods and the hill leading to them, Murder Hill.
17. E.R. Samuel, 'The First Fifty Years', *Three Centuries of Anglo-Jewish History* (Cambridge, 1961), p. 30 (afterwards quoted as E.R. Samuel, 'The First Fifty Years').
18. The full title is of some interest: *An Abstract of all the Statutes made concerning Aliens trading in England, from the first year of King Henry the VII also of the Laws made for securing our Plantation Trade to ourselves, with observations thereon, proving that the Jews (in their practical way of Trade at this time) Break them all, to the great Damage of the King in his Customs, the Merchants in their Trade, the whole Kingdom, and His Majesties Plantations in America in their Staple, together with the Hardships and Difficulties the Author hath already met with, in his endeavoring to find out and Detect the ways and Methods they take to effect it* (afterwards quoted as Hayne, *An Abstract*).
19. E.R. Samuel, 'The First Fifty Years', p. 37.
20. Hayne, *An Abstract*, p. 27.
21. Ibid. p. 28.
22. E.R. Samuel, 'The First Fifty Years', p. 37.
23. *The Exeter Post-Master or The Loyal Mercury*, 25 September 1729. I am indebted to Mr Frank Gent for this reference.
24. Ottolenghe, *An Answer*, p. 5. For the full title of Ottolenghe's book, see Additional Note 2, below, p. 335.
25. Ibid. p. 17.
26. Roth, *Magna Bibliotheca Anglo-Judaica* (1937), p. 258 (afterwards quoted as Roth, *Magna Bibliotheca*). See also L.D. Barnett, *Bevis Marks Records* (Oxford, 1949), p. 82, nos 517, 519; Lucien Woolf, *The Treves family in England* (1896), pp. 3, 5.
27. *TJHSE*, XXII (1970), 149. His name indicates that he was a Sephardi.
28. J.H. Mathews, *A History of St. Ives, Leland . . .* (Elliot Stock, ?Cornwall, 1892), p. 296.
29. Ottolenghe, *An Answer*, p. 17. See also L.P. Gartner, 'Urban History and the Pattern of Provincial Jewish Settlement in Victorian England', *Jewish Journal of Sociology*, 23 (1981), 37.
30. *The London Gazette*, 24 October 1741. He was in Southgate Prison, 11 November 1742. His name is typically Jewish, otherwise there is no evidence to show he was a Jew.
31. Exeter Poor Rate Book, 1752–6, pp. 1, 125.
32. Exeter City Council Act Book, XIV, p. 232.
33. Jewish law forbids exhumation except in very restricted circumstances.
34. A copy of lease and release dated 27 and 28 March 1760 quotes an agreement made between Sarah and John Cummings. The lease is listed in Ply. Syn. Cat. 3, 6.
35. R.N. Worth, *The History of Plymouth* (Plymouth, 2nd Edition 1873), p. 170.
36. They were Joseph Cohen, bankrupt in 1749 (*Gent. Mag.* 1749, p. 430); Solomon Abraham and Jacob Myer Sherrenbeck who were on the conduit water system from 1755 and 1757 respectively (Plymouth Town Rental, 1755,

1757); Sarah and Joseph Jacob Sherrenbeck, who are mentioned in a lease dated 1744 (now in the possession of the Plymouth Congregation).

37. In the author's collection.
38. *Gent. Mag.* 1805, p. 1176.
39. I. Solomon, *Records of my Family* (New York, 1887) (afterwards quoted as Solomon, *Records*), pp. 5–7. For a similar situation in Liverpool, see Margoliouth, *Jews in Great Britain*, III, 110–11.
40. Photostat of Hart family bible in Roth collection.
41. *TJHSE*, XVII (1951), 66.
42. A photo of the clock face appeared in *The Illustrated London News*, 27 February 1971.
43. See above, pp. 47–51.
44. Cf. above, p. 172.
45. There are several Sherrenbecks in Germany: it has not been possible to narrow down the field.
46. Photostat in the Roth Collection.
47. Wm. Schonfield, 'The Josephs of Cornwall', p. 2.
48. Lipman, 'The Plymouth Aliens List', *TJHSE*, VI (1962) (afterwards quoted as Lipman, 'Aliens List'), *passim*.
49. PHC A/c. 1759.
50. They are: David, Levy and Mordecai Abrahams (5, 52, 51); Emanuel Cohen and David Jacob Coppel (58, 9); Eleazer, Moses and Solomon Emdin (4, 15, 6); Aaron and Levi Jacobs (12, 13); Israel Jacobs (38) and a Nathan Jacobs of Dartmouth; Joseph and Mordecai Levy (16, 4); Abraham and Jacob Simon (1, 27); and possibly Simon and Solomon Nathan (54, 44).
51. See PHC A/c. 1815, nos 36, 83; idem, 1821, pp. 16, 21, 28; Lipman, 'Aliens List', nos 11, 42; PHC Bk of Records, p. 1.
52. V.D. Lipman, 'Sephardi and other Jewish immigrants in England in the eighteenth century', *Migration and Settlement* (1971), pp. 47–58. Perhaps the most pathetic of these immigrants was Isaac Penha, a 90-year-old apothecary, who said he was 'flying from the Inquisition, my mother having been burned alive for Judaism'.
53. *Trew. Flying Post*, 19 June 1772 and 20 April 1780. For the name Fiva, cf. *Memorial of Edward Wortley Montague against Abraham Payba . . .* (1752).
54. Duke of Bedford's Estate, William Bray's Account Book.
55. Information from pedigree entered at Herald's College by Masseh Lopes, 20 January 1806.
56. Hyamson, *Sephardim of England*, p. 203. From this Sir Ralph were descended Lord Justice Lopes (Henry Charles Lopes)—later Lord Ludlow, Lord Roborough (Sir Massey Henry Lopes), and the future Viscount Bledislow. The family generously supported Exeter University when it was the University College of the South West.
57. For fuller details of his political career, see W.G. Hoskins, 'A sheaf of modern documents', *Devonshire Studies* (1952), H.P.R. Finberg and W.G. Hoskins, eds, pp. 412–18.
58. Hyamson, *Sephardim of England*, p. 204.
59. Letter to author from the late Dr M. Gordon, Plymouth, 3 September 1971.
60. 5 January 1819. For Ximenes see above, p. 237.
61. A son of this David Lousada became a convert to Christianity and, as the Revd Percy Martindale Lousada, married in 1848 Mary Eliza, the daughter of M. Gutteres of Sidmouth (*Anglo-Jewish Notabilities*, p. 107). Emanuel Baruch

Lousada was married to Rebecca Ximenes (*Anglo-Jewish Notabilities*, p. 127).

62. Possibly a relative of Solomon Sebag (1828–92), English teacher and Hebrew writer, temporary reader at the Bevis Marks Synagogue (*JC*, 6 May 1892). Another namesake arrived in England in 1827 as part of a mission from the Sultan of Morocco (Hyamson, *Sephardim of England*, p. 207). Sebag is Arabic for a painter.
63. Mesh. Nefesh, Ply. A/c. 1808. The Meshivat Nefesh Society was a friendly Society with a social purpose associated with the Plymouth Synagogue (see above, p. 197).
64. S. Rowe, *Plymouth Directory*, 1814. Manasseh Lopes' mother was a Pereira.
65. Apart from the London Sephardi community, the only other Sephardi communities of any permanence were one in Dublin from about 1660 which survived until *c.* 1740, and one which was founded in Manchester in 1872 and which still flourishes. There were also transient Sephardi communities in the eighteenth century, or perhaps just individuals, at Liverpool and Cork (Hyamson, *Sephardim of England*, pp. 146, 358).
66. The chronogram on his tombstone is Deuteronomy, x, 16, 'and you shall circumcise the foreskin of your heart'.
67. Lipman, 'Aliens List', 33. His name is asterisked in the transcript indicating that he arrived in Plymouth between 1798 and 1803, but this is an error.
68. PHC Min. Bk I, p. 47.
69. Lipman, 'Aliens List', 8.
70. Ibid. 4, 16.
71. PHC Min. Bk I, pp. 15, 24, 25.
72. Lipman, 'Aliens List', 9. He died 12 January 1805, and is described on his tombstone as 'from Bialin in Poland' (PHC tomb. B22).
73. Lipman, 'Aliens List', 10.
74. Berlin, PHC tomb. Q24. PHC Min. Bk I, p. 63, see below, pp. 213, 237.
75. 190 miles SW of Warsaw. Valentine was appointed a *shochet* of the Plymouth Congregation (PHC Min. Bk II, p. 63).
76. PHC Min. Bk II, pp. 246–7.
77. Edward Jamilly, 'An Essay on the Georgian Synagogue', (unpublished) (afterwards quoted as Jamilly, 'Georgian Synagogue'), pp. 20, 23. In 1874, it had 54 male and 36 female seats after rebuilding in 1836 (A. Myers, *Jewish Directory for 1874* (1874)).
78. PHC A/c. 1759.
79. Cf. the Exeter Congregation's rule that boys under six and girls under twelve were not allowed in the synagogue for prayers on the High Festivals (EHC Regulations 1825, p. 13).
80. PHC A/c. 1759, *passim*.
81. PHC A/c. 1815.
82. EHC A/c. 1818, *passim*.
83. The decennial census returns, 1841–81, have been used extensively in this book. The difficulties involved in their use are set out in B. Williams, *The making of Manchester Jewry, 1740–1875* (afterwards quoted as Williams, *Manchester Jewry*), pp. 356–7.
84. Figures extracted from 1841 census returns.
85. Ibid.
86. V.D. Lipman, 'The structure of London Jewry in the mid-nineteenth century', *Essays presented to Chief Rabbi Brodie* (1967), p. 267 (afterwards quoted as Lipman, 'The structure of London Jewry').

87. PHC Bk of Records, p. 2b.
88. W. Clegg, *Samuel Harris, a converted Jew. The history and conversion of Shemoel Hirsch, a Polish Jew; containing an account of his early life, of his travels . . .* (Sheffield, 1833), (afterwards quoted as Clegg, *Shemoel Hirsch*).
89. PHC A/c. 1821, *passim*.
90. PHC tomb. B23; PHC Bk of Records, p. 55b.
91. PHC tomb. B26.
92. PHC tomb. A67; PHC Bk of Records, p. 57a. It is not known when he arrived.
93. *JC*, 6 March 1908.
94. PHC A/c. 1821, *passim*. He came with his mother.
95. He married a daughter of a Plymouth Jew called Benjamin Levy (PHC A/c. 1821, pericope *Miketz*).
96. PHC A/c. 1821, *passim*.
97. Cf. the Turkish Rhubarb seller in Exeter, for whom see above, p. 38.
98. PHC A/c. 1821; he paid for wax for a Day of Atonement candle.
99. PHC Bk of Records, p. 56.
100. PHC Bk of Records, p. 10b.
101. PHC Bk of Records, p. 56. The Hebrew name could be read as Busker, indicating an occupational type surname. This last suggestion is somewhat confirmed by the presence of a similar appellation appended to *Meir ben Aaron Israel*, a visitor to Plymouth in 1822 and 1823 (PHC A/c. 1821, pericopes *Balak* and *Vayera*, 1822; *Tazria*, 1823). If the conjectured name Busker is correct it could mean a seller of obscene prints in public houses (E. Partridge, *Dictionary of Slang and Unconventional English*, (1913) (afterwards quoted as Partridge, *Dictionary of Slang*), p. 113). In all probability, however, the surname Busker refers to their (unknown) town of origin.
102. His brother Joshua, an executor of his will, was married to Sarah Ximenes (PCC A. April 1813).
103. Published 1829, Angra. The author has a copy in his collection.
104. At 1 Gibbon Street.
105. Mayhew, *London Labour*, I, 52.
106. *De Sola Pamphlet*, 2, Mocatta Library, London.
107. EHC Meat Tax Book. She bought 2 lbs. of meat. Perhaps she was connected with the Samuel Lopes of 1797, above, p. 34, or possibly it was the Mrs Lopes of Maristow House buying *kosher* meat for a visitor.
108. EHC tomb. 6. She was a daughter of Moses Vita Montefiore and first cousin to Sir Moses Montefiore.
109. EHC Min. Bk I, p. 60.
110. EHC Necrology (at the Jewish Museum, London), 62. Lindo was born in London in 1772 and died there in 1852. He was intimately connected with the Bevis Marks Congregation: in 1838 he founded a society called *Shomere Mishmeret Akodesh* to oppose Reform Judaism (*Jew. Encycl.* s.v. DAVID ABRAHAM LINDO).
111. EHC A/c. 1855, strangers' A/c.
112. Chief Rabbinate Archives, MSS 104.
113. Ibid.
114. PCC Loveday, 298. See also C. Roth, 'Jewish Art and Artists', *Jewish Art*, ed. C. Roth (Tel Aviv, 1961), p. 531. For a similar bequest made by a Philadelphian Jew in the mid-eighteenth century to his home synagogue in Silesia, see E. Wolf and M. Whiteman, *The History of the Jews of Philadelphia* (Philadelphia, 1957), p. 42.

115. W. Ayerst, *The Jews of the Nineteenth Century* (1848), p. 129.
116. PHC tomb. B22, and Lipman, 'Aliens List', 9.
117. The £25 which he left was sent back to his relatives in Poland (PHC Bk of Records, p. 55b).
118. PHC tomb. B23.
119. PHC tomb. B26 and PHC Min. Bk II, p. 159.
120. 1851 census; information from His Honour Mr Justice Marks, Melbourne.
121. M. Brown and J. Samuel, 'The Jews of Bath', *TJHSE*, XXIX (1982), 153.
122. EHC A/c. 1875.
123. Letter from M.L. Dight of Birmingham to *JC*, 15 July 1881.
124. PRO HO 129/12/309.
125. For editorial comment on this, see *JC*, 24 October 1873. Alexander Alexander, the dominant personality in the Exeter Congregation, appears to have been a very difficult person (see EHC Minute Bk 1823, EHC Minute Bk 1860, *passim*; *Western Briton*, 20 February 1846, for examples of his quarrelsome nature).
126. *Ha-Melitz*, XXVI, 155, 25 November 1886.
127. According to Williams, *Manchester Jewry*, p. 270, the proportion of Jews of Russian or Polish birth in Manchester rose from 19 per cent of the total Jewish population there in 1861 to 35 per cent in 1871.
128. *JC*, 20 April 1855. The Jewish prisoners were 'mostly, if not all, natives of Poland'.
129. *Ha-Melitz*, XXVI, 155, 25 November 1886.
130. L.P. Gartner, *The Jewish Immigrant in England* (1960) (afterwards quoted as Gartner, *Jewish Immigrant*), p. 149.
131. Information from the Home Office, London, 1 March 1966.
132. The enumerators who recorded the Census returns had a hard job and often made mistakes when spelling foreign names and places. Sometimes they misheard, sometimes the enumerated themselves were not sure of the spelling, or whether their town of origin was in Russia, Russian-Poland or Poland.
133. Verbal testimony of Mark Woolfson, Bournemouth, to the author, August 1990.
134. A term used to express the comradeship in their new country of emigrants from a common town or district.
135. A. Levy, *Sunderland Jewish Community* (1956), p. 94. For group migrations see also *Jewish Journal of Sociology*, VI (1964), p. 158.
136. S.A. Ascheim, *Brothers and Strangers* (Wisconsin, 1982), pp. 37, 42.
137. Ibid. pp. 43ff.
138. For the general hostility in England against the Russo-Polish immigrant see Gartner, *Jewish Immigrant*, pp. 24–56.
139. Gartner, ibid. p. 214, also suggests that there was a greater rapport between native born Jews and immigrants in the smaller communities.
140. For an account of the power struggle between the new immigrants and the established Jewish community, see Williams, *Manchester Jewry*, pp. 298–317.
141. An estimate based on the number of members in 1883 (PHC A/c. 1883).
142. *Jew. Year Bk* 1906, p. 237.
143. Ibid. 1935, p. 367. Exeter had 37 Jews in 1935.
144. Ibid. 1970; ibid. 1985.
145. See above, p. 31.

146. A Jewish pedlar, down on his luck, making his way to Bridgwater in 1821, was asked if he had friends there. 'No, but there are Jews there and they will help me' (Clegg, *Shemoel Hirsch*, p. 33).
147. Cf. Zender Falmouth, above, p. 32.
148. MS in the writer's collection.
149. Index to PHC A/c. 1759.
150. PHC Bk of Records, p. 1.
151. Ibid. p. 2.
152. Ibid. p. 2.
153. *The County Journal*, 31 May 1729.
154. Could 'Jew' here mean miser or moneylender?
155. Roth, *Great Synagogue*, pp. 50, 65. Roth, *Provincial Jewry*, p. 59, seems to have confused Lemuel Hart (PCC Potter, 72) with this Samuel Hart.
156. Admin. 13 March 1746. He left his books to his kinsman Revd Samuel Hart of Dibford (Diptford) and his property to his daughter Mary. Samuel Hart of the Plymouth Congregation was probably a cousin (cf. the will of Judith, sister of Moses and Samuel Hart, PCC Ducarel, 191, 28 April 1785).
157. *Jew. Encycl.* s.v. PARISH-ALVARS, Elias.
158. *Pigots Directory*, 1823 (afterwards quoted as *Pig. Direct*. 1823). Benjamin Jonas, father of Joseph Jonas, first Jew in Ohio, lived next door.
159. *Gent. Mag.* 1805, 1764.
160. Pet. 24, Royal Institution of Cornwall.
161. Lipman, 'Aliens List', nos 30, 20, 31, 3, 45, 57, 53.
162. *Pig. Direct*. 1823.
163. *Universal British Directory*, 1798.
164. *Pig. Direct*. 1823.
165. Ibid. 1823; D.M. Stirling, *History of Newton Abbot* (Newton Abbot, 1830), p. 176.
166. *Pig. Direct* 1823.
167. Ibid.
168. *Harrison, Harrod Co's Dawlish Directory*, 1862.
169. *Pig. Direct*. 1823; *Whites Direct*. 1850.
170. PHC A/c. 1759, p. 25.
171. Because Aaron's wife was born in 1783.
172. His father was established in Dartmouth by 1786.
173. 1851 census.
174. PHC Bk of Records, p. 56; 1851 census.
175. 1841 census; PHC tomb. B52.
176. 1871 census; PHC tomb. F5.
177. *JC*, 1 September 1989, 'London Extra', pp. 3, 6.
178. Perhaps Jews adopted for use amongst themselves the low cant term for a penny, bosh, on account of their contacts with the poor. See E. Partridge, *A Dictionary of the Underworld* (1961), s.v. POSH. See also H.J. Zimmels, *'Pesakim Uteshuvot Mibet Dino shel R. Shelomo b. Zvi'*, *Sefer Hayovel Tiferet Yisrael* (1967) (afterwards quoted as H.J. Zimmels, *Pesakim Mibet Dino*), p. 225, n. 6.
179. Cf. the movement of London Jews from the East End to Hackney, on to Stamford Hill, then on to Golders Green, and out to Edgware, Bushey and beyond.
180. There were perhaps a hundred Jewish families in Plymouth during this period.
181. They established their own *minyan* about 1815.

182. H.F. Whitfeld, *Plymouth and Devonport in Times of War and Peace* (Plymouth, 1900), p. 261, refers to 'the Hebrews who infested the lanes near the yard'.
183. According to Lipman, 'Aliens List', the brothers Abraham lived at the same address (nos 51 and 52), as did the brothers Jacob (nos 12 and 13), and Levy (nos 4 and 16).
184. To this day most Jewish communities live within fairly easily delineated areas, much as other immigrants of the same ethnic origins tend to settle in certain localities.
185. N. Kokosalakis, *Ethnic Identity and Religion: Tradition and Change in Liverpool Jewry* (Washington, 1982) (afterwards quoted as Kokosalakis, *Ethnic Identity*), p. 148.
186. L.P. Gartner, 'Urban History and the pattern of Jewish settlement in Victorian England', *Jewish Journal of Sociology*, 23 (1981), p. 40.
187. In the *Western Flying Post*, 2 March 1761 he advertised that he had plate, jewels etc. and 'all sorts of watches as cheap as in London'.
188. PCC Dec. A. 1775.
189. Solomon Hart, *Reminiscences*, ed. A. Brodie (1882) (afterwards quoted as Hart, *Reminiscences*), p. 1; PHC Min. Bk II, p. 132, 20 October 1816. See Illustration 6.
190. No. 45.
191. PHC A/c. 1759, pp. 146, 160.
192. EHC A/c. 1827. See Illustrations 1 and 2. They were ancestors of Mr Edgar Samuel, Director of the Jewish Museum, London.
193. *JC*, 24 November 1874. He was born in Prussia in 1822. He married Deborah, daughter of E. Lazarus, in Exeter, 8 August 1849 (*Ex. Flying Post*, 30 August 1849). He moved to London about 1870.
194. EHC A/c. 1855. He lived in Jermyn Street, and presumably worshipped at the Western Synagogue to which he bequeathed a magnificent Torah mantle (A. Barnett, *The Western Synagogue through two centuries* (1961), p. 152).
195. For the important part played by these two men see R. Apple, *The Hampstead Synagogue* (1967).
196. See above, p. 126.
197. Lipman, 'Aliens List', 23, 31, 14.
198. 1851 census; PHC Min. Bk I, p. 73.
199. In 1813, he was resident in Plymouth as a bachelor (List of contributors to the Plymouth Synagogue's Building Fund, 1813); by 1815 he was married (PHC Min. Bk I, p. 104).
200. List of contributors to the Plymouth Synagogue's Restoration Fund, 1805.
201. Lipman, 'Aliens List', 42.
202. PHC Bk of Records, p. 5.
203. Family tree compiled by the Jewish Historical Society of Australia.
204. *JC*, 22 August 1845.
205. *Sherborne and Yeovil Mercury*, 2 December 1799.
206. *JC*, 17 June 1864.
207. He was a son of Samuel Hyman, for whom see Lipman, 'Aliens List', 5.
208. *JC*, 27 January 1865.
209. *Trew. Flying Post*, 29 June 1843.
210. PHC A/c. 1821, *passim*.
211. PHC tomb. A13; PHC Min. Bk II, p. 194.
212. M. Brown and J. Samuel, 'The Jews of Bath', *TJHSE*, XXIX (1982), 140.
213. Ibid. p. 146

214. Circum. Reg. 15; PHC A/c. 1821.
215. Wm. Schonfield, 'The Josephs of Cornwall', a paper given at the Jewish Historical Society of England, 20 December 1938, p. 14.
216. *JC*, 6 March 1908.
217. Her mother, Flora, died there (*JC*, 11 December 1874).
218. PHC A/c. 1821, pp. 34,35. Isaac must have been an old Plymothian or a very important person, because he was called up to the Torah on the Day of Atonement. Hayyim Issacher who kept the account had some difficulty in spelling Manchester which appears as מענסצר, מענצר, מענסץ.
219. PenHC Marriage Register, 9.
220. Letter from Flora Marcuson to Godfrey Simmons, 20 July 1946.
221. PHC Marriage Register, 39. She lived there until her death about 1890 (letter from her grandson Lucien Isaacs to the author, 20 November 1963).
222. Wm. Schonfield, 'The Josephs of Cornwall', p. 10.
223. PenHC Marriage Register, 4, 12, 13, 14, 15. The last married Elias Pearlson who wrote on his circumcision register: 'In Newcastle on Tyne I married Yetta, the daughter of Isaac Bischofswerder, cantor and shochet in Penzance from a worthy family in Germany . . .'
224. PHC Marriage Register, 16, 45, 55, 57, 58, 69, 73, 80, 90, 101, 106, 108.
225. Ibid. 24, 29, 30, 48, 63, 74.
226. Ibid. 12, 35, 70.
227. Ibid. 11, 68.
228. Ibid. 46, 104.
229. H.J. Zimmels, *Pesakim Mibet Dino*, p. 242.
230. Cf. John F. Kennedy, *A Nation of Immigrants* (New York, 1959), pp. 7–10. Letter to the author from Dr I. Grunfeld, 28 February 1963.
231. He died in London 19 February 1846, aged 62, and was buried in Penzance (*VJ*, 13 March 1846; PenHC tomb. 26).
232. Joseph Solomon Ottolenghe was probably the first Jew who had lived in Exeter to go to America. He named his plantation in Georgia, 'Exonia' (letter to the author from Jacob R. Marcus, 7 May 1965). See also, Additional Note 2, p. 336.
233. Their success may have sparked off an Exeter emigration fever in 1832: 'The tide of emigration to the United States has set in powerfully within the district' (*Trew. Flying Post*, 1 March 1832).
234. He was one of the 22 children of Benjamin Jonas and Annie Ezekiel. Benjamin Jonas, a watchmaker, who was almost certainly born in England, lived in Teignmouth, 8 Wellington's Row, next door to Parish the music seller (above, p. 48, n. 158).
235. J.R. Marcus, *Memoirs of American Jews* (Philadelphia, 1955) (afterwards quoted as Marcus, *American Jews*), I, 203.
236. Marcus, *American Jews*, I, 205. The author had a similar experience when he was evacuated to a village in North Wales in 1939. Local people felt his back to discover his wings, thinking that Jews were some form of angel. This was no isolated experience. A twelve-year-old Jewish evacuee from Hackney was examined by her hosts to find her 'horns', whilst another was expected to show her 'pointed ears', supposed to be a Jewish characteristic (*JC*, 1 September 1989, 'London Extra', pp. 3, 5).
237. *JC*, 19 August 1864. This may be an amplification and variation of the preceding anecdote quoted in the text above. See also Lady Magnus, *Outlines of Jewish History* (Philadelphia, 1948), p. 353.

238. Perhaps a son of the Joseph Johnson of Exeter who paid poor rate from 1757 until 1760, and who died in 1748 (*Trew. Flying Post*, June 1784), and almost certainly a relation of Moses and Phineas Johnson who lived in Exeter, Plymouth and Portsmouth.
239. He returned to Exeter where he died in 1884 aged 85 (Ex. tomb. 98). for further details of this man see below, n. 244.
240. See below for a more detailed account of Abraham Jonas.
241. He was *Moses ben Jose Gosport* (PHC Bk of Records, p. 2b). He may have been born in England. He had a shop at Cawsand, opposite Plymouth, in Cornwall. He 'married Sarah, daughter of Benjamin Jonas of Plymouth Dock on Wednesday last' (*Trew. Flying Post*, 2 March 1815). Morris Moses was originally called Moses Moses and there were two men of this name in Plymouth. One was a broker in Pembroke Street (*Ply. Direct. 1812*). Both were members of the Meshivat Nefesh, one, the senior, from 1806–13, the other from 1811–13. In 1813, he gave five guineas to the Plymouth Congregation (PHC Inscrip. 1813).
242. Presumably the one who is entered in EHC A/c. 1820, as due to pay £1 5*s*. 6*d*. but did not do so.
243. He does not appear in either the Plymouth or Exeter records, so presumably he came straight from Teignmouth.
244. On his return to England he received an authorization to slaughter from Rabbi Solomon Hirschell, in anticipation of going back to Cincinnati: 'Jonas Levy from Exeter, Devonshire, living at Cincinnati, State of Ohio, United States of America, acknowledge that I have given my hand to the rules mentioned on the other side. *Jonah ben Menahem*, 15 *Shevat* 5583 [= Monday, 27 January 1823]. Jonas Levy, Dirrect for Philip Symonds, Cincinnati, State of Ohio, U.S.A.' (C. Duschinsky, *The Rabbinate of the Great Synagogue*, (Oxford, 1921), p. 265). The rules were: not to slaughter if a permanent, professional *shochet* was appointed, not to shave with a razor, and not to drink the wine of Gentiles.
245. Marcus, *American Jews*, I, pp. 206, 208.
246. Ibid. I, p. 209. A detailed account of this man follows below.
247. *Proceedings American Jewish Historical Society* (afterwards quoted as *PAJHS*), VIII, 57.
248. PHC A/c. 1815, No. 86.
249. *PAJHS*, XVII, 123–8.
250. *Contra* W. Jessop 'A coat of many colours; a history of the Joseph families of Devon and Cornwall' (unpublished MSS, afterwards quoted as Jessop, 'Joseph Families'), p. 120.
251. PHC Min. Bk II, p. 26, 7 October 1803.
252. PHC Inscrip. 1805; Jessop, 'Joseph Families', p. 120. A brother of Rebecca Myers also emigrated and bought a farm outside Philadelphia.
253. Ibid. He may have learned the beer and wine trade from Lemon Hart in Penzance. This would account for his absence from the Plymouth Congregation records before 1803. After his death, letters of administration were granted to his brother-in-law Nathan Joseph, as attorney for his widow. She died of cholera in 1849 and was buried in the Chestnut Street Cemetery, Cincinnati (ibid).
254. EHC A/c. 1827.
255. He was born in Plymouth, 30 November 1784 (Circum. Reg. 1). His English name was probably Levy Emanuel, Emanuel being his father's surname.

256. PHC A/c. 1819, p. 3 (Hebrew pagination).
257. Letter from Mrs F. Abrahams, 26 August 1963.
258. Solomon, *Records*, pp. 15, 16.
259. Mesh. Nefesh, Ply. A/c. 1795. the West Indies had important Jewish communities from the seventeenth century. Barbados removed all disabilities of the Jews in 1802.
260. Jessop, 'Joseph Families', p. 117.
261. Letter from (?R and P) Phillips to Henry Ezekiel, 23 February 1832, in possession of E. Finestone of Sheffield.
262. *VJ*, 8 December 1843. He was born in Portsmouth where his father went after leaving Plymouth in 1823.
263. *TJHSE*, VII (1912), 197.
264. Tablet in the Exeter Synagogue.
265. Silverstone and his wife were born in Poland, 1807 and 1813. In 1851, their three oldest daughters were Honiton Lace manufacturers and they had eight other children (1851 census).
266. *The Jews in South Africa*, eds G. Saron and L. Hotz (Oxford, 1955), p. 118.
267. Born in Berlin, 1825, the son of Israel Albu. He was probably related to the South African mining magnates Sir George and Leopold Albu (letter from Mr Austen Albu M.P. 9 June 1966).
268. Frank R. Bradlow, 'Sidney Mendelssohn', *Quarterly Bulletin of the South African Library*, 22, 4, June 1968. Revd Meyer's son Sidney produced a monumental bibliography which 'is the foundation of South African culture' (ibid.). See above, p. 209.
269. Witwatersrand Hebrew Congregation Register of Births, no. 276.
270. Verbal testimony of Mrs Phyllis Tucker, Plymouth, great-granddaughter of Morris.
271. Home Office letter to author dated 1 March 1966; letter from Mrs Ruth Goldstein, 1964.
272. PHC Marriage Register, 89. Two brothers became farmers there, their family living in Johannesburg in the 1970s.
273. Plymouth was a port of departure for Australia and many Jews must have passed through. Samuel Pollack of Brompton, returning from Plymouth after saying farewell to his son, slipped from the train and was killed (*JC*, 15 April 1861).
274. BBC Broadcast, December 1949.
275. Jessop, 'Joseph Families', chap. XII, *passim*. Two of their sons became editors, one of the *Gympie Miner*, and the other of his uncles' paper, the *Tamworth News*.
276. Circum. Reg. 75.
277. Jessop, 'Joseph Families', p. 160.
278. Original was in the possession of the late Wilfred Jessop, Chicago.
279. Jessop, 'Joseph Families', chap. XII, quotes an interesting letter Solomon sent shortly before his death to his son Sidney suggesting that he should take over the newspaper.
280. *JC*, 23 September 1853.
281. This man does not appear in the Congregational records.
282. It is difficult to identify him. He might by *Isaac b. Reuben Selig b. Isaac b. Menahem*, born 10 October 1790 (Circum. Reg. 20), but he would be a bit old at 62 to emigrate. He is not likely to be *Isaac b. Abraham* born Bavaria 1803, hardware man of Bilbury Street (*Ply. Direct*. 1850) and

31 Frankfort Street (1851 census), because that man still occupies the latter premises in 1857 (Ply. Town Rental) and is buried in Plymouth in 1872 (PHC tomb. B113). On the other hand Isaac the son of John Isaacs (*Isaac b. Menahem b. Isaac*) seems too young, being only eleven years old in 1853 (1851 census).

283. Not the Mark Levy, widower, general dealer of 3 Westwell Street, son of Moses Levy, who married Ann Rosenthal, widow, on 21 May 1844 (PHC Marriage Register, 10). He died in Guernsey on 23 December 1848 (Death Register, St Peter Port, Guernsey, no. 965). She died December 1851/January 1852 (PHC tomb. B51). His brother-in-law, Joseph Marks of Exeter, emigrated to Australia in 1853 (see below). Another Mark Levy, (*Mordecai b. Abraham b. Issachar Jacob b. Joel*) was still in Plymouth in 1866 (*Kelly's Directory*, 1866, 140 Union Street).
284. Probably the son of Aaron and Lavisa Marks, born 21 May 1834 (Circum. Reg. 75; PHC Bk of Records, p. 10).
285. PHC Marriage Register, 38; M. Gordon, *Van Diemen's Land* (Victoria, 1965), p. 110.
286. Letter from Miss Stone, 19 March 1965.
287. *JC*, 30 November 1855. Charles Marks was elected as an Assessor of the Borough of Plymouth in 1846, and was the first Jew in Plymouth to be elected to municipal office (*VJ*, 13 March 1846).
288. *JC*, 8 July 1853.
289. Letter from His Honour Mr Justice Marks, Melbourne, 1 February 1990, who also supplied the information about the family in Australia.
290. EHC tomb. 33. He was born *c.* 1838 according to a family tree prepared in Australia but more likely to have been born after 1851, because his father writes on the tombstone, 'my youngest son', and he does not appear at home in either the 1841 or 1851 censuses. Possibly the boy had returned to, or perhaps never left, Exeter, and his father had arranged for the stone to be erected.
291. Obituary from unknown source, 12 June 1908, in papers of Godfrey Simmons, Penzance.
292. Information from Flora Marcuson in a letter dated 20 July 1946 to Godfrey Simmons, Penzance.
293. *Collectanea Cornubiensis*, item 85; *Western Briton*, 8 June 1849.
294. Minister of the Exeter Hebrew Congregation, 1840–1853.
295. Leslie D. Davis, *The House of Hoffnung, 1852–1952* (privately printed Australia, 1952). His brother Abraham subsequently took over the running of the firm. Both of them eventually returned to England where they took a prominent part in communal affairs, especially in Liverpool.
296. *Australian Dictionary of Biography* (Melbourne, 1972), iv, 408.
297. Some Jews objected to him accepting an order with which the name 'Christ' was identified, but the Chief Rabbi in London ruled that Hoffnung could accept the decoration because it no longer bore a religious significance (*JC*, 6 March 1908). His sojournings are, perhaps, not untypical: born Poland 1834, lived in Exeter, then Newcastle-on-Tyne, went to London to work, then to New York (1853), then at Quincy, Illinois, back to England, then in Montreal (1858); settled in Liverpool (1866), then in London (1877), and, for a rolling stone, died a rich man.
298. *JC*, 6 March 1908.

299. A little earth from Israel is often put into the coffins of Jews who die in the diaspora so that they too may lie in the soil of the Holy Land.
300. Information from his daughter, Mrs Monty Cohen of Plymouth, and holograph letter in her possession.
301. He joined the Hand-in-Hand Society, 22 October 1876 (H. in H. Min. Bk and A/c. Ply. p. 33).

# NOTES TO PAGES 68–91

## *Chapter Three*

1. Lloyd Gartner, *Jewish Immigrant* p. 171.
2. There is a similar dearth of evidence about the age of migrants from the English countryside to town in the Victorian period. Of 295 male and female migrants to London from English villages, 235 or 85 per cent had been between the ages of fifteen and twenty-five (E.E. Lampard, 'The Urbanizing World', *The Victorian City* (1976), I, pp. 16,17 quoting from Charles Booth, *Life and Labour*, III, 139).
3. Lipman, 'Aliens List'.
4. One was a Solomon Simon, aged 48 when he entered England. He had been in New York, however, for the previous 17 years (Lipman, 'Aliens List', 25). The other was Moses Ephraim who was 39 years old when he arrived in London in 1784. V.D. Lipman gives his date of birth as 1774, (Lipman, 'Aliens List', 28 and see his remarks on p. 189 end), and if correct would make him the youngest immigrant, aged 10. The actual age in the Aliens List is very difficult to read, the clerk in 1798 probably intended to put fifty-four. This coincides better with his age given as 70 in his obituary notice (*Gent. Mag.* 1815, p. 376). Furthermore, Joseph Joseph notes in his Circumcision Register that Rabbi Moses Ephraim died 27 January 1815 after being a teacher 'in my father's house for 9 years and mine for 23 years'. In spite of his precocity, supposedly receiving a Rabbinical Diploma at the age of 8, it is hardly likely that he came to Abraham Joseph as a teacher aged only 10.
5. PRO HO I/85/2732.
6. 1851 census.
7. See above, p. 45.
8. Ibid.
9. Certificate of Nalturalisation, 16 June 1863, No. 4093.
10. These were Israel Myers of King Lane in 1841 and 11 York Lane in 1851; the Revd Myer Stadthagen of 21 Queen Street in both 1841 and 1861. There was also a Jacob Lyons in Plymouth in 1841 at 7 Barrack Street and a namesake in 1851 at 19 Barrack Street, who might, notwithstanding that his wife's name in 1841 was Fanny and in 1851 was Barina, have been the same person.
11. One of the 7 immigrants with a foreign-born wife, Samuel Alexander, only married her legally in November 1851 (PHC Marriage Register, 25). For further details of this man, see above, p. 252.
12. *Twenty-fourth Annual Report of the Registrar-General*, (1861), p. xxx.

13. Joseph Joseph was born in 1766 and died in 1846. His register is on the blank pages at the front and rear of *Sefer Sod Hashem, Brith Hashem* (Amsterdam, 1745). On the front cover is a small silver plate inscribed in Hebrew '*Josepha ben Abraham Plymouth*', and bound into the book is the service, written on parchment, for circumcision in the Synagogue.
14. Pp. 10a–13a, 14a–22b. The Exeter Hebrew Congregation's accounts indicate that there were 10 boys born from 1829 until 1835, but the record is not necessarily complete.
15. A copy is in the possession of Mr Godfrey Simmons. Most *mohelim* kept and keep, with varying reliability, registers of their circumcisions.
16. 74 male births in 32 years gives an average of 2.3 per year.
17. *TJHSE*, XVII (1951), 251. For seven and a half years a section (the larger) of the Portsmouth community brought their own *mohel* from London when his services were needed. Reb Leib's list also includes a number of circumcisions which he performed at Brighton and elsewhere.
18. Lucien Wolf, *The Families of Yates and Samuels* (privately printed 1901), p. 13.
19. When he left in 1823, the next incumbent was appointed *shochet*, reader of the scriptures, *mohel*, and 'cantor in time of need' (PHC Min. Bk II, p. 178).
20. The chronogram on his tombstone was the Hebrew of 'who circumcised children to enter them into the covenant' (Berlin, PHC tomb. 13/2).
21. See *Yoreh Deah*, 260. *Mohelim* will not circumcise a child unless he weighs 6 lbs, (see *Metzudat David* to S. Ganzfried's *Kizzur Shulchan Aruch*, 163).
22. It is instructive to compare this with Elijah Pearlson's 220 circumcisions performed between 1875 and 1906 in Newcastle upon Tyne and Hull of which all but two were on time (MS register in possession of Elias Pearlson, Sunderland).
23. Circum. Reg. 7, 16, 21, 24, 31, 33, 41, 47, 52, 56, 58, 60, 66, 74.
24. Ibid. 3.
25. In the early 1970s, about 15 per cent of Jewish births in the North East of England were by caesarian (information from the Sunderland *mohel*, the late Revd J. Braunold).
26. 1912 is a convenient date at which to stop, just before the First World War and before a number of marriages took place involving parties or their children still (1990) alive.
27. See n. 26, above.
28. Letter to the author from the General Register Office, 2 December 1964.
29. Nineteen of the weddings are mentioned in the local press giving mostly the same information as in the registers, and sometimes supplementing it.
30. 6 and 7 Wm. IV, cap. 85, sec. 2. Eva Lavinia Kinsey Atkins and Emma Boramlagh Hawke who, judging by their names were not Jewesses, and who were nevertheless married under the auspices of the Plymouth and Penzance Congregations, must therefore have been converts to Judaism (PHC Marriage Register, 99; PenHC Marriage Register, 17).
31. Joseph Jacobs, *Jewish Statistics* (1891) (afterwards quoted as Jacobs, *Jewish Statistics*), p. 49.
32. PHC Marriage Register.
33. See Table 18, p. 77. There were between 1.5 and 2.6 Jewish marriages in Plymouth per year during this period.
34. Whitehill, *Bevis Marks Records*, III, p. 5.
35. Sunderland Hebrew Congregation Marriage Registers.

36. V.D. Lipman, *Social History of the Jews of England, 1850–1950* (1954) (afterwards quoted as Lipman, *Social History*), p. 187.
37. *Jew. Year Bk* 1903.
38. S.J. Prais, 'Polarization or Decline?' *Jewish Life in Britain, 1962–1977*, eds S. Lipman and V.D. Lipman (1981), p. 5.
39. PHC Marriage Register, 10, 73.
40. Ibid. 32, 72, 125. Three widowers listed as 'over 21' were Abraham Emdon (no. 5) who was 41, Mark Levy (no. 10) who was 44, and Raphael Harris (no. 55) aged 35.
41. Jacobs, *Jewish Statistics*, p. 51.
42. Figures extracted from the Sunderland Hebrew Congregation Marriage Registers. Figures for the Jewish population of Great Britain as a whole are not available.
43. Cf. Table 22, p. 80.
44. For a dramatic description of the horrors endured by Jewish conscripts in the Russian army in the nineteenth century, see S.M. Dubnow, *History of the Jews in Russia and Poland* (Philadelphia, 1946), II, 24ff.
45. Jacobs, *Jewish Statistics*, p. 53.
46. Ibid. In Sunderland, in 135 marriages there were 6 widows and 3 widowers. Two of the widows, aged 24 and 30, married younger men aged 19 and 21 respectively (Sunderland Hebrew Congregation Marriage Registers).
47. *Twenty-fourth Annual Report of the Registrar-General* (1861), p. iv.
48. Ibid. p. v.
49. Ibid.
50. PHC Marriage Register, 61, 62.
51. A brother of this bride, Edward Marcoso, married in the Plymouth synagogue, six weeks after his sister (PHC Marriage Register, 63).
52. PHC Marriage Register, *passim*.
53. For a full discussion of the possible explanations and for an account of a similar phenomenon in London Sephardi circles, 1837–1901, see Whitehill, *Bevis Marks Records*, III, pp. 2, 3.
54. Hyamson, *Sephardim of England*, p. 300. See also Whitehill, *Bevis Marks Records*, III, p. 5. The *mahamad*, the governing body of the London Sephardim, probably had no jurisdiction over him as he came from North Africa. Moreover, some London Sephardim, Spanish in origin, looked down on their North African counterparts, calling them *forres terros* (= wild beasts).
55. PHC A/c. 1821, pericope *Miketz*, 1822.
56. PHC Marriage Register, 62, 63.
57. *Census of Great Britain in 1851*, (1854), p. 12.
58. Lipman, *Social History*, p. 10.
59. Gartner, *Jewish Immigrant*, p. 166, 'The Jewish home has perhaps received an exaggerated measure of adulation'.
60. PHC Min. Bk I, Regulation 17.
61. See above, p. 237.
62. Probably Samuel Benedict (1740–1821) of Exeter.
63. PCC Loveday 298, 1810.
64. MSS minutes of *Beth Din*, 1805–35, p. 15a (in C. Roth collection).
65. *Judah ben Hayyim Mannheim* who was acquitted of forgery charges (see above, p. 103) may have been related.
66. A more stringent category than bastards resulting from unions prohibited by rabbinic law.

67. PHC Bk of Records, p. 10.
68. Ibid. p. 55b.
69. PHC Min. Bk II, p. 227.
70. PHC A/c. 1821, p. 17.
71. In the Plymouth Congregation's Marriage Register he is Sam Alexander.
72. Chief Rabbinate Archives, 2, under date 8 September 5611. The marriage took place in November (PHC Marriage Register, I, 25, 19 November 1851). His bride was Rachel Bowman, illiterate (see also above, p. 252).
73. Ibid. 8, no date, no number; 9, letter no. 521.
74. Ibid. 9, letter no. 2007, 18 October but no year recorded.
75. Presumably this was the Monsieur Zamoisky who entertained the Exeter public with a series of four demonstrations of Mesmerism at the Globe Hotel Assembly Rooms, Exeter (*Western Daily Mercury*, 23 February 1863).
76. Chief Rabbinate Archives, 9, letter no. 2008.
77. It is not clear why he wanted this assurance.
78. Ibid. letter no. 2069.
79. PHC Min. Bk II, p. 57.
80. = Emanuel Hart.
81. PHC A/c. 1821, p. 107. There could also have been a practical reason for the examination, accidental loss of virginity halved the value of her *ketubah* (= marriage settlement) (Mishnah *Ketubot*, 1 : 3). This perhaps, was the reason for a complaint laid before the court of Chief Rabbi Hirschell in 1821 by a husband who claimed his wife was not a virgin on their marriage. She admitted the fact but denied it was due to misconduct but that when she was a child she slipped on some stones and ruptured her hymen (Minutes of *Beth Din*, 1805–35, p. 38a, in C. Roth collection).
82. MS Register in possession of Eias Pearlson, Sunderland. The Jewish Association for the Protection of Girls and Women maintained correspondents in provincial communities. In Plymouth in 1925 they were Ald. M. Fredman and L. Robins and there were 21 subscribers to its funds. L. Robins continued to act on his own until the Second World War.
83. Quoted by J. Rumney, 'The Anglo-Jewish Community, Some Aspects of its Social and Economic Development', *JC*, Supplement, June 1936 (afterwards quoted as Rumney, 'Social Development'), p. 135.
84. Francis Abell, *Prisoners of War in Britain* (1914), p. 257. Gartner, *Jewish Immigrant*, p. 108, quotes *The Polish Yidel* of 1884, which declared that anti-Semitism in England was partly caused by the misdeeds of individual Jews in, *inter alia*, obscene publishing.
85. *Gent. Mag.* IV (1734), p. 215.
86. He was not tried at the Quarter Sessions, for which records survive. The Assize records are at the Public Record Office, London, but little has survived of this circuit and nothing which throws further light on the case.
87. *Dorchester and Sherborne Journal*, 5 April 1805. It is possible that Hymes was not Jewish, but he might be Solomon Hymes (Lipman, 'Aliens List', 22, but aged 65?), or a namesake who figures in PHC A/c. 1821.
88. There were actually eight buried in the Jewish cemetery as Mrs Franco, though married to a Jew (see above, p. 227) was not a Jewess.
89. PHC Bk of Records, p. 56a. See Illustration 8.
90. This last phrase was added in Hebrew to avoid the theological implication of 1832 *Anno Domini*.

91. PenHC tomb. 25. There is some mistake in the tombstone inscription, as 7 November 1832 was a Wednesday. Jewish law prohibits burials on the Sabbath. If this burial did take place on a Saturday, it may have been after the Sabbath, or perhaps an exception was made in the plague emergency.
92. PenHC tomb. 38, but according to Joseph Joseph's Circumcision Register, p. 13, it was 'Thursday night, that is Friday, 9 November'.
93. EHC A/c. 1827, p. 125.
94. White's *Devonshire Directory*, pp. 634, 639.
95. N. Longmate, *King Cholera* (1966) (afterwards quoted as Longmate, *King Cholera*), p. 132.
96. Quoted at length in the *Hebrew Observer*, 28 January 1853.
97. Longmate, *King Cholera*, pp. 12, 14.
98. Above, p. 52.
99. A *Moses ben Judah Woolf* who had not been in England long enough to acquire an established English name.
100. Thos. Shapter, *The History of the Cholera in Exeter in 1832* (1849) (afterwards quoted as Shapter, *Cholera*).
101. Shapter, *Cholera*, p. 55.
102. Ibid.
103. EHC A/c. 1827, p. 125; Shapter, *Cholera*, p. 169.
104. Shapter, *Cholera*, p. 254.
105. Ibid. p. 55.
106. Ibid. p. 169.
107. Ibid. p. 170. After the cholera epidemic of 1848 Parliament passed the *Metropolitan Interment Act, 1850*, which compelled the cemeteries of religious groups to conform to secular santiary standards. The Act was replaced in 1852 by one which was more acceptable to the Jewish community (A. Gilam, 'The Burial Grounds Controversy between Anglo-Jewry and the Victorian Board of Health, 1850', *Jewish Social Studies*, XLV (1983), 147–56).
108. Longmate, *King Cholera*, p. 131.
109. Ibid. p. 4.
110. Graetz, *History of the Jews*, IV, 101–2; V, 728.
111. Longmate, *King Cholera*, p. 114.
112. London *Beth Din Minutes*, p. 63a (in C. Roth collection). If the children were to be brought up as Jews, a conversion was necessary, as in Jewish law a child takes the religion of its mother, and Mrs Franco was not a Jewess. One of the girls, Rebecca, who was fourteen years old when her parents died, later married Woolf Emden, her marriage on 16 August 1837 being the first recorded under the new Act in the Plymouth Congregation's Marriage Register. Her sister Bluma, eight years younger, was still living with Rebecca at the time of the 1851 census.
113. 1851 census. She was born in Dartmouth, 1810, and died in Plymouth in 1890, (PHC tomb. F5). She was supported by the Jewish Hand-in-Hand Society, Plymouth, for the last 20 or so years of her life (Hand-in-Hand A/cs, p. 36).
114. 1861 census.

# NOTES TO PAGES 92–125

## *Chapter Four*

1. C. Roth, *The Jewish Contribution to Civilization* (New York, 1940), p. 33.
2. Roth, 'Jews in England', p 254.
3. Ibid. pp. 228, 290. For a general discussion of Jewish occupations in England see Roth, 'Jews in England', pp. 227–31; Gartner, *Jewish Immigrant*, pp. 57–99; V.D. Lipman, 'Trends in Anglo-Jewish occupations', *Jewish Journal of Sociology*, II (1960).
4. Ottolenghe, *An Answer*, pp. 5, 17. See also above, p. 30.
5. Solomon, *Records*, pp. 5–7. See also above, p. 32.
6. Roth, *Provincial Jewry*, p. 22.
7. See above, p. 32.
8. Ply. Syn. Cat. 3,6.
9. *Gent. Mag.* September 1749, p. 430.
10. Ibid. October 1764, p. 499; December 1765, p. 525; December 1768, p. 591.
11. *Ex. Pocket Journal*, 1791.
12. His trade card is preserved at the Victoria and Albert Museum, London.
13. *Sherborne and Yeovil Mercury*, 2 December 1799.
14. *Ex. Pocket Journal*, 1796. There were two silversmiths, an engraver who sold a variety of goods, a pawnbroker, and a stationer.
15. *Sherborne Mercury and Western Flying Post*, 31 March 1783.
16. However, the verb 'to jew' had the connotation, from 1845, of 'to cheat or overreach' (*Oxford English Dictionary*). For the boycott of Jewish traders at Limerick in 1904, and attacks on Jewish miners at Dowlais see C.H. Emanuel, *A Century and a Half of Jewish History* (1910), pp. 160, 161. There were also anti-Jewish riots in South Wales in 1911 when troops were called out by Sir Winston Churchill, the then Home Secretary (C. Roth, 'The Anglo-Jewish Community in the context of World Jewry', *Jewish Life in Modern Britain*, eds J. Gould, S. Esh (1964), p. 99.
17. A. Arnold, 'Apprentices of Great Britain, 1710–73', *MJHSE*, VII (1970), pp. 145–57. Arnold lists 23 apprentices in the South West with Jewish sounding names, but only these four, and a fifth apprenticed to a Jewish master, can be positively identified as Jews. The references which follow are those given in Arnold's article.
18. 54/222/1762.
19. Another Moses Moses of Taunton (51/215/1752) was probably not Jewish.
20. 21/97/1757.
21. 56/174/1767.

22. 51/213/1753. See PHC Min. Bk I, p. 23 for *Abraham ben Solomon* carpenter, in 1779.
23. 26/43/1769. He was probably Solomon Solomon of Falmouth, who in 1813 was assessed for poor rate 12*s*. on his house and 2*s*. for his stock. In 1819, administration of his estate was granted to his widow Betsey, the value of his estate being under £1,500.
24. 50/68/1772. Nathan (Lipman, 'Aliens List', 44) was born in 1740 in Germany, and came to Plymouth soon after landing in 1756.
25. 15/173/1738. His name suggests that he was a Sephardi.
26. 15/68/1737.
27. 54/212/1762.
28. 55/31/1764. He might be the grandfather of Abraham Abrahams, watchmaker of Morley Place, Plymouth, who was born in 1794.
29. 56/189/1768.
30. 54/83/1761.
31. 54/82/1761.
32. *Trew. Flying Post*, 18 December 1806.
33. Ibid.
34. Ibid. 5 February 1784.
35. Journeyman describes him well. He was born in Dieppe about 1728. He was conscripted in 1747 and served in the French Army. He then worked in Vienna, Berlin, Wesel, Cologne, Nimegen, The Hague, Amsterdam, Rotterdam, The Briel, and Helvoetsluis. He came to London in 1758 and worked in Swansea, Ilfracombe, Drogheda, Dublin, Newry, Fishguard, Pembroke, Swansea again, and then for a year in Plymouth and then, in 1768 or 1769, Redruth.
36. Pet. 24 at the Royal Institution of Cornwall.
37. Ibid. This working week was not onerous, even for those days (see E.P. Thompson, *The Making of the English Working Class* (Gollancz, 1963), p. 357, quoted in J.G. Rule, 'Some social aspects of the Cornish industrial revolution', *Industry and Society in the South-West* (Exeter, 1970), p. 71.
38. *Trew. Flying Post*, 30 October 1828; 17 December 1835.
39. PHC Min. Bk I, p. 23. There is, however, a possibility that the descriptions 'carpentria' and 'goldshtikker' appended to these names are surnames.
40. St Glewias Parish Registers, 15 February 1760.
41. *Western Flying Post*, 1 July 1776.
42. Lipman, 'Aliens List', see above, p. 32.
43. Ibid. nos 17, 18, 20, 21, 22, 23, 24, 30, 31, 34, 35, 38, 39, 41, 42, 43, 44, 45, 50, 51, 56, 57; and in the *Universal British Directory*, Aaron Aaron, Samuel Cohin, Emanuel Hart, and Joseph Joseph who were probably all born in England.
44. Lipman, 'Aliens List', nos 36, 49, 52, 55, 37, 27.
45. Ibid. nos 32, 47, 6, 10, 11, 53; and in the *Universal British Directory*, Abraham Jacobs, Levi Levi, and Abel Alexander.
46. Lipman, 'Aliens List', nos 1, 4, 5, 7, 12, 13, 14, 25, 26, 29, 40, 48, 54.
47. Ibid. 2, 3, 19.
48. Ibid. 16, 33, 28.
49. Clegg, *Shemoel Hirsch*, p. 24.
50. Mayhew, *London Labour*, I, 338.
51. He was Samuel Joyful Lazarus and his licence was issued at Falmouth, 31 July 1843 and is preserved at the Jewish Museum, London.
52. Clegg, *Shemoel Hirsch*, p. 22.

53. Ibid. p. 24.
54. Ibid. p. 28.
55. Ibid. p. 30.
56. Ibid. p. 35.
57. PHC A/c. 1821, pericope *Hayyei Sarah*.
58. Clegg, *Shemoel Hirsch*, p. 38.
59. Ibid. p. 45.
60. Ibid. p. 48.
61. Ibid. p. 55.
62. Ibid. p. 63. Few Jews appear to have become servants, and if they did then they served fellow Jews. Moses Isaac was servant to Mordecai Abraham of Plymouth in 1773, and robbed him (*Trew. Flying Post*, 24 December 1773, see below, n. 114); Rebecca Israel was a domestic servant in the Exeter home of Rebecca and Henry Rothschild and their six children (1861 census).
63. C.P. Moritz, *Journeys of a German in England in 1782* (1965), *passim*.
64. See above, p. 29.
65. *Gent. Mag.* 1760, p. 43.
66. The *Gentleman's Magazine* gives his name as Sherrenbeare.
67. Whitfeld, *Plymouth and Devonport in Peace and War*, p. 280. This murder whose motive was robbery is to be distinguished from incidents of anti-Semitic hooliganism then current. At Monmouth Assizes in 1769, for example, one Prosser was convicted for tying up a Jew in front of a large fire and stuffing hot bacon down his throat (*Annual Register*, 1769, p. 92). Following an incident in 1776 when a Jew was assaulted and his chin greased with pork the *Gentleman's Magazine* criticized 'a pleasantry that was becoming frequent' (*Gent. Mag.* 1776, p. 189).
68. The present Newton Abbot.
69. *Trew. Flying Post*, 4 December 1767, 25 March 1768.
70. G.R. Pullman, *The Book of the Axe* (1875), p. 587. The late Mr B. Emdon drew my attention to this reference. Up(s)hay is a parish of Axminster.
71. *Trew. Flying Post*, 28 December 1814.
72. Lipman, 'Aliens List', 2, 3, 19.
73. One of them was an Alexander Sunder. It is tempting to identify him with Sender Alexander, tailor in Cambridge St, Plymouth in 1841, though Sender was apparently born in Devon (1841 census) whereas Sunder was an alien.
74. Of the forty-six Jewish hawkers and pedlars in Manchester on Census Night, 1841, only six settled in that town (Williams, *Manchester Jewry*, p. 119).
75. Other than a nine-year-old namesake who, according to the 1851 census, was the son of Aaron Israel, Jeweller, 8 Synagogue Place, Exeter.
76. Uncatalogued papers at Devon Record Office. I am indebted to Mr Frank Gent, Crediton, for this information.
77. Mayhew, *London Labour*, I, 338.
78. J.F. Mortlock, *Experiences of a Convict*, eds G.A. Wilkes and A.G. Mitchell (Sydney, 1965), p. 129.
79. EHC A/c. 1818.
80. See above, p. 293, n. 51.
81. EHC Marriage Register, 15.
82. Uncatalogued papers at Devon Record Office. I am indebted to Mr Frank Gent, Crediton, for this information.
83. EHC Marriage Register, 27.
84. Wm. Schonfield, 'The Josephs of Cornwall', pp. 11–13; see above, p. 106.

85. *Western Flying Post*, 2 March 1761.
86. The original is at the Victoria and Albert Museum, London.
87. *Trew. Flying Post*, 13 March 1800, records a barometer made by him. For a similar advertisement by Joseph Abrahams, see *Trew. Flying Post*, 17 January 1799.
88. PRO HCA 2/235.
89. J.R. Leifchild, *Cornwall: Its Mines and Miners* (1862), p. 287.
90. G. Green, 'Anglo-Jewish trading connections with officers and seamen of the Royal Navy, 1740–1820', *TJHSE*, XXIX (1988) (afterwards quoted as Green, 'Royal Navy'), 97. The following account is very largely based on this paper.
91. Holograph letter in the possession of the late W. Jessop, Chicago. The signature is a little uncertain, Jessop reading Hilliart instead of William.
92. Holograph letter in the possession of the late W. Jessop, Chicago.
93. *Catalogue of Anglo-Jewish Art and History*, item 433.
94. *Trew. Flying Post*, 27 November 1794. For a similar encomium in the *Liverpool Mercury*, 7 October 1825, about a prominent Liverpool Jew, Samuel Yates, see Kokosalakis, *Ethnic Identity*, p. 64.
95. Green, 'Royal Navy', p. 101.
96. *Trew. Flying Post*, 27 November 1794.
97. Bail bonds for navy agents which gave full name, domicile, and profession of all agents, and similar details of two bondsmen were preserved at the PRO (HCA 30, Miscellanea), but were destroyed before the author could see them. Some of the agents were men of high social standing, e.g. Lt Col Tucker, secretary to the Admiralty; and all had to swear that they were worth more than £5,000 (HCA 30/14, 15).
98. Green, 'Royal Navy', p. 115, n. 33.
99. There are 32 Jewish and 21 non-Jewish navy agents in Devon and Cornwall listed in John Murray's *Navy List*, 1819, and only 7 Jews in 1825.
100. This trust was not always repaid by the sailors. An E. Moss of Chatham (probably Elias Moss, of Portsmouth and later of Plymouth) advanced £100 to a boatswain, and wanted to receive the sailor's wages direct from the Admiralty. The man, however, had deserted his ship, got no pay, and Moss lost his money (PRO Adm. I, 4440/243).
101. Green, 'Royal Navy', p. 107.
102. Ibid. p. 110.
103. Slops were ready made clothing and other furnishings supplied to seamen from the ship's stores; hence ready made, cheap, or inferior garments generally.
104. The original was in the possession of the late W. Jessop, Chicago.
105. Green, 'Royal Navy', p. 123.
106. PHC A/c. 1814, p. 240.
107. Green, 'Royal Navy', pp. 123–5.
108. Martin Valentine 'one of the "Cheap John" fraternity' died at an inn at Camelford, June 1844 (H.E. Douch, *Old Cornish Inns* (Truro, 1966), p. 43). The Jew pedlar's retort to a dissatisfied customer whose razor did not shave is a classic: 'You gave but the price of the handle—I made you a present of the blade' (*Universal Songster* (1825), I, 108).
109. = Jew, by rhyming slang, perhaps from the Dutch *smouse* = news [brought by itinerant] Jews.
110. *The Universal Songster*, I, 262.
111. Ibid. 230.

112. Ibid. II, 72.
113. *Trew. Flying Post*, 16 August 1827. The Special Jury sitting with the Lord Chief Justice found that Levy had known that the customer was insane when he dealt with him.
114. Ibid. 24 December 1773. Abraham came to England in 1766 (Aliens Register no. 51), and was a successful silversmith when he died in 1811 (PCC 153 Oxford).
115. Roth, *Jews in England*, p. 105.
116. Partridge, *Dictionary of Slang*, s.v. AARON, IKEY MO. Ikey Mo is short for Isaac Moses.
117. Such perks are currently known as 'rabbits'. One Recorder told the author that it was almost impossible to empanel a jury in Plymouth which would convict a dockyard labourer of pilfering.
118. Whitfield, *Plymouth and Devonport in Times of Peace and War* p. 261.
119. Information from F.W. Moses of Sydney in a letter dated 1 November 1978, based on the trial transcript, the shipping indent and J.S. Levi, G.F.J. Bergman, *Australian Genesis* (1974) (afterwards quoted as *Australian Genesis*), pp. 162, 163.
120. *Plymouth, Devonport and Stonehouse Herald*, 9 July 1853; *Western Times*, 24 February 1884.
121. Lipman, 'Aliens List', 35.
122. PHC Min. Bk I, p. 23. Only 4 out of the 42 contributors gave more.
123. Ibid. II, p. 174.
124. Ibid.
125. *Bailey's Directory*, 1783.
126. *Gent. Mag.* 1794, p. 1156.
127. Plymouth Town Rental.
128. PCC 91, Newcastle.
129. *Trew. Flying Post*, 29 December 1794.
130. Plymouth Town Rental, 1806
131. Plymouth General Rental, 1817.
132. Green, 'Royal Navy', p. 124.
133. PHC Min. Bk II, p. 145.
134. PHC Burial Ground A/c. 1820, p. 4.
135. PHC Min. Bk II, p. 222.
136. No relative of the Joseph family of Plymouth just mentioned.
137. Wm. Schonfield, 'The Josephs of Cornwall', pp. 11–13.
138. Levi Benjamin (*Ply. Direct.* 1822).
139. Aaron Lyons, (*Ply. Direct.* 1822), and *Zvi ben Judah Lyons* (PHC Burial Ground A/c. 1820, p. 78).
140. Elizabeth Benjamin, if she was a Jewess (*Ply. Direct* 1822).
141. The son of Libche Truro (fragment in PHC Min. Bk II).
142. Eliza Emanuel (*Ply. Direct.* 1812), Fanny Lyons (*Ply. Direct.* 1822), Julia Marks (*Ply. Direct.* 1836).
143. Sander Alexander (1841 census), Eleazer Brock (PHC Marriage Register, 2).
144. Samuel Benjamin (*Ply. Direct.* 1812), W. Benjamin (*Ply. Direct.* 1814).
145. *Judah Zvi ben Solomon*, known as Zalman Boxmaker, who died in 1821. His effects were sold up for £17 *6s. 9d.*, the burial and other expenses were £5 *6s. 3d.* The balance of £12 was given to Abraham Emanuel of Dock in settlement of a loan he had made to Zalman.
146. There were 24 in all in this category, of whom 16 were probably self-employed whilst the remainder were employees.

147. 1804–87. He invented an eye shield for the Earl of Carnarvon, and wrote *A treatise on the nature of vision, etc.* (London, 1833) and *Observations on the preservation of sight* (Exeter, 1837). In 1845 he presented an address to the Queen Dowager 'in grateful acknowledgement of his late Majesty's patronage of him in his [Alexander's] boyhood' (*VJ*, 29 August 1845).
148. Chief Rabbi Hirschell considered the announcement of Harris' appointment of sufficient importance to warrant entering it into his daybook (Roth, *Great Synagogue*, p. 141). Both Alexander and Harris were warrant holders to King William IV.
149. Born in Plymouth 1811, died there 1860.
150. PRO LC 5/243. The others were two embroiderers in London, two jewellers in Brighton and Portsmouth, and a quill and pen manufacturer in Edinburgh.
151. Marion Lochhead, *The Victorian Household*, pp. 30–1, quoted in V.D. Lipman, 'The Structure of London Jewry', p. 256, n. 13.
152. *JC*, 5 July 1867.
153. In 1857 he owned 27 properties with a gross value of £411 per annum (Plymouth Town Rentals).
154. His charitable donations in 1852 noted in the *Jewish Chronicle* alone amounted to £70.
155. This class would almost certainly have some families on relief, but there is no evidence to show how many.
156. For the sake of comparison it may be mentioned that in Manchester one Jewish merchant had seven servants, including a butler; two had five; two had four and one had three (Williams, *Manchester Jewry*, p. 87).
157. *Census of Great Britain in 1851* (1854), p. 72.
158. See above, pp. 93–4.
159. 1841 census.
160. 1851 census.
161. 1861 census.
162. See above, p. 40.
163. See above, p. 108.
164. See above, p. 108.
165. PHC Marriage Register, 32, 42, 43.
166. Ibid. 2, 7, 40, 43, 44, 79(2), 80, 81, 82, 83, 119.
167. Ibid. 42, 107, 113, 119, 123, 139.
168. Ibid. 2, 32, 36, 47, 48(2), 51, 57, 77, 78, 82, 88, 108, 109(2), 110, 111(2), 113, 114(3), 117, 118(2), 120, 121(2), 125, 127, 137, 140, 143(2), 150(3), 154.
169. *Encyclopaedia Brittanica*, (1951), 18, p. 947.
170. Gartner, *Jewish Immigrant*, p. 59. It is related of three Russian immigrant brothers in Plymouth at the end of the nineteenth century that they went away peddling from Monday to Friday, taking loaves of bread and herring for their week's food.
171. *JC*, 27 May 1853.
172. *JC*, 22 July 1853.
173. *JC*, 27 May 1853.
174. *JC*, 11 November 1853.
175. See Illustration 9 for the letterhead of his firm, Abrahams and Sons, 39 Catherine Street, Devonport. His sons adopted his first name as their surname, calling themselves 'Abrahams'. His grandson Reginald Lewis, father of Gabrielle and Andrea, was the photographic correspondent for the *Daily Mirror* in the South West.

176. 1871 census. He was the great-grandfather of Mrs Ruth Overs, Netanya.
177. The term was apparently first used in 1870, (*Oxford English Dictionary*).
178. His address was 9 Union Street, Plymouth (PHC Marriage Register, 93).
179. *Eyre's Plymouth Directory*, 1896, p. 387.
180. Personal recollection of his widow. See Illustration 9.
181. See Illustration 9 for letterhead.
182. The following two paragraphs are taken from R. Burman, 'Jewish Women and the Household Economy in Manchester, *c.* 1890–1920', *The Making of Modern Anglo-Jewry*, ed. D. Cesarani, (Oxford, 1990) (afterwards quoted as Burman, 'Jewish Women').
183. Burman, 'Jewish Women', p. 56.
184. Ibid, p. 57. Moses Mordecai prohibited his wife from taking Mr Bendix as a lodger (above, p. 83).
185. See above, p. 126.
186. e.g. Fanny Rosenberg, general dealer and Agnes Moses, haberdasher.
187. e.g. Phoebe Levi, dealer in toys; Mathilda Lazarus, haberdasher; Hannah Silverstone, dealer in shells; Zipporah Levy, pawnbroker; Anne Franks, slopseller.
188. e.g. Catherine and Amelia Ezekiel who carried on the business of their father after the death of their brother Ezekiel A. Ezekiel (*Trew. Flying Post*, 25 December 1806).
189. But, however, see S. Alexander, 'Women's Work in 19th Century London', *The Rights and Wrongs of Women*, A. Oakley and J. Mitchell (1976).
190. See above, p. 116.
191. Harry Bischofswerder, a member of the Penzance Congregation, owned the Wheal Helena Mine near Marazion, Cornwall, for a brief while in 1891 and 1892 (*JC*, 15 January 1892).
192. S. Aris, *The Jews in Business* (1970), p. 232.
193. Ibid. p. 233.
194. Ibid. p. 234.
195. The surname appears as 'Freedman' in the 1871 census and in PHC Marriage Register, I, 79 (1882).
196. PHC Marriage Register, I, 80.
197. Ibid. 88.
198. PHC tomb. C15; Q19.
199. PHC Marriage Register, I, 79.
200. Ibid. 89. He now calls himself Fredman.
201. Ibid. 97.
202. G.E. Harfield, *A commercial directory of the Jews of the United Kingdom* (1894), s.v. PLYMOUTH, DEVONPORT.
203. *Eyre's Plymouth Directory, 1896.*
204. See note 202, above.
205. Letter to the author from Mr David Maxwell, Dublin, 8 February 1989.
206. PHC Marriage Register, I, 2.
207. Ibid. 40.
208. Their estate enabled the city of Plymouth to erect a block of sheltered flats on the Barbican.
209. PHC tomb. P27.
210. Cyril Stein is currently chairman of the public company, Ladbrokes.
211. Letter to the author from Mr Jack Smith, Haifa, 6 March 1989.
212. *Calendar of Probate and Letters of Administration*, published annually by the Registrar-General.

# NOTES TO PAGES 126–135

## *Chapter Five*

1. Unlike the situation in the modern State of Israel where whole communities from the diaspora sometimes settle in one organized move.
2. Roth, *Provincial Jewry*, p. 14; *Australian Genesis*, p. 218.
3. Ply. Syn. Cat. 3,6. This information is not given in the catalogue, but is set out in the lease itself.
4. Ibid. 3,5. C.W. Bracken, *A History of Plymouth* (Plymouth, 2nd edn 1934), p. 272, writes that the Hoe Jewish Cemetery was opened in 1748, which fits very well with this account.
5. Ply. Syn. Cat. 4,3.
6. Ibid. 5,3.
7. They were Abraham Emanuel of Plymouth Dock and Michael Nathan of Plymouth, shopkeepers, Benjamin Levy of Plymouth, optician, and John Saunders of Plymouth, gentleman.
8. For the same reason the property owned by Sarah Sherrenbeck and the Plymouth Congregation was held in trust for them by some prominent non-Jewish Plymothian.
9. Roth, *Jews in England*, p. 171.
10. Ply. Syn. Cat. 4,3.
11. Ibid. 5,3.
12. Ibid. 6,1.
13. Exeter City Council Act Book, XIV, p. 232. The yard was previously let to one Chas. Tanner at 6*s*. 8*d*. per annum, but was in ruins when the Jews took it over (Warden of the Poor Rentals, Michaelmas, 1756). The plot was let for a maximum of 99 years. When the last of the three 'lives' who were mentioned in the lease died, the lease ended.
14. A Book of Maps of all lands and tenements belonging to the Chamber of Exon. (*c*. 1770), Map 5 (Exeter City Archives).
15. Information in the companion to the Book of Maps referred to in the previous note.
16. Ibid.
17. Ibid. Roth, *Provincial Jewry*, p. 60, is referring to this extension when he writes that the cemetery was acquired in 1807. Roth knew of Jenkins, *History of Exeter* (1806) which already mentions the cemetery, and this accounts for Roth's parenthetic note that the cemetery of 1807 was perhaps not the first.
18. MSS 104, Chief Rabbinate Archives, p. 37.
19. It was also used until the 1960s by the Torquay Congregation, which now has its own cemetery.

20. W. Warn, *Directory and Guide for Falmouth and Penryn*, 1864, p. 45.
21. *64th Annual Report of the Board of Deputies*, (1915), p. 45.
22. FHC tomb. 2, 3, 4.
23. FHC tomb. 1, but the date appears to be 5555 (= 1795).
24. PHC tomb. 58a.
25. C. Roth, 'Penzance', *JC Supplement*, May 1933.
26. From 1805, Jews resident in Devonport prayed from time to time in various rooms hired there for that purpose, to spare them the long walk into Plymouth. The last such miniature synagogue at 66 Chapel Street, was founded in 1907, and destroyed in the Blitz of 1941. See above, p. 130.
27. A non-Jew.
28. Ply. Syn. Cat. 1,2. The foundation stone reads
    קדש לה' / הבית הקדוש והנכבד הזה /
    נתיסד ונבנה בשנת בואו נשתחוה נכרעה ונברכה לפני ה' / לפ"ק
    (Holy to the Lord. This holy and honourable house was founded and built in the year, 'Come let us worship, bow down and bless before the Lord'). The verse is from Psalms, xcv, 6, with slight changes, and the chronogram gives 5522 (= 1761/1762).
29. Ply. Syn. Cat. 2,1.
30. Ibid. 1,3; 1,4.
31. Ibid. 1,6; PHC Min. Bk II, p. 229.
32. PHC Min. Bk II, 53ff.; D. Black, *The Plymouth Synagogue* (Plymouth, 1961), p. 10.
33. The silver box, in the shape of a four leaf clover worked in fine filigree, is not hallmarked, and was probably made by a local Jewish silversmith. It is still used by the Plymouth Congregation.
34. A non-Jew.
35. Reference in MSS at Exeter City Library, 48/12/7/18b. According to *Billing's Devonshire Directory* (Birmingham, 1857), p. 29, the synagogue of 1764 was built on the site of an earlier one. As there is no mention in the lease of an earlier one it is not likely that there was an earlier synagogue.
36. Letter of C.A. Mansell, 410 South Lincoln Street, Martinsville, Indiana, USA, dated 17 July 1929 to the Postmaster, Exeter, found at the Devon Record Office, Exeter, by Mr R. Sweetland.
37. W. Warn, *Directory and Guide for Falmouth and Penryn*, 1864.
38. This and later conveyances are at J. Jewill Hill, solicitors, Penzance.
39. There is a foundation stone with an all Hebrew inscription on the front of the Plymouth synagogue, but this may not have been its original position. The original stone has crumbled away and its recent replacement has a number of minor errors.
40. Portsmouth and Edinburgh were approached by lanes off a street, Dublin and King's Lynn through a yard at the rear of another property (E. Jamilly, 'An essay on the Georgian Synagogue' (not published) (afterwards quoted as Jamilly, 'Georgian Synagogue'), p. 14.
41. Jamilly, 'Georgian Synagogue', p. 14.
42. The inscription spells out the reference to this law: *Orah Hayyim*, Section 94.
43. Cf. Hyamson, *Sephardim of England*, p. 78, for Avis, the Quaker builder of Bevis Marks who would not accept a profit on his work.
44. Jamilly, 'Georgian Synagogue', p. 8.
45. Ibid. p. 35. But note *JC*, 9 June 1854, when apparently a new ark was made.

46. See J. Godfrey-Gilbert, 'Proposed Restoration of the Plymouth Synagogue, 1964' (afterwards quoted as Godfrey-Gilbert, 'Plymouth Synagogue Restoration, 1964'), p. 2.
47. The *bimah* (central dais).
48. The ark.
49. These have disappeared.
50. These have now been silver plated.
51. These were later withdrawn from the mortgage as they were the private property of individuals (Ply. Syn. Cat. 2,3).
52. Ply. Syn. Cat. 2,3.
53. Godfrey-Gilbert, 'Plymouth Synagogue Restoration, 1964', p. 2. See Illustration 10.
54. Jamilly, 'Georgian Synagogue', pp. 36, 37. The inscriptions on the ark are the Ten Commandments and בשנת אשתחוה אל היכל קדשך ביראתך (I will worship towards Thy holy ark in fear of Thee). It is the second half of Psalms, v, 8, and the chronogram gives 5522 (= 1761/1762). See Illustration 11.
55. Godfrey-Gilbert, 'Plymouth Synagogue Restoration, 1964', p. 1.
56. The gallery in the Exeter Synagogue runs round the south and west walls.
57. Godfrey-Gilbert, 'Plymouth Synagogue Restoration, 1964', p. 2.
58. PHC Min. Bk I, p. 42.
59. I Kings, vii, 21. Frontispieces to sacred Jewish books invariably show such stylized columns.
60. PHC Min. Bk II, p. 64.
61. Ibid. p. 65.
62. Tablets of commemoration hanging in the Plymouth synagogue. An Order of Service (the Hebrew hand-written) for the consecration of the synagogue on 21 February 1864 survives.
63. *Western Morning News*, 28 February, 1910. See Illustration 12 for the title page of the Order of Service.
64. The restoration cost some £10,000. See Illustration 13 for a photograph of the Torah Scrolls being taken into the synagogue.
65. *JC*, 9 June 1854.
66. *Jew. Year Bk*, 1930. The cost was underwritten by the Hoffnung family.
67. A *mikveh* is like a miniature swimming pool, constructed to strict specifications of Jewish law, in which married Jewesses must immerse themselves after their monthly periods before resuming marital relationships.
68. MSS 104 Chief Rabbinate Archives, pp. 37, 47, 135, 139.
69. Ibid. p. 37.
70. PHC Min. Bk II, p. 220.
71. Ibid. p. 218.
72. MSS 104 Chief Rabbinate Archives, p. 139.
73. Copies of correspondence and agreement in the author's collection.
74. Since the reconstruction of the Congregation's house in 1975 there has been no *mikveh* in Plymouth.
75. MSS 104 Chief Rabbinate Archives, p. 48.
76. See above, p. 249.

# NOTES TO PAGES 136–171

## *Chapter Six*

1. See V.D. Lipman, 'Synagogal Organization in Anglo-Jewry', *The Jewish Journal of Sociology*, I (1959) (afterwards quoted as Lipman, 'Synagogal Organization'), 80–93.
2. Lipman, 'Synagogal Organization', p. 85.
3. See *Laws and Regulations of the Plymouth Hebrew Congregation* (1835) (afterwards quoted as *PHC Regulations, 1835*), p. 2. Seatholders were sometimes called *orchim* in Plymouth (PHC Min. Bk I, p. 10).
4. *PHC Regulations, 1835*, p. 9. In Manchester they were called Free Members (Williams, *Manchester Jewry*, p. 54).
5. In 1789 it was found necessary to specify that poor vestry members had equal rights with the rich (PHC Min. Bk I, p 49).
6. The term is transliterated in the Ashkenasi pronunciation, which was used by the immigrants.
7. PHC Min. Bk I, regulation 14, p. 5.
8. Ibid. regulation 13. The amounts payable for vestry membership did not vary much until the category was abolished in 1945. The London synagogues charged five or ten guineas (Lipman, 'Synagogal Organization', p. 81).
9. EHC Regulations, 1833, no. 15.
10. Lipman, 'Synagogal Organization', p. 87.
11. Roth MSS 205, PenHC Revised Regulations, 1844, no. 7.
12. See Glossary for these and the following Hebrew terms.
13. *PHC Regulations, 1835*, p. 9, of which Roth MSS 205, PenHC Revised Regulations, 1844, nos. 31–8 are a word-for-word repetition. In Manchester in 1840, the Free Members had reserved burial plots in the Pendleton cemetery (Williams, *Manchester Jewry*, p. 88).
14. PHC Min. Bk I, p. 9. In Exeter he could keep his *Hezkat HaKehillah* rights by paying an annual retainer of five shillings (EHC Regulations, 1833, no. 16).
15. PHC A/c. 1815.
16. *PHC Regulations, 1835*, p. 1. The London Congregations had much the same organization on a more elaborate scale consonant with their larger numbers (Lipman, 'Synagogal Organization', pp. 81–3).
17. There does not appear to have been the rivalry between the 'old', established, section of the community and the up and coming aspiring newcomers, leading to pressures to found new synagogues as in Manchester, London, Liverpool and Newcastle (Williams, *Manchester Jewry*, pp. 135–63; D. Cesarani, 'The Transformation of Communal Authority in Anglo-Jewry,

1914–1940', *The Making of Modern Anglo-Jewry*, ed. D. Cesarani, pp. 6, 115; Kokosalakis, *Ethnic Identity*, p. 68; G.D. Guttentag, 'The Beginnings of the Newcastle Jewish Community', *TJHSE*, XXV (1977), pp. 1–25).

18. *PHC Regulations, 1835*, p. 2.
19. Ibid. p. 3.
20. Palm, citron, myrtles, and willows required for the festival of Tabernacles.
21. Ibid. pp. 4, 5.
22. Ibid. p. 2.
23. It is customary for the beadle to hold out a charity box at the cemetery, crying the verse, 'But charity delivers from death' (Proverbs 10:2) in Hebrew.
24. Ibid. pp. 5, 6.
25. EHC Regulations, 1833, no. 6.
26. Ibid. no. 14.
27. Ibid. no. 1.
28. PHC Min. Bk I, p. 17.
29. Ibid. p. 17.
30. Ibid. p. 5. Cf. the special prayer offered on Sabbaths in all synagogues for 'those who give lamps for lighting and wine for *kiddush* and *havdalah*'. Only by chance did the author discover that Mr Jack Cohen's family has defrayed the expense of wine for the Plymouth synagogue for the past century.
31. *PHC Regulations, 1835*, p. 7.
32. PHC Min. Bk I, p. 5. A similar rule was in force in London (Lipman, 'Synagogal Organization', p. 84) and in other provincial communities, e.g. Bristol (*Rules for regulating the congregation of the Old Synagogue, Bristol* (1838), p. 4.
33. Ibid. p. 45.
34. Lipman, 'Synagogal Organization', p. 85.
35. Mr Lionel Aloof recollects these being kept in a small room reserved for that purpose.
36. London and some provincial Congregations were moving towards a more democratic pattern in the second half of the eighteenth century (Lipman, 'Synagogal Organization', p. 86).
37. See Gartner, *Jewish Immigrant*, pp. 190, 215. See also Williams, *Manchester Jewry*, p. 182. Even the appointment of Rabbi Dr Schiller-Szinessy to the Manchester community in 1851 was subject to the agreed proviso that 'he would not assume for himself any decision on rabbinical questions' (Williams, *Manchester Jewry*, p. 188). For an account of Manchester Jewry's struggle to appoint its own independent Rabbi in the mid-nineteenth century, see Williams, *Manchester Jewry*, pp. 209–20, 234–7. Orthodox, small, independent *hevrot* in Manchester and London began to appoint *dayan*-type rabbis towards the end of the nineteenth century (see B. Williams, 'East and West', *The Making of Modern Anglo-Jewry*, ed. D. Cesarani, (Oxford, 1990), p. 17. In Liverpool, no rabbi had any special power in the running of the community; as late as the 1920s, the authority of the communal Rav, Rabbi S.J. Rabbinowitz, was not recognized by the Old Hebrew Congregation (Kokosalakis, *Ethnic Identity*, pp. 69, 147).
38. See Gartner, *Jewish Immigrant*, pp. 24, 195.
39. The Chief Rabbi was virtually the only person in England with Congregational responsibilities who was called by the title of Rabbi.
40. Cf. the usurpation of rabbinical function by the *Kohol* of Plymouth in 1819 which decided, for example, who were the obligatory *aliyot* (PHC Min. Bk I, p. 90).

41. S. Sharot, *Judaism: A Sociology* (1976), p. 72, quoted by Kokosalakis, *Ethnic Identity*, p. 78.
42. PHC Bk of Records, p. 29.
43. Roth, *Provincial Jewry*, p. 92.
44. PHC Min. Bk I, p. 92.
45. Roth MSS 205, Regulation 79.
46. This prayer is currently recited in many synagogues belonging to the United Synagogue, London.
47. *Jonathan ben Nathan* and Moses Myers do not figure in the authoritative list of Chief Rabbis given by C. Roth in the *Festival Prayer Book for Tabernacles* published by Routledge, London, 1951, though he mentions Moses Myers in his article on the Chief Rabbinate, *JC*, 31 July 1931.
48. PHC Min. Bk I, p. 1.
49. Ibid. p. 3.
50. PHC Min. Bk II, p. 23.
51. Ibid. p. 55.
52. Chief Rabbi from 1758–64.
53. C. Roth, 'The Chief Rabbinate in England', *Essays presented to J.H. Hertz*, ed. I. Epstein, E. Levine, C. Roth (1942), p. 374.
54. PHC Min. Bk II, p. 55.
55. Ibid. p. 48.
56. Ibid. p. 65.
57. *VJ*, 16 August 1844.
58. The Exeter Congregation's necrology is in the Jewish Museum, London. See Illustration 14.
59. *VJ*, 16 August 1844.
60. EHC Regulations, 1833, no. 31.
61. Ibid p. 3.
62. Rabbi Moses Ephraim (Lipman, 'Aliens List', 28) who was in Plymouth from 1780–1815 was a tutor in a private family, the Josephs, but had no rabbinical position.
63. PHC Min. Bk II, p. 16.
64. Ibid. p. 21.
65. Besides his *Midrash Phineas* (which has a useful subscribers' list) *Phineas* also wrote *Sefer Kinoteha DePhineas* (Berlin, 1788) which has an introduction with some autobiographical details.
66. The Revd Dr M. Berlin, minister in Plymouth from 1896–1906, had a Rabbinical Diploma, but as was common at that time, did not style himself Rabbi (information from his son, Mr B. Berlin). He was a considerable scholar, the Revd S. Singer pays tribute to him in 1891 for valuable assistance which he gave in preparing the first edition of the *Authorised Daily Prayer Book*.
67. The prohibition of any musical accompaniment makes the cantor (and choir) almost a necessity when there are several hundred worshippers present.
68. *PHC Regulations, 1835*, no. 99.
69. See below for the functions of the *shochet*.
70. PHC Min. Bk II, p. 178.
71. Cf. ibid. p. 115.
72. He was Levi Benjamin, one of the teachers of Leoni, the master of Braham (*Annual Register*, 29 March 1829). He had the most powerful voice in the kingdom (*Gent. Mag.* 1829, p. 380).
73. EHC Regulations, 1833, no. 45.

74. *PHC Regulations, 1835*, no. 84.
75. PHC Min. Bk I, p. 60. The *Kregil* was the white bibs such as are worn nowadays by barristers and clergymen.
76. EHC Regulations, 1823, no. 22. *Biff* is probably a Yiddish form of 'bib', cf. the previous note.
77. Letter from Mr Aloof's son, Lionel, 19 January 1989.
78. The use of surplice or gown by the minister when he preached in church had been a matter of bitter controversy in Elizabethan and Stuart times, and there were still rumblings in the nineteenth century (J. Thurmer, 'The Nineteenth Century: The Church of England', *Unity and Variety* (Exeter, 1991), pp. 121–2).
79. *PHC Regulations, 1835*, no. 85.
80. No written records of the Falmouth Congregation appear to have survived.
81. Ibid. no. 84; PenHC Regulations, 1844, no. 64; EHC Regulations 1833, no. 45.
82. *PHC Regulations, 1835*, nos. 91, 93.
83. Ibid. Nos. 91, 92.
84. PHC Min. Bk II, p. 107.
85. *PHC Regulations, 1835*, nos. 100, 101.
86. Ibid. no. 96.
87. The author pays a filial debt in giving this description of a beadle's activities—his father was the much beloved beadle of the Golders Green Synagogue, London, for more than thirty years.
88. PHC Min. Bk II, p. 24.
89. Less in the case of a *shochet* who slaughters only fowls.
90. Abbreviated in Hebrew as שו"ב. These initial letters, when placed after a practitioner's Hebrew name, have given rise to the surname Shoob.
91. Above, p. 144.
92. Duschinsky, *Rabbinate*, p. 264. Presumably these particular prohibitions, the one Biblical, the other Rabbinic, were widely disregarded at the time.
93. PHC Min. Bk II, p. 95. The following Saturday night was Purim, and he was no doubt expected to be back on Friday. Apparently the return journey could be done in a week.
94. Penzance Minutes, Roth MSS 271, p. 82.
95. Roth MSS 205, no. 69.
96. PHC Min. Bk I, p. 60.
97. PenHC Min. Bk 1843, p. 6.
98. See Genesis, xxxii, 33.
99. PHC Min. Bk I, p. 60. For the obligation to water meat within three days of slaughtering (unless soaked in the meantime) see S. Ganzfried, *Code of Jewish Law* (New York, 1927), XXXVI, 27.
100. Cf. above, p. 283, n. 244.
101. *PHC Regulations, 1835*, no. 141.
102. PHC Min. Bk II, p. 41.
103. Ibid. p. 96. It works out to about £78 per annum.
104. Ibid. pp. 8, 20, 133.
105. Ibid. p. 121.
106. Ibid.
107. Ibid. pp. 34, 37.
108. Ibid. p. 171. As the fee was 3*d.* for a *pair* of fowls, did Jewish housewives then generally use two chickens at a time?
109. A. Shoshan, *Man and Animal* (Jerusalem, 1963), p. 124.

110. Such was the tradition until *kosher* meat ceased to be sold in Plymouth under the aegis of the Congregation about 1964. According to the Exeter Meat Tax Book, 1828, the Revd Moses Levy received 12 lbs. of meat each week. The unvarying amount indicates a perquisite of a regular allocation.
111. EHC Min. Bk 1838–45.
112. PenHC Min. Bk 1843–63.
113. *JC*, 10 February 1865.
114. PHC Annual Balance Sheet, 1884.
115. PenHC Min. Bk 1863–92.
116. Will of E.A. Ezekiel, Admin. January 1807, in Archdeaconry Court of Exeter. The original was destroyed but a transcript by Michael Adler was given to Cecil Roth and a copy is now in the possession of the author. Ezekiel had no children to say *kaddish* for him.
117. PCC *Crickitt*, 415, 1811. Rachael Benjamin, died 1817, left £1 to Moses Solomon to say prayers for twelve months, and £1 to 'Mr Isacher to learn for 1 month' (Devon Record Office, Wills, B652). See above, p. 162 for further details.
118. PHC Min. Bk II, p. 171.
119. Even the Chief Rabbi in London at this period had secondary sources of income (Duschinsky, *Rabbinate*, pp. 238, 243).
120. Plymouth Library Archives, Worth, 285. He occupied a house in Basket Street, Plymouth from before 1797 to 1800.
121. He signed a lease for conduit water for a house in Butler Street, Plymouth, in 1815, and is described as a slop-seller.
122. PHC Min. Bk II, p. 78. He was engaged by the Plymouth Congregation as a *shochet* on 18 August 1811 (PHC Min. Bk II, p. 63), and was murdered at Fowey in November 1811 whilst buying up golden guineas for paper money (*Royal Cornwall Gazette*, 30 November 1811, 28 March 1812, 3 April 1812).
123. Holograph letter to him from Abraham Joseph II in the possession of Godfrey Simmons of Penzance. He was minister in Penzance from 1812 until 1854.
124. When the Penzance Congregation's official was absent from duty, for example, the committee wrote to him that 'such conduct . . . was not becoming a Reverend' (Roth MSS, 204).
125. EHC Min. Bk 1838, p. 72. He was the Revd Michael L. Green, relation of Revd A.L. Green. In the 1841 census he is described as a clothier and in July 1841 he married Rosetta, eldest daughter of M. Davis of Exeter (*Trew. Flying Post*, 1 July 1841).
126. EHC Minute Book II, meeting held 21 December 1851.
127. PHC Min. Bk II, p. 39.
128. Ibid. p. 68.
129. John Hatchard, *The Predictions and Promises of God respecting Israel* (Plymouth, 1825), Appendix by M.S. Alexander.
130. *JC*, 7 August 1863. Similarly, the reader of the Norwich Congregation, who was paid £1 a week in 1864, advertised, 'Hebrew, Chaldaic, German lessons . . . Testimonials from English and Continental Universities' (*Norwich Argus*, 8 October 1864).
131. Revd Elias Pearlson kept a record of his income as a *mohel* in Hull. In 1883, he did 31 circumcisions which brought him in £8 1*s.* 6*d.*; in 1884, 25 and received £7 9*s.*; 31 in 1886 brought him £4. 16*s.*; and from 77 in the eighteen

months ended December 1888 he made £9 4*s.*; (original MSS in possession of E. Pearlson, Sunderland).

132. A Jew may not possess *hametz*, leaven, on Passover. If he has leavened food, say a bottle of whisky, which he does not wish to throw or give away, then he sells it to a non-Jew with the option of buying it back after Passover. Such sales are usually made through a rabbi, and a small payment is made to him for his trouble.
133. PHC Min. Bk II, p. 15.
134. 1841 census.
135. 1851 census, 1861 census.
136. PCC 1862/414.
137. Jessop, 'Joseph Families', p. 141, Family Trees W.5, W.6.
138. Letter to the author from the Revd J.A. Thurmer, then Lazenby Chaplain, University of Exeter, 24 January 1973.
139. Mr Allan Brockett of the University of Exeter kindly gave this and the following information on the salaries of dissenting ministers in Exeter, in a letter to the author, 13 January 1973.
140. In the vicinity of Exeter between 1837 and 1844 the average farm labourer's wage was only 7*s.* 6*d.* a week, though with substantial perquisites (*Devonshire Studies*, W.G. Hoskins and H.P.R. Finberg (1935), p. 428).
141. Revd Raphael of Birmingham received £180 per annum in 1845 just for acting as headmaster of the Jewish School there, and this was besides his stipend as Minister of the Congregation. Of the other teachers at that school, two received £80 per annum and one £60 (Chief Rabbinate Archives, MSS 104, p. 5). The *Hazan* (reader/minister) of the Bevis Marks Congregation, London, was paid £250 per annum in 1860 (Hyamson, *Sephardim*, p. 364).
142. He had lucrative sidelines and wealthy sons, and had sufficient income to remain in Penzance even when he was not employed by the Congregation.
143. *Gent. Mag.* 1837, p. 99.
144. Only the names of six of the Falmouth Congregation's ministers are known and they are too few to draw any conclusions from them. They were: (a) *Samuel ben Samuel HaLevi*, died 22 March 1814; (b) *Moses ben Hayyim*, died 24 October 1830; (c) Joseph Benedict Rintel, *c.* 1832–49; (d) N. Lipman; (e) –. Herman, 1860; (f) Samuel Orler, *c.* 1875–*c.* 1880 (A.M. Jacob, 'The Jews of Falmouth', *TJHSE*, XVII (1949), p. 66, n. 2, for the first four; Chief Rabbinate Archives, 7, 18 April 1860 for the fifth; letter from Lord Mishcon to the author, 10 March 1993, for the sixth).
145. This seems to be the point of a letter Chief Rabbi Adler wrote to the Exeter Congregation in 1859: 'It will be difficult to find an Englishman with all the qualities you require. *Let me know the salary*' (Chief Rabbinate Archives, 5, 1 April 1859).
146. *VJ*, 25 November 1842. Levy does not seem to have taken up the Anglo-Jewish ministry.
147. Letter in the possession of Mr Godfrey Simmons, Penzance, quoted by C. Roth, 'Penzance', *JC Supplement*, June 1933, p. ii.
148. PenHC Min. Bk 1839, pp. 86, 91; PenHC Min. Bk 1843–61, p. 28.
149. See Roth MSS 204, 31 July 1864 and 9 April 1865.
150. Chief Rabbinate Archives, 2, letter no. 7129.
151. Ibid. letter no. 7131.
152. Ibid. letter no. 7150.
153. Ibid. letter no. 7159.

154. Ibid. letter no. 7579.
155. Ibid. 7, 18 April 1860.
156. Ibid. 4, letter no. 127. He died in office 21 April 1862 aged 58 (PHC tomb. B76). Perhaps he was already suffering from ill-health and consequently was less able to restrain himself.
157. Ibid.
158. Ibid. 3, letter no. 8944.
159. For examples at Sheffield, see *JC*, 30 December 1859 and 14 September 1860; for Sunderland, see A.B. Levy, *The 200-year-old New Synagogue* (1960), pp. 19, 20; Gartner, *Jewish Immigrant*, pp. 190, 206.
160. PHC Min. Bk II, p. 8. The widow of Moses, the Plymouth Congregation's beadle, was allowed 'all the days of her life whilst she lives in our community two shillings every week, also she is to live rent free . . .' (PHC Min. Bk II, p. 15).
161. EHC Min. Bk 1838 and 1861.
162. Hardly a word of complaint against Myer Mendelssohn was registered in the Exeter Congregation Minute Books, 1854–67. As there were bitter remarks against some of his predecessors and successors, 'Mendelssohn must have been a bit of a saint' (Frank R. Bradlow, 'Sidney Mendelssohn, a short biography', *Quarterly Bulletin of the South African Library*, 22, 4, June 1968, p. 106).
163. PHC Min. Bk II, p. 177.
164. In France and Germany the Rabbis' wages were paid by the state from the nineteenth century onwards. In England, a Provincial Ministers' Fund was set up in 1884 to encourage smaller congregations to engage English-speaking ministers. It was never of great importance. The United Synagogue, and to a lesser extent, the Federation of Synagogues, in London, make loans to newly formed communities to enable them to build synagogues.
165. J. Sherrenbeck and Abraham Joseph loaned their own Scrolls to the Plymouth Congregation, and eventually bequeathed them to it. In 1782, 17 *Baalei Batim* of the Plymouth Congregation gave sums ranging from five shillings to two guineas and totalling £18 4*s*. for the purchase of a Scroll (a new Scroll in the 1990s costs some £12,000). At the same time, Abraham Joseph gave £1 3*s*. to the Congregation to purchase two pairs of phylacteries, which would presumably have been for use in the synagogue by worshippers who could not afford to buy their own, and twelve *mezuzot*, possibly for resale to members of the Congregation (PHC Min. Bk I, p. 28).
166. Examples of this type of expenditure and appeals or levies will be given in the ensuing pages.
167. PHC A/c. 1759, *passim*.
168. PHC A/c. 1883, *passim*.
169. PHC Min. Bk I. Regulations nos 2, 58.
170. Ibid. Regulation 12.
171. Ibid. Regulation 60.
172. e.g. Roth MSS 205; PenHC Regulations, 1823, nos 79, 81, 84; EHC Regulations, 1833, nos 22, 36. An unusual feature in Exeter was the obligation to offer on a 'Good Sabbath' (i.e. one on which two Scrolls are taken out) and the Sabbath of Comfort (*Sabbath Nahamu*) but not on ordinary Sabbaths.
173. PHC A/c. 1759.
174. Ibid.
175. They were Joseph Sherrenbeck and *Mordecai ben Yehiel* (PHC A/c. 1759, pp. 1, 5).

176. PHC A/c. 1883–86, *passim*.
177. In the author's collection.
178. PHC A/c. 1913, Balance Sheet in the author's collection.
179. PHC Min. Bk I, p. 76.
180. PHC A/c. 1759, *passim*. The Exeter Congregation's Meat Tax Book survives at the Jewish Museum, London.
181. For example, E.A. Ezekiel of Exeter, died 1807 (a copy of the will is in the author's collection, the original was destroyed); Abraham Joseph of Plymouth, died 1794, PCC Newcastle, 91; Moses Mordecai of Exeter, died 1809, PCC Loveday 298; Revd Myer Stadthagen of Plymouth, died 1862 (Pr. 1862).
182. PCC Crickitt, 415. He died in 1811.
183. Balance Sheets of the Plymouth Congregation in the author's collection.
184. *PHC Regulations, 1835*, nos 32, 119, 120.
185. Ibid. no. 107.
186. Ibid. no. 107.
187. Ibid. no. 133.
188. Ibid. no. 101.
189. EHC A/c. 1827–33, pp. 5, 6, 7, 9, 19, 28, 42, 103, 110, 113, 114, 130, 150, 162, 165, 172.
190. PHC A/c. 1850–3, *sub* H. Hyman and Mr Levy.
191. *PHC Regulations, 1835*, no. 52.
192. Roth MSS 205, nos. 39, 40.
193. EHC Regulations, 1833, no. 15. Sons of vestry members only had to pay 10*s*. 6*d*.
194. For Canterbury's synagogue building fund, see *JC*, 16 October 1846; for Gt Yarmouth's, *JC*, 4 February 1848; for Dover's, *JC*, 8 November 1861.
195. Ply. Syn. Cat. 2, 2.
196. Ibid. 2, 3.
197. Ibid.
198. Receipts for mortgage payments kept with the Plymouth Congregation's leases.
199. Ply. Syn. Cat. 2, 4. A silver shield, 7 inches by 5 inches hung prominently in the synagogue, inscribed 'and all the work was finished, 4 *Sivan* 5544 [= May 1784]', with nine names of vestry members, probably commemorates this event. There was also a payment of £300 being part of £450 outstanding on 19 April 1786. This may have been for monies owing on the purchase of the cemetery or perhaps a Synagogue house (Ply. Syn. Cat. 3, 6).
200. PHC A/c. 1759, p. 1 (Hebrew pagination). These sums are recorded as *Matnat Yad* i.e. donations at *Yizkor* on these festivals.
201. Ibid.
202. Ibid. p. 5 (Hebrew pagination).
203. Ibid.
204. A beautiful handwritten commemorative inscription with details of donors' names and the amount they gave hangs in the Plymouth Synagogue. The information which follows is taken from this inscription.
205. This inscription also hangs in the Plymouth Synagogue. It is a copy made by Joseph Goldston in 1900 of the original destroyed by damp.
206. This appeal, too, was recorded on an inscription which hangs in the Plymouth Synagogue.
207. PHC Min. Bk II, pp. 31, 33.

208. I.e. the Chief Rabbi in London.
209. PHC A/c. 1814–26, p. 240. This account was written by Hayyim Issachar. See above, p. 103.
210. See above, p. 151.
211. *Royal Cornwall Gazette*, 30 November 1811, 28 March 1812, 3 April 1812.
212. PHC Min. Bk II, p. 71, foot of page.
213. Ibid. top of page.
214. Ibid. At least one member strenuously objected to the proposal, insulted the committee and was disciplined with an 18*s*. fine (PHC Min. Bk II, p. 72).
215. Ibid.
216. The threat to imprison on default of payments seems to have emanated from the Mayor. Apparently the vestry here resolves to aid any of its members who does not pay so that he should not be sent to prison.
217. PHC Min. Bk I, p. 22.
218. Ibid. II, p. 98.
219. The chronogram, מעשה צדקה שלום, is based on Isaiah 32:17, and gives 1779.
220. PHC Min. Bk I, p. 69. The chronogram is, יהי שלום בחילך שלוה בארמנותיך (Psalms 122:7) and gives 1796.
221. Wertheimer printed rule books for a number of Congregations about this time.
222. Original in the author's collection.
223. Facsimile in the author's collection.
224. Roth MSS 205.
225. Chief Rabbinate Archives, MSS 104.
226. For the Hebrew terms see Glossary.
227. PHC Min. Bk I, Regulation 3. For the Hebrew terms see Glossary.
228. Ibid. Regulation 4.
229. Ibid. Regulation 8.
230. Ibid. Regulation 12.
231. Ibid. Regulation 19.
232. Ibid. Regulation 20.
233. Ibid. Regulation 21.
234. Ibid. Regulation 22; worshippers just returned from a long journey were exempt from the fine.
235. Ibid. Regulation 23; see also Roth, *Great Synagogue*, p. 67.
236. Ibid. Regulation 30; A *minyan* is a quorum for prayers.
237. Ibid. Regulation 46. Unruly behaviour in the Synagogue is traditional on this day.
238. Ibid. Regulation 9.
239. Ibid. Regulation 40.
240. Ibid. Regulation 56.
241. See Roth, *Great Synagogue*, pp. 68, 120, 122.
242. From the New Moon *Av* until the Fast of the Ninth of *Av*.
243. PHC Min. Bk I, p. 41.

# NOTES TO PAGES 172–191

## *Chapter Seven*

1. See *PHC Regulations, 1835*, no. 2; Roth MSS 205, no. 2; and J.H. Hertz, *Authorised Daily Prayer Book* (1947), pp. xxii, 401; Roth, *Great Synagogue*, p. 67.
2. S. Singer, *Authorised Daily Prayer Book* (1947) (afterwards quoted as *Authorised Daily Prayer Book*), pp. 166, 223.
3. The only event of a *Hasidic* nature that has been noted was the annual pilgrimage until the early 1920s of Eleazer Orgel to the court of his *Rebbe*, the rabbi of Belz (letter from Orgel's great-grandson, A.P. Rose, 23 October 1989).
4. *Authorised Daily Prayer Book*, pp. 142–8.
5. This is a typically German custom.
6. *Authorised Daily Prayer Book*, p. 224.
7. A.I. Sperling, *Sefer Taamei Minhagim* (Jerusalem, 1957), p. 302.
8. PHC Min. Bk I, Regulation 22.
9. There is a short Reading of the Law on these days, and observant Jews make a special effort to attend service on them.
10. It was the practice in Plymouth in the 1960s to hold evening services for members having *yahrzeit*.
11. A minor festival in the Jewish calendar.
12. PCC Crickitt, 415.
13. *PHC Regulations, 1835*, no. 2.
14. The eve of the New Moon has a special service known as *Yom Kippur Katan*.
15. I.e. when a member had *yahrzeit*.
16. From Passover to Pentecost.
17. Lipman, *Social History*, p. 186.
18. *JC*, 26 September 1856.
19. *Trew. Flying Post*, 26 July 1821.
20. *JC*, 3 January 1862.
21. PHC A/c. 1886, p. 344.
22. Chief Rabbinate Archives, 5, no date or letter number.
23. A copy of his Order of Service was signed by J. Sanger.
24. Original Order of Service in the possession of Godfrey Simmons.
25. *JC*, 9 June 1854.
26. 25 May 1815.
27. *Trew. Flying Post*, 8 June 1815.
28. *Trew. Flying Post*, 22 June 1815.
29. J. Harris, *The Royal Devon and Exeter Hospital* (1922), p. 72.

30. Samuel Pepys made the point in his diary: 'But, Lord! to see the disorder, laughing, sporting, no attention, but confusion in all their service, more like brutes . . .' (A. Cohen, *Anglo-Jewish Scrapbook* (1943), p. 275.
31. Roth, *Great Synagogue*, p. 251.
32. Ibid. p. 258. Adler's suggestions were considered in Penzance and the sale of *Aliyot*, as suggested by him, was discontinued (PenHC Minutes, 1843–61, p. 12).
33. PHC Min. Bk I, Regulation no. 21.
34. Ibid. no. 34.
35. Ibid. no. 46.
36. EHC Regulations, 1823, no. 9. The licence certified the building as one used for a religious purpose, but was not a prerequisite for holding religious services.
37. EHC Regulations, 1823, no. 18. See Illustration 15.
38. Ibid.
39. In some Congregations it is still customary to fix an apple and a lighted candle in it to the top of the flags the children carry round the synagogue on this day.
40. Ibid. no. 48.
41. It being customary to drown out Haman's name each time the reader mentions it.
42. EHC Regulations, 1823, no. 48.
43. Mayhew, *London Labour*, I, p. 430.
44. In the author's collection.
45. EHC Min. Bk 1848, p. 7.
46. Kokosalakis, *Ethnic Identity*, p. 78.
47. *PHC Regulations, 1835*, nos. 110, 111.
48. The prayer book was written and illuminated by one Sender, beadle in Portsmouth, at the behest of Joseph Joseph. For a resumé of the laws of precedence as practised in Plymouth, see Additional Note 3, p. 336.
49. PHC Min. Bk II, p. 216.
50. PHC Regulations, 1835, no. 46. Roth MSS 205, no. 33 spells out the same rights for Penzance members. For Exeter, see EHC Regulations, 1833, no. 21.
51. Religious majority.
52. A thanksgiving blessing recited after escape from danger, after overseas travel or after recovering from childbirth or serious illness.
53. EHC Regulations, 1833, no. 21, end.
54. Known as *shlishi*.
55. Known as *shishi*.
56. *Eins fur acharon*, as it was known.
57. *JC*, 13 July 1855.
58. *JC*, 26 September 1855.
59. *JC*, 12 September 1856.
60. *JC*, 26 September 1856.
61. *Plymouth Journal*, 4 September 1856.
62. *JC*, 14 November 1856.
63. The *Plymouth Herald*'s report of the following incident and strictures are quoted at length in the *JC*, 15 October 1858.
64. PHC Min. Bk I, Regulations, no. 19
65. Roth, *Provincial Jewry*, p. 62.
66. PHC Min Bk II, p. 21.

67. Ibid. p. 25.
68. Ibid. p. 33.
69. Ibid. p. 95.
70. Ibid. p. 172.
71. Ibid. p. 227.
72. PHC A/c. 1884, p. 224.
73. EHC Regulations, 1833, no. 30.
74. PHC A/c. 1884, p. 224.
75. EHC Regulations, 1833, no. 30.
76. PHC A/c. 1821, *passim*; PHC A/c. 1886, p. 344.
77. e.g. PHC A/c. 1883, p. 112; PHC A/c. 1887, p. 121.
78. See *Authorised Daily Prayer Book*, pp. 294–300.
79. EHC A/c. 1830, *passim*.
80. I.e. the *lulav* (= palm branch).
81. PHC A/c. 1884, p. 224.
82. *VJ*, 22 October 1845.
83. EHC Regulations 1823, no. 25.
84. PHC Meshivat Nefesh A/c. p. 23.
85. PHC A/c. 1884, p. 224. See Illustration 16 for its present *succah*.
86. EHC Regulations, 1833, no. 30; PHC Meshivat Nefesh A/c., *passim*.
87. Cf. Roth, *Great Synagogue*, plate 49, opp. p. 212.
88. The cushion and inventory (about 1925) have now disappeared.
89. PCC 1862/414. The Jewish Historical Society of England has a set of silver circumcision instruments hallmarked Exeter and London 1771, 1794, 1821.
90. *Jewish Encyclopaedia*, IV, p. 99, where an account of the opposition to *metzitzah* by mouth is given.
91. See *Encyclopaedia Judaica*, V, p. 288.
92. She was a daughter of Abraham Joseph I and married Nathan Joseph.
93. The binder, or wimpel, is in the author's collection. The Hebrew text is
אברהם בן נתן נטע כ"ץ
נולד למ"ט ביום ה' כ"ח כסלו תקס"ב לפ"ק, ה' יגדלהו לתורה לחופה ולמעשים טובים א"ס.
94. On the thirtieth day after the natural birth of a firstborn male child the father is obliged to redeem him at a ceremony called *Pidyon HaBen*. Some Congregations kept platters, especially decorated and inscribed, upon which to place the baby during this ceremony. There is no record of such a platter in the South-West Congregations.
95. PHC Min. Bk I, p. 45.
96. See above, pp. 192, 193.
97. PHC A/c. 1821, p. 127.
98. Ibid. p. 127.
99. In Poland, where it was customary to recite the *haftarah* silently, a boy under bar mitzvah was often called to read the portion.
100. *JC*, 16 May 1890.
101. A.I. Sperling, *Sefer Taamei Minhagim*, p. 408.
102. PHC Marriage Registers, I, nos 1–68.
103. Ibid. I, nos 69–154; II, nos 1–3.
104. PenHC Marriage Register.
105. EHC Marriage Register, *passim*.
106. See Deuteronomy, xxv, 5–10.
107. Facsimile of original in author's collection.
108. Isaac Mosely, born 12 May 1821, died 12 July 1850.

109. She was Ellen or Eleanor, the youngest daughter of Henry (son of Abraham) Ezekiel (*Exeter Flying Post*, 29 June 1843).
110. He was Abraham Mosely of 30 Park Street, Bristol, born 19 June 1817.
111. He was Benjamin Jonas, originally of Plymouth, then Exeter, and then lived in Teignmouth. He was the father of Abraham and Joseph Jonas, the first Jews in Ohio.
112. He married Georgiana, second daughter of Henry Ezekiel, on 6 March 1839 (EHC Marriage Register, no. 4).
113. EHC Marriage Register, no. 11.
114. Chief Rabbinate Archives, 9, letter no. 1141.
115. Corresponding perhaps to the middle middle-class of Gentile society of that time.
116. *JC*, 30 September 1853.
117. *JC*, 1 September 1854.
118. EHC Regulations, 1823, no. 40.
119. EHC A/c. 1832 p. 127.
120. PHC Min. Bk I, Regulations, no. 29.
121. Devon Record Office, Wills, B652, E351.
122. *Vachers* as they are known in Yiddish. See Rule Book of the *Bikkur Cholim*, Plymouth, rule no. 2.
123. Ganzfried, *Code of Jewish Law*, 198, 3.
124. It is worth noting that Revd Stadthagen was buried on the *last day of Passover*, burial on the eighth day of a Festival being permitted by Jewish law, but unusual.
125. Lyon Joseph was one of these five, he died in Bath on Tuesday, 7 May 1825, and was buried in Plymouth five days later (PHC tomb. A13).
126. See above, p. 136ff.
127. Chief Rabbinate Archives, 34, 13 April 1890.
128. An undated roster for three nights (about 1830) has survived (stuck to PHC Min. bk II, p. 180). It includes for the first night 'a Christian being Friday'. He was needed to keep the fire up and trim the lamp, the Jews being unable to do this on their Sabbath.
129. *PHC Regulations, 1835*, no. 132.
130. EHC Regulations, 1833, 52(7).
131. Baring-Gould, *Devonshire Characters* (1908), p. 407.
132. A garment worn by observant Jews under their jacket or shirt with 'fringes' (cf. Numbers, xv, 37–41) at each corner. Hirsch's assertion is arrant nonsense.
133. Clegg, *Shemoel Hirsch*, p. 41. He was paid one pound for the four nights (p. 43).
134. Rule Book of the *Bikkur Cholim*, p. 2.
135. EHC Regulations, 1823, no. 44; repeated in 1833, no. 52 (9).
136. See Singer, *Authorised Daily Prayer Book*, p. 421.
137. This is permitted by Jewish law, provided six handbreadths of earth interpose (Ganzfried, *Code of Jewish Law*, 199, 5).
138. PHC Min. Bk II, p. 195.
139. Ganzfried, *Code of Jewish Law*, 199, 6.
140. EHC Min. Bk 1838, p. 56.
141. Jessop, 'Joseph Families', p. 126. Samuel Ralph died 17 March 1867, aged 64, and is buried on a raised plot separated from the rest of the cemetery (Plym. tomb. A133).
142. EHC Regulations 1823, no. 41, and similarly in the 1833 Regulations.

143. PHC Min. Bk II, p. 243.
144. Second edition, London, 1861, p. 212. See also Ganzfried, *Code of Jewish Law*, 199, 17.
145. PHC Min. Bk II, p. 253.
146. The arithmetic leaves something to be desired.
147. PHC Min. Bk II, p. 252.
148. Ibid. p. 105. For similar cases, see ibid. pp. 121, 167.
149. Chief Rabbinate Archives, 4, letter nos 758, 811, 827.
150. PHC Min. Bk II, p. 68, and Berlin, S7. He is, perhaps, the Naftali Benjamin who came to Plymouth in 1745 (see Lipman, 'Aliens List', no. 32). In the Plymouth Congregation a dead member's seat is left vacant, and a notice to that effect is placed on the seat, for at least one year. It is not known when this custom began.
151. Devon Record Office, Wills, B652, E353.
152. Extract of will kindly supplied by Lucien Isaacs, a descendant, in a letter dated 20 November 1963.

# NOTES TO PAGES 192–205

## *Chapter Eight*

1. Sidney and Beatrice Webb, *English Local Government: English Poor Law History: The Old Poor Law* (1927) (afterwards quoted as Webb, *Poor Law*), p. 1.
2. Ibid. p. 60.
3. Ibid. p. 399.
4. Ibid.
5. Ibid. p. 400.
6. J.R. McCulloch, *Principles of Political Economy* (1852), pp. 400, 407, quoted in Webb, *Poor Law*, p. 405.
7. Kokosalakis, *Ethnic Identity*, p. 86, comes to the same conclusion.
8. B. Kirkman Gray, *A History of English Philanthropy* (1905), (afterwards quoted as Gray, *English Philanthropy*), pp. 261–2.
9. Ibid. pp. 233–5.
10. Ibid. p. 235.
11. Rumney, *Anglo-Jewish Development*.
12. PHC Min. Bk II, p. 96.
13. *Annual Register*, 29 March 1829.
14. PHC Min. Bk II, p. 199.
15. Ibid. p. 15.
16. PHC Min. Bk II, pp. 25, 68.
17. Ibid. p. 133.
18. Ibid.
19. Ibid. p. 199.
20. Ibid. pp. 31, 33.
21. PHC A/c. 1814–26, pp. 240, 211. See also above, p. 165.
22. Original in the author's collection. For full text see above, p. 226.
24. PHC Min. Bk II, p. 203.
25. Ibid.
26. John Sykes, *Local Records of Northumberland, Durham* (Newcastle/Tyne, 1866), II, p. 215.
27. PHC A/c. 1883, pp. 101–4.
28. See above, p. 160.
29. PHC A/c. 1883–90, p. 417, which gives the names of the different boxes.
30. PHC Min. Bk II, p. 48.
31. PHC Regulations, 1779, no. 45.
32. An institution known in Yiddish as *pletten*.
33. Ibid. no. 44.
34. *PHC Regulations, 1835*, no. 22.

35. Ibid. no. 13.
36. EHC Regulations, 1833, no 23.
37. Report at the Jewish Board of Deputies, London, 1841.
38. Kokosalakis, *Ethnic Identity*, p. 91.
39. PHC A/c. 1883, p. 107.
40. Jews in Exeter were paying poor rate from 1752; a Mrs Sarah Abrahams paying 1½*d*. weekly, and Mr Ezekiel 2*d*. from 1756 (Exeter Corporation Poor Rate Book, 1752–6, pp. 1, 125). See also above, p. 30.
41. Quoted in Rumney, *Anglo-Jewish Development*.
42. In 1705, 50 Lutheran refugees from Catholic persecution came from the Palatinate to England where they received one shilling a day from the Queen. Their welcome was followed four years later by a further 10,000 Palatines, who were lodged in tents on Blackheath, but their arrival led to much opposition (Gray, *English Philanthropy*, p. 155).
43. PHC Min. Bk II, p. 127.
44. Gray, *English Philanthropy*, pp. 79–81.
45. Ibid. p. 244.
46. ח"ק משיבת נפש. For a society of similar name founded in London in 1779, see L. Wolf, *The Meshebat Nephesh* (1897). This, the oldest Ashkenasi charity in England, was founded to alleviate distress following food riots (*Jew. Year Bk*, 1902, p. 94).
47. Meshivat Nefesh A/c. p. 1.
48. Ibid. *passim*.
49. Ibid. p. 34. This rule book does not appear to have survived.
50. Ibid. for years 1799, 1802, 1805, 1807, 1810. There are about twelve and a half Jewish months to the solar year.
51. Ibid. *passim*.
52. The London Meshebat Nephesh allocated its funds entirely to the poor.
53. Meshivat Nefesh A/c. *passim*.
54. Meshivat Nefesh A/c. *passim*.
55. Ibid. p. 33.
56. Ibid. p. 127. The Liverpool Congregation established a Jewish Philanthropic Society in 1811. It, too, had a dinner in December, 'equivalent to a grand Christmas party', to which non-Jewish civil dignitaries were invited (Kokosalakis, *Ethnic Identity*, p. 50).
57. Ibid. pp. 59, 67.
58. Ibid. p. 80.
59. Ibid. pp. 95, 99.
60. Devon Record Office, Moger, Testamentary Causes, Series II, pp. 2124, 2127.
61. Gray, *English Philanthropy*, p. 243.
62. *JC*, 2 January 1852.
63. It probably merged into the United Jewish Hand-in-Hand Benevolent Society, Plymouth, Stonehouse and Devonport, founded 1861 (see above, p. 200).
64. *JC*, 2 January 1852.
65. Ibid. They were not celebrating Christmas, but using the opportunity of a public holiday to meet together.
66. *JC*, 17 December 1852.
67. Cf. above, p. 196, n. 37.
68. *Suggestions to the Jews for Improvement in Reference to their Charities, Education and General Improvement, By a Jew* (1844), p. 32.

69. Rumney, 'Social Development'; Kokosalakis, *Ethnic Identity*, p. 87.
70. Proverbs, xix, 17. The date is derived from this chronogram.
71. He signed, 'The small one *Samuel ben Menahem Men[del] KZ*'. He came from Cheltenham and settled in Plymouth in the early 1820s (PHC Burial Ground A/c. 1820, p. 98). He died 27 April 1860, and the translation of the inscription on his tombstone reads: An honourable and faithful man . . . working loving kindness with the poor and hastening to prayer, evening, morning and noon' (Ply. Tomb. B71).
72. After Job, xix, 23.
73. EHC A/c. 1827–30, p. 82.
74. The Society's 1865 balance sheet (in author's collection) was its fourth.
75. Hand-in-Hand Min. Bk p. 26.
76. Ibid. p. 2.
77. Ibid. *passim*.
78. Ibid. accounts for years 1875–7, for example.
79. Out of the monthly payment of 1*s*. 3*d*. to the Sunbury Friendly Society in 1773, 3*d*. had to be spent on beer (Gray, *English Philanthropy*, p. 244). Many Friendly Societies were attached to public houses, which naturally encouraged them.
80. Hand-in-Hand Min. Bk p. 8.
81. Ibid.
82. Hand-in-Hand Min. Bk p. 43.
83. See Webb, *Poor Law*, pp. 376–95. About 1821, there were at least 60,000 poor in England continually circulating up and down the country at the public expense.
84. EHC Min. Bk I, p. 116.
85. Hand-in-Hand Min. Bk pp. 16, 68.
86. Gray, *English Philanthropy*, p. 34.
87. Clegg, *Shemoel Hirsch*, p. 32.
88. Ibid. p. 24.
89. Ibid. p. 35.
90. Ibid. p. 38.
91. Ibid. p. 40.
92. Hand-in-Hand Min. Bk p. 46.
93. Cf. Plymouth Hebrew Board of Guardians, *1st Annual Report* (Plymouth, 1909), p. 5 (in the author's collection).
94. Letter to the author from the Curator, Jewish Museum, London, 12 May 1964.
95. *Jew. Year Bk*, 1935, p. 244.
96. *Transactions of Plymouth Institute*, VI, p. 82.
97. *Jew. Year Bk*, 1935, p. 244.
98. Her husband had been Treasurer of the United Jewish Hand-in-Hand Society.
99. *JC*, 15 August 1890.
100. Ibid. The Ladies Benevolent, as it is known in the Plymouth Congregation, still functions.
101. Exeter Corporation Poor Rate Book, 1752–6, pp. 1, 125.
102. It is a *mitzvah* to do so (*Tur, Yoreh Deah*, 251, 1).
103. *Trew. Flying Post*, 22 June 1815.
104. J. Harris, *The Royal Devon and Exeter Hospital* (1922), p. 72.
105. *JC*, 20 May 1864.
106. Information from his family.

107. *Trew. Flying Post*, 23 February 1887.
108. *Transactions Plymouth Institute*, VI, p. 82.
109. An alternative name for Tregurtha Downs Mine near Marazion. It was the last survivor of a brief upsurge of interest in mining in 1881, and was closed in 1895.
110. *Royal Cornwall Gazette*, 21 January 1892.
111. Roth, MSS 271; Roth MSS 273.
112. Devon Record Office, Wills, C794.
113. *JC*, 16 June 1854.
114. Chief Rabbinate Archives, 9, letter nos 749, 1206.
115. Ibid. letter no. 1205.
116. Letter in the author's collection.
117. The Jewish year is given as 5635 but this was probably an error for 5636.
118. Gray, *English Philanthropy*, pp. 261–2.
119. Ibid. p. 57, n. 3.
120. Cf. Jacob Nathan's 'strong opposition on religious principles' to a Jewish child being admitted to the non-Jewish Deaf and Dumb Asylum, Kent Road, London (*JC*, 6 January 1865, letter of L. Hyman).
121. *JC*, 14 September 1860.
122. Francis Abel, *Prisoners of War in Britain, 1756–1815* (1914), p. 257.
123. PRO Adm. 103/268, no. 1859.
124. Henry Cohen, *Settlement of the Jews in Texas* (American Jewish Historical Society Publication, no. 2, 1894), p. 8.
125. Letter to the author, dated 16 February 1965.
126. Francis Abel, *Prisoners of War in Britain, 1756–1815* (1914), p. 257; Basil Thomson, *The Story of Dartmoor Prison* (1907), pp. 66, 69, 146.
127. Chief Rabbinate Archives, 3, letter no. 9008.
128. *JC*, 20 April 1855.
129. Chief Rabbinate Archives, 4, letter nos 147, 203.
130. Ibid. 7, 3 September 1861.
131. Ibid. 7, 11 March 1861.
132. Ibid. 7, letter no. 6846. *Matzot* and food for Passover are still sent to the prison by the Visitation Committee, United Synagogue, London, and for many years the members of the Plymouth Congregation donated money to supplement the prisoners' rations at Passover and the High Holydays.
133. Chief Rabbinate Archives, 11, 7 August 1868.
134. Ibid.

# NOTES TO PAGES 206–220

## *Chapter Nine*

1. In a letter to the author, 14 February 1964.
2. He kept a shoe and patten warehouse in Exeter from 1796 until his death in 1834 (*Exeter Pocket Journal*, 1796–1834; EHC tomb. 28).
3. Roth, *Provincial Jewry*, p. 61.
4. L. Cohen, *Sacred Truths* (Exeter, 1808), postscript.
5. Ibid.
6. L. Cohen, *A New System of Astronomy* (1825)
7. Ibid. p. 129.
8. Ibid. p. 142.
9. See Illustration 7.
10. *Trew. Flying Post*, 23 February 1887.
11. Printed in Exeter by W.C. Pollard and published in London, 1833.
12. Published in Exeter, 1837.
13. *Trew. Flying Post*, 30 January 1834.
14. 24 June 1865, no. 1695 at the Patent Office, London.
15. H. Joseph, *An account of the extraordinary discovery of fossil animal remains at Oreston* (1859).
16. *JC*, 18 March 1859; 13 May 1859.
17. Original in possession of the late Wilfred Jessop, Chicago. See also B. Susser 'Voyages to Australia', *Plymouth Western Morning News*, 15 August 1965.
18. Mary and Ann Cawrse, daughters of Henry and Anne Cawrse were baptized at St Neot on 2 June 1793 (information from H.L. Douch, Royal Institution of Cornwall, 19 February 1966).
19. *Trew. Flying Post*, 14 January 1813.
20. *Autobiography and Journal of B. R. Haydon*, ed. M. Elwin (1950), pp. 457, 554, 555. Eric George, *The Life and Death of B. R. Haydon* (1948), p. 196.
21. *Autobiography and Journal of B. R. Haydon*, ed. M. Elwin, p. 469.
22. Ibid. p. 483.
23. *MJHSE*, IV, p. 110.
24. See above, p. 237.
25. She moved in poetic circles. The second Mrs Robert Southey met her in Teignmouth in 1833 (*TJHSE*, XVI (1945), 140).
26. *JC*, September 1947.
27. *TJHSE*, XI (1927), 178; letters to the author 26 January 1965, 25 December 1965 from the late Mrs Beth-Zion Abrahams author of (as yet) unpublished biography of Amy Levy.

28. His was one of the three lives on which the lease of the Exeter Congregation's cemetery was secured in 1803 when he was seventeen years old (Book of Maps and Lands belonging to the Chamber of Exon. Map 5, no. 10).
29. His house was assessed at 5*s.* and stock at 1*s.* 6*d.* for poor rate, 1812–15, in Falmouth.
30. *Jewish Encyclopaedia*, V, p. 318.
31. Ibid.
32. Roth MSS 276.
33. *Jewish Encyclopaedia*, V, p. 318.
34. The following account is taken from F.R. Bradlow, 'Sidney Mendelssohn Collection', *Jewish Affairs* (Johannesburg), May 1965, pp. 12–17, see above p. ??.
35. 2 Vols. London, 1910.
36. I.D. Colvin, 'In Memoriam; Sidney Mendelssohn', *The African World*, LX (778), 294, 6 October 1917.
37. J. Leftwich, *Israel Zangwill* (1957), p. 77.
38. A. Rubens 'The Daniel Family', *TJHSE*, XVIII (1953) (afterwards quoted as Rubens, 'The Daniel Family'), p. 105.
39. Rubens, 'The Daniel Family', p. 106.
40. He married her on 20 November 1798 (Catalogue of Permanent Collection of Paintings in the possession of the Bath Corporation, s.v. DANIEL, (JOSEPH) (afterwards quoted as Cat. Paintings, Bath).
41. PCC Pitt 919, his sister's declaration; *contra* Rubens, 'The Daniel Family', p. 106, who credits Joseph with illegitimate children in Bristol.
42. PHC Min. Bk I, p. 23.
43. Rubens, 'The Daniel Family', p. 106. Hart died in December 1838 aged 73 (*Gent. Mag.*, 1839, p. 105).
44. Cat. Paintings, Bath according to which Joseph was born at Bath and christened at St Michael's Church, 21 May 1758. (According to Felix Farley's *Bristol Journal*, 3 September 1803, he died aged 43 in 1803, so must have been born in 1760—a slight, but not necessarily fatal discrepancy.)
45. PHC Min. Bk I, p. 47. It appears to be for a חילוק, which probably means 'a dispute'.
46. Under the heading of 'Plymouth'.
47. For portrait see Roth, *Provincial Jewry*, opp. p. 92; and Rubens 'The Daniel Family', plate 11, after p. 108. For Ephraim, see above, p. 287, n. 4.
48. See Rubens, 'The Daniel Family', plate 15, after p. 108.
49. Rubens, 'The Daniel Family', p. 107.
50. Rubens, 'The Daniel Family', pp. 107, 108.
51. PCC Pitt 919.
52. Lipman 'Aliens List', no. 49, according to which he settled in Plymouth in 1760. Samuel Hart, however, is said to have been born in Plymouth in 1755 (*Gent. Mag.*, 1839, p. 105).
53. PHC Min. Bk II, p. 49, 23 October 1808. He left £10 to the Synagogue to be remembered after his death at *Matnat Yad*.
54. See above.
55. Alfred Rubens, 'Early Anglo-Jewish Artists', *TJHSE*, XIV (1937), 120 (afterwards referred to as Ruben's 'Early Anglo-Jewish Artists').
56. PHC A/c. 1759, p. 49. He died 3 November 1809.
57. PHC Min. Bk II, p. 81.
58. Ibid. p. 150.

59. *Autobiography and Journals of B. R. Haydon*, ed. Malcolm Elwin, p. xx.
60. Illustration 17.
61. Then aged 21.
62. Then aged 62, a hard life had aged him.
63. Haydon was then in the Fleet Prison for debt.
64. Eric George, *Life and Death of B. R. Haydon*, p. 186.
65. *Autobiography and Journals of B. R. Haydon*, ed. Malcolm Elwin, p. 283.
66. Information from Mr H.L. Douch, Royal Institution of Cornwall. See also above, p. 207.
67. *Memoirs and Letters of Charles Boner with Letters of Mary Russell Mitford* (1871), p. 99, letter dated July 1846.
68. Ibid. p. 69, 10 September 1849.
69. *The Diary of B. R. Haydon*, ed. W.B. Pope (Harvard, 1963), II, p. 351, no. 6.
70. See his obituary in the *Hampshire Repository*, II, 20 November 1790, quoted by C. Roth in *TJHSE*, XIII (1936), 177.
71. She died in Exeter in June 1806, aged 70. (See his obituary, *Exeter Flying Post*, 17 June 1806, quoted in Rubens, 'Early Anglo-Jewish Artists', p. 104, no. 52b.)
72. Cf. above, pp. 222, 228.
73. *Devonshire Freeholder* 12 July 1822.
74. Rubens, 'Early Anglo-Jewish Artists', p. 104 quoting G. Pycroft, *Art in Devonshire* (Exeter, 1883), p. 45. For his trade card, see Rubens 'Further Notes on early Anglo-Jewish Artists', *TJHSE*, XVIII (1953), plate 19, after p. 108.
75. Quoted above, p. 99.
76. *Trew. Flying Post*, 5 February 1784.
77. Rubens, 'Early Anglo-Jewish Artists', p. 104.
78. Ibid. p. 127.
79. *Notes and Queries*, Series II, vol. viii, p. 494.
80. Opie as a lad in Cornwall, painted 'An Old Jew' which he showed to George III, and which helped to establish his reputation (Catalogue, *John Opie (1761–1807), Exhibition* (1962), p. 5).
81. Catalogue of Engraved British Portraits in the British Museum.
82. *Exeter Flying Post*, 18 December 1806. *The Dictionary of National Biography*, s.v. EZEKIEL, SOLOMON makes Solomon Ezekiel of Penzance the son of this Ezekiel Abraham Ezekiel, but the last sentence of the obituary taken from the discourse indicates that E.A. Ezekiel died a bachelor (cf. too, Rubens, 'Early Anglo-Jewish Artists', top p. 106).
83. Rubens, 'Early Anglo-Jewish Artists', p. 105, no. 58.
84. Cf. above, p. 201.
85. This account of Mordecai is mainly derived from Rubens, 'Early Anglo-Jewish Artists', p. 103.
86. Rubens, 'Early Anglo-Jewish Artists', p. 125.
87. Together with Abraham Ezekiel and his son Ezekiel, and Samuel Jonas.
88. See *Report of Charity Commissioners on the Endowed Charities of Exeter* (Exeter, 1904), p. 250.
89. *Anglo-Jewish Historical Exhibition Catalogue* (1888), item 761.
90. PCC Loveday, 298. See also above, p. 83.
91. Solomon, *Records of my Family*, p. 13.
92. *Royal Cornwall Gazette*, 18 December 1825; FHC tomb. 18.
93. See Illustration 6.
94. See above, p. 250 for details of his education.

95. Hart here translates the Hebrew expression *tsa'ar leba'alei hayyim*, the injunction against cruelty to animals, which in Yiddish and popular Jewish speech has the wider connotation of suffering imposed on the defenceless.
96. Hart, *Reminiscences*, p. 8.
97. Rubens, 'Early Anglo-Jewish Artists', pp. 120, 121.
98. Hart, *Reminiscences*, p. 15.
99. *JC*, 17 June 1881.
100. See Declaration of Trust, dated 17 November 1881, in the archives of the Plymouth Congregation.
101. Some of the ritual silver of the Falmouth Congregation is deposited at the Jewish Museum, London.
102. EHC tomb. 55.
103. It has an Exeter silver mark with maker's initials 'S.L.' (= Simon Levy).
104. Exeter silver mark, maker's initials 'S.H.' (= ?Samuel Hart). See Illustration 18.
105. Cf. above, p. 31.
106. = Simon Levy.
107. Cf. above, p. 309, n. 199.
108. *Lech L'cha*, Genesis, xii.
109. = cup used to hold wine whilst reciting a prayer of sanctification at the onset of Sabbaths and Festivals.

# NOTES TO PAGES 221–274

## *Chapter Ten*

1. For a full discussion of the terms 'ethno-centric', 'acculturated' and 'assimilated', much beloved by modern sociologist students of the Jewish scene, see E. Krausz, 'Concepts and Theoretical Models for Anglo-Jewish Sociology', *Jewish Life in Britain, 1962–1977*, S.L. Lipman and V.D. Lipman, eds (New York, 1981). For a discussion of the interaction of the processes of assimilation and acculturation on one another, see M. Freedman, 'Jews in the Society of Britain', *A Minority in Britain* (1955), pp. 227–42.
2. It has been cogently argued that more subtle and sophisticated concepts than those of ethnocentrism, acculturation and assimilation must be used to describe the changes which took place in the Jewish Diaspora (Kokosalakis, *Ethnic Identity*, p. 10).
3. See T.M. Endelman, *Radical Assimilation in English Jewish History, 1656–1945*, (Indiana University Press, 1990) (afterwards quoted as Endelman, *Radical Assimilation*), p. 4.
4. M. Freedman loc. cit. mentions the terms which a Jew uses when describing his neighbours in neutral and unemotional way, as: 'either "Christians", "Gentiles", or the somewhat ethnocentric "non-Jews". When (particularly immigrant) Jews couch their polite reference to the great majority around them in the simple word "English" they are documenting the first approach in the adjustment of an ethnic and religious group to a predominantly secular society'.
5. Leviticus, xix, 27.
6. Jessop, 'Joseph Families', p. 71.
7. Roth, *Provincial Jewry*, p. 92.
8. Ibid. p. 72.
9. *Trew. Flying Post*, 17 January 1799.
10. Portrait in Jewish Museum, London; photo of cameo in the author's collection. See Illustration 4.
11. Photographs of portraits of three men reputed to be her sons were kindly given to the author by the late Dr Richard D. Barnett of the British Museum.
12. Portrait in the Jewish Museum, London. See Illustration 1.
13. Photograph of his miniature given to the author by one of his descendants, Miss Allegra Dawe. None of his fourteen children and hundreds of descendants is known to have died as a Jew. See Illustration 5.
14. Solomon, *Records of my Family*, p. 12.
15. Ibid. p. 7.

16. *Jewish Encyclopaedia*, XII, p. 519, s.v. WIG. Well into the twentieth century acculturated Jews regarded the *sheitel* as a 'filthy, horrible and disgusting tradition, backward and long outdated' (R. Livshin, 'The Acculturation of the Children of Immigrant Jews in Manchester, 1890–1930', *The Making of Modern Anglo-Jewry*, ed. D. Cesarani, (afterwards quoted as Livshin, 'The Acculturation of the Children of Immigrant Jews') p. 89, quoting *JC* 22 February and 7 March 1924).
17. Portrait in possession of the late Dr Richard D. Barnett, British Museum. See Illustration 3.
18. Jessop, 'Joseph Families', p. 133.
19. Clegg, *Shemoel Hirsch*, p. 24.
20. *Plymouth Weekly Journal*, 9 June 1825. Wolfe was kept in prison until the lad was out of danger and until he found the money to pay Hitchcock's expenses and loss of time.
21. A. Ruppin, *The Jews of Today*, p. 148 quoted in Abraham Cohen, *An Anglo-Jewish Scrapbook* (1943), p. 254.
22. Frances Abel, *Prisoners of War in Britain, 1756–1815* (1914), p. 223.
23. Ibid. p. 233. Williams, *Manchester Jewry*, p. 9, quotes a graphic letter of 1798 which describes the abuse which a Jew might expect in a London theatre.
24. Ottolenghe, *An Answer*, p. 18.
25. Jessop, 'Joseph Families', p. 158.
26. Ibid. p. 157.
27. Ibid.
28. Ibid. p. 139.
29. Ibid. p. 131.
30. *Devonia*, May 1908, VIII, p. 103.
31. Adler MSS 4160, p. 259, at the Jewish Theological Seminary of America.
32. *Devonia*, May 1908, VIII, p. 103.
33. It is unwise, however, to judge too much from portraits because painters used to have pictures painted in advance and paint in the face when they found a customer (R. Henriques, *Marcus Samuel* (1960), p. 18).
34. Lecture by Miss Natalie Rothstein, Assistant Keeper, Textile Department, Victoria and Albert Museum, reported in *Anglo-Jewish Art and History* (Jewish Historical Society of England, 1967), no. 2.
35. *The Universal Songster* (1825), I, 408.
36. Cf. the nineteenth century low catch-phrase 'a long nose is a ladies liking': length of the male nose being held to denote corresponding length elsewhere (Partridge, *A Dictionary of Slang and Unconventional English*, s.v. NOSE).
37. Plymouth City Archives, Worth, w.367, p. 37. To walk in the garden = to ply as a prostitute.
38. Mayhew, *London Labour*, I, 452. Turkish rhubarb was used as a mild laxative.
39. Letter in the author's collection.
40. Original in the possession of F. Ashe Lincoln, QC, London.
41. Cohen, *Anglo-Jewish Scrapbook*, p. 236.
42. PCC Loveday, 298.
43. County Record Office, Truro; the will was proved 8 June 1807.
44. Original in author's collection. He was asking for memorial prayers to be recited for his late wife.
45. His son's portrait of him exhibited in 1826 describes him as 'Professor of the Hebrew language' (see above, p. 216).
46. PHC Min. Bk II.

47. 'Waste book in bookkeeping, book in which rough preliminary entries are made' (*Concise Oxford Dictionary* (Oxford, 1949), s.v. WASTE).
48. The earliest surviving cash book of the Liverpool Congregation dated 1806 is in Hebrew, after that date all other records are in English (Kokosalakis, *Ethnic Identity*, p. 59).
49. EHC tomb. 31; 1841 census; Book of Maps of all lands and tenements belonging to the Chamber of Exon. Map 5, no. 10 (in Exeter City Archives).
50. Copy in the author's collection.
51. EHC Min. Bk I, p. 57.
52. See above, p. 226.
53. Original in the author's collection. H. Ralph was probably Hannah Ralph, born in Plymouth in 1768 (PRO HO 107/1879/287/2C/p.514) and was probably née Nathan, and widow of Lewis Ralph who had died a few months earlier (*Trew. Flying Post*, 5 February 1824). She was presumably applying not only to live in the house but also to act as caretaker and *mikveh* attendant. She was unsuccessful, the situation and house going to Solomon Emden and his wife (PHC Min. Bk II, p. 183). For letters written in English by the beadle and a poor Jew to the Liverpool Congregation in 1836, see Kokosalakis, *Ethnic Identity*, p. 61. The beadle's letter may have been written for him by a well-educated person.
54. Meshivat Nefesh A/c. p. 4. Note, too, the secular form of date.
55. *PHC Regulations, 1835*, no. 96. In the Sephardi Synagogue in Lauderdale Road, London, announcements are still made in Spanish, and the Sephardim left Spain in 1492!
56. On the reverse side of the tombstone of Lyon Joseph, died June 1825 (PHC tomb. A13).
57. Sarah, wife of Moses Jacob of Redruth (FHC tomb. 28).
58. Aaron Selig, died 18 July 1841 (PenHC tomb. 11) and Solomon Levy, 'a native of Exeter', died 20 August 1841, aged 56 (PenHC tomb. 32).
59. Nancy, wife of Moses Lazarus (EHC tomb. 52).
60. Letter to the present writer from Stone's granddaughter.
61. *Melech* (Hebrew) = king; *feld* (German) = field.
62. The bookplate of the Holcenberg Collection, endowed by his daughters at the Plymouth Public Library, plays on this name.
63. There was a well-known Jewish family called Johnson in Devon (see *Catalogue of an Exhibition of Jewish art and history*, item 443), but there is no evidence to suggest that their name was derived from an original Jonas.
64. Family tradition recounted to the author. Benjamin Goetz, first rabbi of the Liverpool Congregation, adopted the name of Yates, a well-known local name (Kokosalakis, *Ethnic Identity*, pp. 47, 67).
65. EHC A/c. 1827, pp. 97, 158, 191, 256; 1851 census, 1861; EHC Marriage Register, 24; EHC tomb. 91.
66. R. Livshin, 'The Acculturation of the Children of Immigrant Jews', p. 82.
67. Jacob compared Benjamin to a wolf (Genesis, xlix, 27).
68. *Hirsch* is the Yiddish for the Hebrew *Zvi*, a hart.
69. Jacob compared Issacher to an ass which 'bears' a burden (Genesis, xlix, 14).
70. Jacob compared Judah to a lion (Genesis, xlix, 9).
71. Bunam was a medieval French form for *bon homme*, itself a translation of the Hebrew.
72. PHC Min. Bk II, pp. 54, 80, 133, 141.
73. Solomon, *Records of my Family*, p. 6.

74. Mayhew, *London Labour*, I, 452. See above, p. 324.
75. PHC A/c. 1821, p. 17.
76. Chief Rabbinate Archives, 7, March 1860.
77. The Jewish Sabbath is kept from just before sunset on Friday until just after sunset on Saturday.
78. Mayhew, *London Labour*, I, 53.
79. *Trew. Flying Post*, 8 August 1844.
80. H. Miles Brown, *Cornish Clocks and Clockmakers* (Dawlish, 1961) (afterwards quoted as Brown, *Cornish Clocks*), p. 70.
81. *Devonshire Freeholder*, 5 November 1825. A Plymouth woman kept her beer shop open on Sundays during Divine Service and 11 men were found drinking there; also fruit shops were open. The mayor initiated prosecutions.
82. *JC*, 26 August 1853. When a Manchester Jew claimed in 1850 that his business depended on his Sunday trade, the magistrate declared: 'If you choose to shut up on Saturday, in obedience to your own religion, you must close on Sunday for ours' (Williams, *Manchester Jewry*, p. 203).
83. Better known as The Board of Deputies.
84. C.H.L. Emanuel, *A century and a half of Jewish History* (1910), pp. 68, 76, 85, 87 *et passim*.
85. *JC*, 22 February 1861. Three Jews and 10 non-Jews were involved.
86. EHC Min. Bk I, 16 January 1853.
87. EHC Min. Bk I, 20 February 1854. For similar problems in the wider community see *Responsa of Hatan Sofer* (1912), no. 28, written in 1873, and Hatam Sofer to *Orach Hayyim*, 15.
88. Chief Rabbinate Archives, 5, letter 3460.
89. Adler, MSS 4160, p. 259, at the Jewish Theological Seminary of America. For the doctrine of 'the law of the land' see *Baba Kamma*, 113a.
90. PRO HO 129/11/286–7.
91. PRO HO 129/11/282.
92. Partridge, *Dictionary of Slang*, s.v. JEW.
93. Letter to the present writer from Mrs H. Conick, 29 September 1962.
94. EHC Min. Bk 1838, p. 51. No manufactured bath or pool, other than a properly constructed *mikveh* can meet the requirements of Jewish law regarding the monthly immersion of the Jewish woman.
95. Chief Rabbinate Archives, MSS 104.
96. Chief Rabbinate Archives, MSS 104, p. 37.
97. Chief Rabbinate Archives, 7, letter dated 17 May 1860. Cf. E. Krausz, *Leeds Jewry* (Cambridge, 1964), p. 110.
98. PHC Min. Bk I, Regulation 17.
99. Genealogical notes prepared by Mr Geoffrey White (in the Cecil Roth Collection). Almost any Jewish family tree of the nineteenth century in England such as those quoted in Jessop, 'Joseph Families' gives similar results.
100. The author met a number of non-Jews in Devon in the mid-1960s who claimed to have had a Jewish grandparent. One non-Jewish woman seeing candles alight on a Friday evening in the home of Mrs Rose Owen of Plymouth, remarked that at the beginning of the twentieth century her own mother also used to light candles on a Friday night not knowing why, but because 'it was traditional in our family'. For a similar incident concerning a Marrano, see C. Roth, *Gleanings* (New York, 1967), p. 139, n. 81.
101. *Annual Register*, 26 March 1831.

102. W.G. Hoskins, 'A sheaf of Modern Documents', *Devonshire Studies*, W.G. Hoskins and H.P.R. Finberg, eds, (1952) p. 413.
103. Information supplied by the then Lazenby Chaplain, University of Exeter, the Revd J.A. Thurmer, in a letter dated 18 May 1973.
104. *Trew. Flying Post*, 14 January 1813.
105. *The Diary of B. R. Haydon*, ed. W.B. Pope, II, 94. See also above, pp. 213, 217.
106. *TJHSE*, XVII (1953), 105ff. Cf. above, p. 209.
107. PCC Pitt 919, probate 30 June 1806. His two brothers each left four 'natural and lawful' children, but presumably these were legitimate.
108. His letter of application is quoted verbatim, above, p. 227.
109. Jessop, 'Joseph Families', p. 126, though he was buried on the 'high ground' (PHC tomb. A133).
110. Endelman, *Radical Assimilation*, p. 18.
111. The traditional patronymic of a convert to Judaism.
112. EHC tomb. 41.
113. He died in 1829 and lies in the next grave, EHC tomb. 40.
114. EHC Min. Bk I, p. 38. Cf. Hyamson, *Sephardim*, p. 358, the first convert converted by the Sephardi authorities in England was in 1877. It had been traditionally believed that a condition of the Jews' re-admittance to England was the refusal of proselytes in England itself. The Ashkenasim soon broke the convention, though probably only when it was convenient for the party to go abroad (cf. Hyamson, *Sephardim*, p. 390, n. 6).
115. MSS Minutes of London Beth Din, 1805–35, p. 10b (in the Cecil Roth Collection).
116. *A List of Converts to Judaism in the City of London, 1809–1816*, ed. Barnett A. Elzas (New York, 1911).
117. Joseph Joseph's Circum. Reg. no. 2.
118. PHC A/c. 1821, p. 39.
119. MSS Minutes of London Beth Din, 1805–35, p. 63a (in the Cecil Roth Collection).
120. PHC Book of Records, p. 21a. In 1874, a Mrs Pinner received 10*s*. from the Hand-in-Hand Society.
121. PHC Book of Records, p. 53.
122. PHC Marriage Register, no. 99.
123. e.g. Joseph Ottolenghe, Menasseh Lopes.
124. St Glewias Parish Registers. Possibly this is the source of the 'most curious accusation' that Isaac Bing, known as Levy Isaacs, Secretary of the Great Synagogue, London, married a Gentile in Ireland (see Roth, *Great Synagogue*, p. 193).
125. S.S. Levin, 'The changing pattern of Jewish Education', *A Century of Jewish Life* (1970), p. 58.
126. The following account, including direct quotations, is based on A.M.W. Stirling, *The Ways of Yesterday* (1930).
127. See report in *VJ*, 7 January 1842, and above, p. 241.
128. Endelman, *Radical Assimilation*, p. 27.
129. See above, pp. 30, 224; and Additional Note 2, p. 335.
130. Williams, *Manchester Jewry*, p. 27.
131. Or Schönlanke.
132. *Plymouth Journal*, 23 June 1825.
133. *Trew. Flying Post*, 8 December 1825.
134. *Plymouth and Devonport Weekly Journal*, 23 June 1825.

135. John Hatchard, *The Predictions and Promises of God respecting Israel* (Plymouth, 1825), pp. 37–40.
136. Her maiden name was Deborah Levy, and she was the daughter of a Plymouth Jew.
137. *Plymouth and Devonport Weekly Journal,* 17 November 1825. A convert's embarrassed family might well leave town. Barnet and Jane Lyons settled in Plymouth after their daughter, Esther, was enticed from her family in Cardiff and converted to Christianity in 1868. The parents sued her abductors, a Baptist minister and his wife, and the case became a *cause célèbre* (*JC,* 1868 and 1869, *passim*); R. Woolfe, 'The abduction of Esther Lyons', *Cajex,* ii (1952), 14–23, quoted in *MJHSE,* XI (1979), pp. 67, 71.
138. *The Jewish Expositor* (1826), XI (1826), p. 78.
139. *The Cambrian,* 25 June 1825. The Revd M.H. Malits kindly provided this reference.
140. *Gent. Mag.* 27 April 1847, p. 675.
141. *VJ,* 7 January 1842.
142. J.H. Newman, *Apologia pro Vita Sua,* (1964), pp. 146–152.
143. Roth, *Magna-Bibliotheca,* p. 292. Other editions were published in 1835 and 1840.
144. EHC A/c. 1822–1825, at the entry of Hyman Levy Davis Isaac.
145. *Trew. Flying Post,* 23 May 1832.
146. C.S. Hawtrey, *A summary account of the origin, proceedings, and success of the London Society for Promoting Christianity among the Jews* (1826), p. 25.
147. A.M.W. Stirling, *The Ways of Yesterday,* p. 137.
148. Endelman, *Radical Assimilation,* p. 49.
149. St Mary's, Truro, Baptism Register, p. 654.
150. St Andrew's, Plymouth, Parish Baptism Register, 1825.
151. *Plymouth and Devonport Weekly Journal,* 24 March 1825.
152. *The Cambrian,* 20 November 1830. The Revd M.H. Malits kindly provided this reference.
153. *West Briton and Cornwall Advertiser,* 12 November 1830.
154. Roth, *Jews in England,* p. 226.
155. PHC Book of Records, p. 53.
156. There is ample evidence of conversions being effected in London under Solomon Hirschell until 1835, see above, p. 237.
157. Underlined in the original. A century later, the Anglo-Jewish community's attitude has hardly changed. Congregations and communities display a marked reluctance to collect demographic mateiral or conduct in-depth surveys, 'for what would the "fascists" do with the results?' (B.A. Kosmin, 'The case for the local perspective in the study of contemporary British Jewry', *Jewish Life in Britain, 1962–1977* (New York, 1981), pp. 83–94). The politics of Anglo-Jewry are still considered to be a taboo subject (G. Alderman, *The Jewish Community in British Politics* (Oxford, 1983), p. vii). A current (1990) series of British Telecom advertisements portraying the actress, Maureen Lipman, as a 'typical' Jewish housewife leaves many Jews with a distinct feeling of unease.
158. *JC,* 27 July 1860.
159. Roth, *Magna-Bibliotheca,* pp. 256–68.
160. See above, p. 206.
161. L. Cohen, *Sacred Truths* (Exeter, 1808), p. ii.

162. The first edition may be found at the Mocatta Library, the second at Jews' College, London.
163. Published in New York, 1850.
164. London, 1815. See C. Roth, 'Educational Abuses and Reforms in Hanoverian England', *Essays and Portraits in Anglo-Jewish History* (Philadelphia, 1962), p. 228.
165. London, 1817. Roth, *Magna-Bibliotheca*, p. 265.
166. Endelman, *Radical Assimilation*, p. 52, quotes a number of examples. These provisos were generally struck down by the courts in the twentieth century (Re Moss's Trusts [1945] 1All ER 207); but see Re Tuck [1976] 1All ER 545, where the condition against intermarriage was upheld because the bride's status was subject to clarification by the Chief Rabbi.
167. Endelman, *Radical Assimilation*, p. 52.
168. Devon Record Office, Wills, L492.
169. Verbal tradition in the family recounted to the author.
170. He was born in Mezeritz in 1728 and landed at Harwich in 1748. He was beadle to the Plymouth Congregation in 1778 (PHC Min. Bk I, p. 47). He did not come to Plymouth between 1798 and 1803 (*contra* Lipman, 'Aliens List', no. 33).
171. PHC Min. Bk I, p. 28.
172. Ibid. p. 10.
173. Ibid. p. 39.
174. In the Yiddish idiom the teacher learns with his pupil.
175. PHC Min. Bk II, p. 68.
176. Ibid. p. 65. The west wall is at the back of the synagogue.
177. EHC Min. Bk I, under date 26 October 1851.
178. Ibid. at 19 December 1852.
179. Ibid. at 24 October 1853.
180. See above, p. 228.
181. Moses Ephraim was his tutor.
182. See his obituary in *JC*, 29 May 1868.
183. Cf. the Penzance Hebrew Society for the Promotion of Religious Knowledge which also met on Friday nights (Roth MSS 276).
184. Original in possession of Mr Edgar Samuels, London.
185. Chaim Rabin has pointed out that there was a group of Hebraists in London in the early 1840s who constituted a Hebrew-speaking circle (Chaim Rabin, *Ivrit Meduberet Lifnei 125 Shanah* (Jerusalem, 1964).
186. PHC Min. Bk II, p. 39.
187. Ibid p. 171.
188. Quoted in P.L.S. Quinn, 'The Jewish Schooling Systems of London, 1656–1956' (unpubd PhD Thesis, University of London, 1958) (afterwards quoted as Quinn, 'Jewish Schooling'), p. 382.
189. Chief Rabbinate Archives, MSS 104.
190. For example, Chief Rabbinate Archives, 11, letter nos 7811, 7796; 12, pp. 501, 503; 13, p. 181.
191. *JC*, 17 January 1862.
192. See *Transactions of Plymouth Institute*, vol. 6, p. 82.
193. A. Myers, *Jewish Directory for 1874* (1874).
194. Photocopies were made by the late B.H. Emdon.
195. Again the magic age of *bar mitzvah* beyond which formal Jewish education ceased.
196. Revd Dr M. Berlin, minister of the Congregation, 1896–1906.

197. *Jew. Year Bk*, 1906.
198. Letter to the author 19 January 1989.
199. Hart, *Reminiscences*, p. 7.
200. His father married (a) Leila, sister of Jacob Jacobs, and (b) Hannah, widow of Jacob Jacobs, daughter of Hyam Barnett of Gloucester.
201. See R. Livshin, 'The Acculturation of the Children of Immigrant Jews in Manchester, 1890–1930', *The Making of Modern Anglo-Jewry*, ed. D. Cesarani (Oxford, 1990), pp. 79–94.
202. *JC*, 24 June 1881.
203. *JC*, 17 June 1881.
204. Quoted in Quinn, 'Jewish Schooling', p. 249.
205. Cyrus Redding, *Illustrated Itinerary of the County of Cornwall* (1842), p. 245.
206. Quinn, 'Jewish Schooling', p. 249.
207. Chief Rabbinate Archives, 2, letter dated 8 September 1851.
208. PRO HO 107/1879/287/2C/p.688.
209. PHC Marriage Register, 32.
210. Ibid. 26.
211. Ibid. 56. These were members of the Roseman and Fredman families which dominated the affairs of the Plymouth Congregation for nearly a century after 1870. See above, pp. 121–2.
212. Ibid. 57.
213. PRO HO 50/43.
214. Ibid.
215. Ibid.
216. Ibid. 22 June 1798.
217. PRO HO 50/66, 64: 20 13/530, 536, 553, 554, 349.
218. He sent a copy of his notes to Prof. Cecil Roth on 3 March 1955.
219. A blank was left in the original notes for the name to be filled in.
220. C. Thomas, 'Cornish Volunteers in the Eighteenth Century', *Devon and Cornwall Notes and Queries*, XXVIII (1959), 11–12. Mr Godfrey Simmons drew my attention to this article.
221. W.G. Hoskins, *The Exeter Militia List, 1803* (1972).
222. *JC*, 29 September 1899. His lineal descendants currently trade in Teignmouth.
223. On one day in April 1793 every man on the streets of Plymouth was impressed (*Sherborne Mercury and Western Flying Post*, 29 April 1793).
224. J.R. Hutchinson, *The Press Gang*, p. 244.
225. Green 'Royal Navy', p. 108.
226. *Catalogue Anglo-Jewish Art Exhibition, 1956*, item 330. Cf. also item 440.
227. D. Ben-Gurion, Letters to Paula (1971), p. 31.
228. PHC Marriage Register, II, 8.
229. PHC Marriage Register, II, 6.
230. PHC tomb. O22. The Judeans later became the Haganah, which in turn became the Israel Defence Force, the Israeli army. Alf Lithman of London recalled in 1990 that his brother John enlisted by falsifying his age and that he died as the result of some accident.
231. PHC tomb. O23.
232. PHC tomb. O17.
233. PHC tomb. D23.
234. Her gravestone has the Hebrew equivalent of GRHDS and this is the only indication that she was of Jewish origin. See *The Plymouth Hebrew Congregation's Digest*, December 1986, p. 14.

235. Both D.L. Maxwell and M. Overs later became Honorary Officers of the Congregation.
236. Ruth Bloom is descended from the Ullman's, a Jewish family with its roots in eighteenth-century Devon.
237. Letter to the author from his nephew, Jack Smith, Haifa, 6 March 1989.
238. G.A. Alderman, *The Jewish Community in British Politics* (Oxford, 1983) (afterwards quoted as Alderman, *The Jewish Community in British Politics*).
239. Ibid. p. 6.
240. Ibid. p. 12. For Lopes, see above, pp. 227, 238; for a full account of his political career and chicanery, see W.G. Hoskins, 'A sheaf of modern documents', *Devonshire Studies*, eds W.G. Hoskins and H.P.R. Finberg (1952).
241. Alderman, *The Jewish Community in British Politics*, pp. 10–17, discusses the reason for the swings of the Jewish vote.
242. Ibid. p. 26.
243. *Plymouth and Devonport Journal*, 2 April 1857, quoted by Alderman, *The Jewish Community in British Politics*.
244. Roth, *Jews in England*, p. 255. The earliest known instance of a Jew *de facto* serving elected public office in England occurred in Brighton in 1822 (*TJHSE*, XII (1928), 46).
245. For his biography and intellectual achievements, see above, p. 245.
246. *VJ*, 9 May 1845.
247. He and his wife Kitty were born in Portsea, Hants, in 1784 and 1788, respectively. He arrived in Plymouth before 1810 (Meshivat Nefesh A/c.), became a vestry member of the Congregation in 1812 (PHC Min. Bk II, p. 73). In 1822, he was a pawnbroker at 62 North Corner Street, and a wholesale slopseller at 48 Queen Street, Plymouth Dock (*Tapp's Plymouth Directory*, 1822). In 1850 he had a Fancy Depot at 15 Catherine Street, Devonport (PRO HO 107/1881/289/1/p.46).
248. *Brindley's Plymouth Directory*, 1830, p. 119.
249. G. Alderman, *London Jewry and London Politics, 1889–1986*, (1989), p. 3.
250. Born in Portsmouth in 1801 and settled in Plymouth about 1821.
251. *VJ*, 13 March 1846.
252. See above, pp. 202–3.
253. *JC*, 7 November 1862; Plymouth Town Council Minutes.
254. He was born in Plymouth in 1799 (the son of Eliezer Emden I who was born in Amsterdam 1764), was in London 1786–94, Portsmouth 1794–98, Plymouth 1798 to his death in 1844 (Lipman, 'Aliens List', no. 14), and died in May 1872.
255. Devonport Council Minutes, D3/AB5. PHC tomb. B112.
256. Devonport Council Minutes, D3/AB6.
257. *JC*, 2 March 1900.
258. Born Plymouth 1851, a pawnbroker at 36 Edgecumbe Street at the time of his marriage in September 1875 to Eliza, daughter of Markes Levy.
259. *JC*, 7 November 1890.
260. See Illustration 19.
261. See Illustration 20.
262. *The Exeter Evening Post*, 1 March 1887; *Trew. Flying Post*, 23 February 1887.
263. Alderman, *The Jewish Community in British Politics*, p. 31.
264. Letter from A. Alexander to *JC*, 9 March 1860. Overt anti-Semitism in England never seems to have played any part in the development or decline of the Congregations of the South West.

265. *JC*, 20 February 1863.
266. *JC*, 13 July 1877.
267. *Jew. Year Bk*, 1901, p. 321. His name is variously spelled as Mendel, Mendl and Mendle. He campaigned for a swimming pool in Plymouth, 'an urgent need'. The pool was built in 1965!
268. Alderman, *The Jewish Community in British Politics*, p. 43.
269. L.M. Goldman, *The History of the Jews in New Zealand* (Wellington, 1858), p. 176.
270. Alderman, *The Jewish Community in British Politics*, p. 69.
271. *Jew. Year Bk*, 1901.
272. For a corrective of this view, see Alderman, *The Jewish Community in British Politics*, pp. 16–30.
273. Quoted from *Daily Western Mercury* in *JC*, 15 November 1861.
274. M.A. Shepherd, 'Cheltenham Jews in the nineteenth century', *Jewish Journal of Sociology*, 21 (1979), 125–133, explains the decline of the Cheltenham Congregation by saying that it was too small and too badly led by its religious and lay leaders to attract new immigrants of good quality. This probably explains the Exeter Congregation's failure to attract sufficient East European Jewish immigrants in the 1880s and an abortive attempt in 1965 to encourage Jewish families to settle in Plymouth. Shepherd admits 'the paradox of a languishing community in the midst of prospering late Victorian Cheltenham may never be satisfactorily explained'.

# NOTES TO PAGES 263–264

## *Epilogue*

1. H. Whitfeld, *Plymouth and Devonport in Peace and War*, p. 280.
2. H.R. Coulthard, *The Story of an Ancient Parish, Breage with Germol* (Penzance, 1913), p. 151.
3. In the parish of St George and St John, in the municipal ward of Petrock near to or off Smythen St (PRO HO RG9/1399). It is not mentioned in the 1838 valuation of Exeter, it appears in the 1871 House Inspection Register but is not marked by name on the 1876 Ordnance Survey 25 inch map (information from Mrs Rowe, Exeter Librarian, 20 December 1967).
4. PRO HO RG9/1398.
5. C. Redding, *Illustrated Itinerary of the County of Cornwall*, p. 165. A well called Gulfwell or the Hebrew Brook at the seat of the Harris family was once attended, according to Borlase, by a Sybil whose death in his time was recent. The well was considered oracular and was consulted for the purpose of recovering lost cattle or stolen goods.
6. It is situated at the seaward end of Lansallos Street, Polperro. See above, p. 273, n. 195.
7. The halibut, *hippoglossus vulgaris*, called Jews' fish because it was favoured in their diet (F.W.P. Jago, *Glossary of the Cornish Dialect* (Truro, 1882), p. 195).
8. Black field beetle which exudes reddish froth. Children hold it in their hand and say, 'Jew! Jew! Spit blood' (M.A. Courtney, *Cornish feasts and folklore*, p. 61).
9. This saying is still current according to the caretaker of the Royal Institution of Cornwall (information from H.C. Douch, 28 July 1964). Wright, *The English Dialect Dictionary*, s.v. JEW, understands the phrase to mean, 'meaner than a Jew', shave = miser, niggardly. It is more likely, alas, that the phrase refers to the cruel joke of setting fire to the beards of Jewish pedlars. 'It was the fiendish custom to set fire to the beards of condemned [Marranos] before the pyre was lighted, so as to increase their sufferings. This they called, "Shaving the New Christians"' (C. Roth, *A History of the Marranos* (New York, 1932), p. 135).
10. Roth, *Provincial Jewry*, p. 91.
11. The Holcenberg Collection, Plymouth City Library. See Illustration 20.
12. Primarily the bequests of Jacob Nathan, see above, p. 203.
13. Brett Lance Herman of Toronto.

# ADDITIONAL NOTES

(The figures in parentheses at the end of each additional note refer to the pages of the text)

1. 'And whereas our borough of Marghas-Jew is an ancient borough, and was once a trading town and of great note, (charter of 13 June, 27 Elizabeth I.)' quoted in T.S. Duncombe, *The Jews of England* (1866), p. 27. But cf. F.M. Muller 'Are there Jews in Cornwall', *Chips from a German workshop* (1880), III, pp. 229–329 who asserts that Market-Jew is a corruption of *Marghas-Jovis* (= Thursday-market). However, 'Jew' was an opprobrious epithet (Graetz, *The History of the Jews*, V, p. 293, 'the scornful nickname of Jew' and *VJ*, 16 February 1844, p. 84, 'because of the remains of ancient prejudice against the title "Jew", the *Athenaeum* calls us "Hebrews"'). It is therefore unlikely that the Cornish would have saddled themselves with such a name unless there had been good reason. (—p. 265, n. 11.)

2. They were:
   (i) Joseph Ottolenghe, An Answer to two papers lately published by Gabriel Treves, a Jew of the City of Exeter. The one intituled, A Vindication of the proceedings of Gabriel Treves against Joseph Solomon Ottolenghe, now a prisoner in Southgate, Exon; the other is intituled, An Advertisement, wherein is contain'd the said Joseph Ottolenghe's vindication of himself, against the aspersions cast on him in the said papers. Together with an account of his conversion from the Jewish to the Christian religion. And also of the hardships which he hath suffered from the said Gabriel Treves, his uncle, etc., since his conversion. (?, 1735).
   (ii) Gabriel Treves, the Ethiopian's skin not changed, nor the leper's spots washed out. Proved in a reply to a Pamphlet pretended to be penned by Joseph Ottolenghe, intituled, An Answer to Two papers lately published by Gabriel Treves etc. By the abused Gabriel Treves. (Exeter, 1736).
   (iii) Lewis Stephens, Discourse preach'd before the inhabitants of the Parish of St. Petrock in Exeter, on Sunday the 6th of July, 1735: occasioned by their delivering Joseph Ottolenghe, a poor convert Jew, out of South-gate Prison; into which he was cast by a Jew, after his conversion to Christianity, etc, (Exeter, 1735).
   (iv) Lewis Stephens,Excellencies of the kindness of Onesiphorus to St. Paul, when he was a prisoner in Rome. Exemplified in a discourse preach'd before the inhabitants of the Parish of St. Petrock in Exeter, on Sunday the 6th of July, 1735: occasioned by their delivering Joseph Ottolenghe, a poor convert Jew, out of South-gate Prison; into which he was cast by a Jew, after his conversion to Christianity, etc. (Exeter, 1735).

There is a copy of *An Answer . . .* in the Mocatta Library, London. There was a copy of *The Ethiopian's Skin* in the Davidson Collection at the Plymouth Institution, but it was destroyed during World War II. It is not listed in C. Roth, *Magna Bibliotheca Anglo-Judaica* (1937).

Ottolenghe's subsequent career in America is of some interest. The trustees of the Society for Establishing the Colony of Georgia as a silk producing colony gave a sample of Georgian silk in 1741 to 'Sampson Levi the Jew, to take to England' (H. Huhner, 'The Jews of Georgia in Colonial Times', *Proceedings of the American Jewish Historical Society*, X, p. 90). In 1751 the Trustees sent Ottolenghe from England to superintend the silk industry, and in 1753 he was given 300 acres of land. He was consulted by the Governor and Assembly of Georgia at first in connection with his office as superintendent of the silk industry and later on other matters (ibid.). Sampson Levi's name suggests that he might have been the grandfather of a Sampson Levi, born Newton Abbot, 1802. (—p. 30, n. 26.)

3. The order of *Kaddish* precedence in the Plymouth Congregation as laid down in its MSS *Siddur*, 1805:
   (a) Definition:
      (i) *Tefillah Kaddish* after *Alenu*, *Anim Zemirot*, *Barechi Nafshi* and *Perek* (afterwards referred to as *T.K.*).
      (ii) *Tehillim Kaddish* after Psalms, Five Megillot, and study (afterwards referred to as *Ps.K.*).
   (b) Friday night after Psalm 92: *Yahrzeit* and mourners ballot.
   (c) There are 3 stages of mourners: 7 day, 30 day, 12 months from death. 7 day / minor (i.e. under 13 years) and Yahrzeit:– the minor says 2 *Ps.K.*'s and *Yahrzeit* 1 *Ps.K.* But a 30 day and 12 month are completely displaced by *Yahrzeit* each *Ps.K.*
   (d) *Yarhzeit* member says every *T.K.* and displaces non-member *Yahrzeit*, and non-member may say one *Ps.K.* without having to ballot.
   (e) *Yahrzeit* member and 30 day member:– *Yahrzeit* has only one *T.K.* after *Maariv* or *Shacharit* and rest to 30 day member.
   (f) *Yahrzeit* member and 30 day non-member:– 2 *T.K.*'s to member and 1 *T.K.* to non-member.
   (g) Several *Yahrzeit*'s:– each has one *Kaddish* and displace 30 day.
   (h) *Yahrzeit* member and *Yahrzeit* non-member and 30 day member:– each have one *T.K.*
   (i) Non-member 30 day and member 12 month:– share all *T.K.*'s.
   (j) Equal ranking:– cast lots.
   (k) Non-member *Yahrzeit* may say one *T.K.* in presence of 30 day or 12 month members but does not join in ballot for *Ps.K.*
   (l) *Yahrzeit* has precedence over mourners.
   (m) A married or bachelor teacher [of Torah] employed by several members OR one who studies at a *Yeshivah* OR a bachelor servant of a member is treated as a member.
   But a married teacher employed by one member or a married servant is treated as a non-member.
   (n) The day a 12 month finishes he says every *Kaddish* unless a *Yahrzeit* or 30 day is present. (—p. 179, n. 48.)

# SELECTED BIBLIOGRAPHY

## *Unprinted*


Archives at:

Board of Deputies of British Jews, London
Chief Rabbi's Office, London
Exeter City Library
Jewish Museum, London
Plymouth City Library, Plymouth
Plymouth Hebrew Congregation, Plymouth
Public Record Offices, London, Truro
Roth Collection, Hebrew University, Jerusalem
Royal Institution of Cornwall, Truro
Somerset House, London
Susser Collection, Jerusalem
Victoria and Albert Museum, London

Tombstone inscriptions at the Exeter, Falmouth, Penzance and Plymouth (2) Jewish cemeteries.

Inscriptions in the Exeter and Plymouth synagogues.


## *Printed*

PRIMARY


*Anglo-Jewish Notabilities* (1949)

*Bevis Marks Records*, vols I, II, ed. R. Barnett (Oxford, 1940, 1949); vol. III, ed. G.H. Whitehill (1973).

*Bibliotheca Cornubiensis* (1874)

Board of Deputies of British Jews, *Annual Reports*

*Calendar of the Plea Rolls of the Exchequer of the Jews*, vols I, II, ed. J.M. Rigg (1905); vol. III, ed. H. Jenkinson (1929); vol. IV, ed. H.G. Richardson (1972)

*Catalogue of an Exhibition of Anglo-Jewish Art and History* (1956)

W. Glegg, *Samuel Harris, a converted Jew* (Sheffield, 1833)

L. Cohen, *Sacred Truths* (Exeter, 1808)

*Collectanea Cornubiensis* (Truro, 1890)

Directories

*Bailey's, 1783*

*Universal British Directory, 1798*
H. Woolcombe, *Picture of Plymouth, 1812* (Plymouth)
S. Rowe, *Plymouth Directory, 1814* (Plymouth)
*Tapperell's Plymouth Directory, 1822* (Plymouth)
*Pigot's Directory, 1823*
*Brindley's Plymouth Directory, 1830* (Devonport)
*Flintoff's Plymouth Directory, 1844* (Plymouth)
*White's Devonshire Directory, 1850* (Sheffield)
*Kelly's Post Office Directory, 1856*
*Billing's Devonshire Directory, 1857* (Birmingham)
*Harrison, Harrod Directory for Devon and Cornwall, 1862*
W. Warn, *Directory and Guide for Falmouth and Penryn, 1864*
*Eyre's Plymouth, Devonport and Stonehouse Directory, 1896*
*Jewish Year Book*, 1896–
S.A. Hart, *Reminiscences* (1882)
Journals
*Annual Register*, 1760–1827
*Exeter Pocket Journal*, 1791–1863 (Exeter)
*Gentleman's Magazine*, 1731–1833
*Navy List*, 1780–1870
*Notes and Queries*, 1900– (Exeter)
R.P. Lehmann, *Anglo-Jewish Bibliography* (1973)
——, *Nova Bibliotheca Anglo-Judaica* (1961)
*A List of Converts to Judaism in the City of London, 1809–16*, ed. B.A. Elzas (New York, 1911)
H. Mayhew, *London Labour and the London Poor* (1851)
*Memoirs and Letters of Charles Boner with Letters of Mary Russel Mitford* (1871)
Newspapers
*The Cambrian* (Swansea, 1813)
*Daily Western Mercury* (Plymouth, 1860–)
*Felix Farley's Bristol Journal* (Bristol, 1776–1853)
*HaMelitz* (Odessa, 1860–1903)
*Hebrew Observer* (1853–4)
*Jewish Chronicle* (1841–)
*London Gazette*, (1665–)
*Plymouth and Devonport Weekly Journal* (Plymouth, 1818–25)
*Royal Cornwall Gazette* (Truro, 1811–)
*Sherborne and Yeovil Mercury* (Sherborne, 1746)
*Trewman's Exeter Flying Post* (Exeter, 1768–)
*Voice of Jacob* (1841–8)
*West Briton and Cornwall Advertiser* (Truro, 1811–)
*Western Flying Post* (Sherborne, 1829–67)
*Western Morning News* (Plymouth, 1860–)
J. Ottolenge, *An Answer . . .* (?, 1735)
*Plymouth Hebrew Congregation Regulations, 1834*
C. Roth, *Magna Bibliotheca Anglo-Judaica* (1937)
*Select Pleas, Starrs and other Records from the Rolls of the Exchequer of the Jews, 1220–84*, ed. J.M. Rigg (1902)
I. Solomon, *Records of my Family* (New York, 1887)
*Starrs and Jewish Charters*, vols. I, II, III, IV (1930–2)
R.N. Worth, *Calendar of the Plymouth Municipal Records* (Plymouth, 1893)

SECONDARY

Articles

M. Adler, 'The Medieval Jews of Exeter', Report and *Transactions of the Devonshire Association*, LXIII (1931)
S. Applebaum, 'Were there Jews in Roman Britain?' *TJHSE*, XVII (1950)
A. Arnold, 'Apprentices of Great Britain, 1710–73', *MJHSE*, VII (1970)
J. Bannister, 'Jews in Cornwall', *Journal of the Royal Institution of Cornwall*, II (1867)
A.M. Jacob, 'The Jews of Falmouth 1740–1860', *TJHSE*, XVII (1949)
V.D. Lipman, 'Sephardi and other Jewish Immigrants in the Eighteenth Century', *Migration and Settlement* (1971)
——, 'A survey of Anglo-Jewry in 1851', *TJHSE*, XVII (1953)
——, 'Synagogal organisation in Anglo-Jewry', *The Jewish Journal of Sociology*, I (1959)
C. Roth, 'Penzance: The Decline and Fall of an Anglo-Jewish Community', *Jewish Chronicle Supplements* (May, June, 1931)
A. Rubens, 'The Daniel Family', *TJHSE*, XVIII (1953)
——, 'Early Anglo-Jewish Artists', *TJHSE*, XIV (1937)
E.R. Samuel, 'The First Fifty Years, *Three Centuries of Anglo-Jewish History* (Cambridge, 1961)
B. Susser, 'Social Acclimatization of Jews in Eighteenth- and Nineteenth-Century Devon', *Exeter Papers in Economic History, no. 3* (Exeter, 1970)

BOOKS

I. Abrahams, *Jewish Life in the Middle Ages* (2nd edition, ed. C. Roth, 1932)
M. Adler, *Jews of Medieval England* (1939)
G. Alderman, *The Jewish Community in British Politics* (Oxford, 1983)
*Anglo-Norman Custumal*, eds J.W. Schopp, R.C. Easterling (1925)
C.W. Bracken, *A History of Plymouth* (Plymouth, 2nd edn 1934)
H.M. Brown, *Cornish Clocks and Clockmakers* (Dawlish, 1961)
D. Cesarani, ed. *The Making of Modern Anglo-Jewry* (Oxford, 1990)
A. Cohen, *An Anglo-Jewish Scrapbook* (1943)
M.A. Courtney, *Cornish Feasts and Folklore* (Penzance, 1913)
C. Ellis, *Historical Survey of Torquay* (Torquay, 1930)
T.M. Endelman, *Radical Assimilation in English Jewish History, 1656–1945* (Indiana, 1990)
L.P. Gartner, *The Jewish Immigrant in England* (1960)
L.M. Goldman, *The History of the Jews in New Zealand* (Wellington, 1958)
B. Kirkman Gray, *A History of English Philanthropy* (1905)
G.L. Green, *The Royal Navy and Anglo-Jewry, 1740–1820* (1989)
W.G. Hoskins, *Devon* (1954)
A.M. Hyamson, *The Sephardim of England* (1951)
J. Jacobs, *Jewish Statistics* (1891)
——, *Jews of Angevin England* (1893)
Jewish Historical Society of England, *Miscellanies*, part 1 (1925), in progress
——, *Transactions*, vol. 1 (1895), in progress
J. Katz, *Exclusiveness and Tolerance* (Oxford, 1961)
N. Kokosalakis, *Ethnic Identity and Religion* (Washington, 1982)
J.R. Leifchild, *Cornwall: Its Mines and Miners* (1862)
J.S. Levi and G.F.J. Bergman, *Australian Genesis* (1974)

A. Levy, *History of the Sunderland Jewish Community* (1956)
V.D. Lipman, *Social History of the Jews in England, 1850–1950* (1954)
N. Longmate, *King Cholera* (1966)
J.R. Marcus, *Memoirs of American Jews* (Philadelphia, 1955)
T. May, *An Economic and Social History of Britain, 1760–1970* (1987)
*A Minority in Britain*, ed. M. Freedman (1955)
G. Pycroft, *Art in Devonshire* (Exeter, 1888)
H.R. Richardson, *The English Jewry under Angevin Kings* (1960)
C. Roth, *The Great Synagogue, London, 1690–1940* (1950)
——, *A History of the Jews in England* (3rd edn, Oxford, 1964)
——, *The Rise of Provincial Jewry* (1950)
G. Saron and L. Hotz, *The Jews in South Africa* (Oxford, 1955)
Thos. Shapter, *The History of the Cholera in Exeter* (1849)
A.M.W. Stirling, *The Ways of Yesterday* (1930)
*Three Centuries of Anglo-Jewish History*, ed. V.D. Lipman (Cambridge, 1961)
Sidney and Beatrice Webb, *English Local Government: English Poor Law History* (1927)
H.F. Whitfeld, *Plymouth and Devonport in times of Peace and War* (Plymouth, 1900)
B. Williams, *The Making of Manchester Jewry, 1740–1875* (Manchester, 1976)

# INDEX

NOTE: Some names are variously spelled in different sources: variants are noted at the head of the first entry.